ELECTROANALYTICAL CHEMISTRY
A Series of Advances
VOLUME 24

ELECTROANALYTICAL CHEMISTRY
A Series of Advances
VOLUME 24

edited by
Allen J. Bard
and
Cynthia Zoski

CRC Press
Taylor & Francis Group
Boca Raton London New York

CRC Press is an imprint of the
Taylor & Francis Group, an **informa** business

CRC Press
Taylor & Francis Group
6000 Broken Sound Parkway NW, Suite 300
Boca Raton, FL 33487-2742

First issued in paperback 2017

© 2012 by Taylor & Francis Group, LLC
CRC Press is an imprint of Taylor & Francis Group, an Informa business

No claim to original U.S. Government works

ISBN-13: 978-1-4398-3750-4 (hbk)
ISBN-13: 978-1-138-11176-9 (pbk)

Visit the Taylor & Francis Web site at
http://www.taylorandfrancis.com

and the CRC Press Web site at
http://www.crcpress.com

Contents

Introduction to the Series

This series is designed to provide authoritative reviews in the field of modern electroanalytical chemistry defined in its broadest sense. Coverage is comprehensive and critical. Enough space is devoted to each chapter of each volume to include derivations of fundamental equations, detailed descriptions of apparatus and techniques, and complete discussions of important articles, so that the chapters may be useful without repeated reference to the periodical literature. Chapters vary in length and subject area. Some are reviews of recent developments and applications of well-established techniques, whereas others contain discussion of the background and problems in areas still being investigated extensively and in which many statements may still be tentative. Finally, chapters on techniques generally outside the scope of electroanalytical chemistry, but which can be applied fruitfully to electrochemical problems, are included.

Electroanalytical chemists and others are concerned not only with the application of new and classical techniques to analytical problems, but also with the fundamental theoretical principles upon which these techniques are based. Electroanalytical techniques are proving useful in such diverse fields as electroorganic synthesis, fuel cell studies, and radical ion formation, as well as with such problems as the kinetics and mechanisms of electrode reactions, and the effects of electrode surface phenomena, adsorption, and the electrical double layer on electrode reactions.

It is hoped that the series is proving useful to the specialist and nonspecialist alike—that it provides a background and a starting point for graduate students undertaking research in the areas mentioned, and that it also proves valuable to practicing analytical chemists interested in learning about and applying electroanalytical techniques. Furthermore, electrochemists and industrial chemists with problems of electrosynthesis, electroplating, corrosion, and fuel cells, as well as other chemists wishing to apply electrochemical techniques to chemical problems, may find useful material in these volumes.

Allen J. Bard
Cynthia G. Zoski

Contributors to Volume 24

Eric Bakker
Department of Inorganic,
 Analytical and Applied
 Chemistry
University of Geneva
Geneva, Switzerland

Victor Climent
Instituto de Electroquímica
Universidad de Alicante
Alicante, Spain

Juan M. Feliu
Instituto de Electroquímica
Universidad de Alicante
Alicante, Spain

Mark B. Jensen
Department of Chemistry
Concordia College
Moorhead, Minnesota

Carol Korzeniewski
Department of Chemistry and
 Biochemistry
Texas Tech University
Lubbock, Texas

Ernö Pretsch
Institute of Biogeochemistry and
 Pollutant Dynamics
ETH Zurich
Zurich, Switzerland

Joaquín Rodríguez-López
Department of Chemistry and
 Biochemistry
The University of Texas at Austin
Austin, Texas

Dennis E. Tallman
Department of Coatings and
 Polymeric Materials

and

Department of Chemistry and
 Biochemistry
North Dakota State University
Fargo, North Dakota

Contents of Other Series Volumes

1 Advances in Potentiometry

Eric Bakker and Ernö Pretsch

CONTENTS

1.1 INTRODUCTION

After more than one century, the history of potentiometry is long. After early applications of glass electrodes, electrodes of the first and second kind, and those based on ion exchangers, a real breakthrough was due to the invention of liquid membrane electrodes in 1967. Spectacular developments during the 1970s and 1980s were followed by a short period of declining interest. This changed radically about 15 years ago when again a series of major developments were achieved. These are the topic of this chapter. As an introduction, a brief summary of the early history is given (cf. Table 1.1).

The potential between two aqueous phases containing ions was first described by the 25-year-old Nernst in 1889 [1]. One year earlier, in connection with his studies on diffusion, he already derived a related equation describing the potential difference between two locations in a solution having different concentrations of a salt [2]. In his extended study of 1889, he additionally treated concentration cells and electrodes of the first and second kind. Later, he investigated transport phenomena between two aqueous solutions separated by phenol as an organic phase [3]. Using the same system, his coworker Riesenfeld studied the potentiometric response in terms of the concentrations of the two aqueous phases [4]. These contributions may be regarded as the first studies on a kind of interface between two immiscible electrolyte solutions (ITIES) and liquid membranes. The invention of ion-selective electrodes (ISEs) is usually attributed to Cremer, whose research was focused on the understanding of the electromotive behavior of tissues. In the extensive theoretical part of his paper on the glass electrode in 1906 [5], he discussed different concentration cells and stated that the membrane potential is the sum of phase-boundary potentials and diffusion potential (a finding usually attributed to Teorell and Meyers and Sievers, see below). He made his first experiments with the system described by Nernst [1] and Riesenfeld [4], but then he found glass more promising. The electric conductivity of glass had already been described by Helmholtz in 1881 [6]. For his studies, Cremer selected glass also because high-temperature experiments by Meyer in 1890 indicated that Na^+ is the conducting species, i.e., glass must be permselective [7]. Cremer's electrodes were glass flasks with a diameter of 6–10 cm and a minimal wall thickness of <0.02 mm. He found that acids added to the neutral solution on one side of the flask induced reproducible potential changes (230 mV by 10^{-1} M H_2SO_4, and 190 mV by 10^{-2} M H_2SO_4). The potential difference between diluted solutions of sulfuric acid on one side and of NaOH on the other was 550 mV [5]. A few years later, Haber and Klemensiewicz in their paper "Über elektrische Phasengrenzkräfte" ("About electrical phase-boundary forces") presented detailed investigations on glass bulb electrodes including pH titration curves and an account of the working mechanism, which, in essence, is still valid today, more than 100 years later [8]. The authors claimed that within the surface layer of the soaked glass, the H^+ and OH^- concentrations do not depend on the acidity or alkalinity of the external aqueous solution [9].

TABLE 1.1
Selection of Historical Landmark Papers in the Field of Potentiometry

Year	Author(s)	Contribution
1889	Nernst [1]	Deduction of the Nernst equation and its application to electrodes of the first and second kind and to concentration cells
1902	Riesenfeld [4]	Potentiometric response between two aqueous phases separated by phenol
1906	Cremer [5]	Potential of glass membrane depends on the acidity of the aqueous solutions on either side
1909	Haber and Klemensiewicz [8]	Full account of the response of glass electrodes and potentiometric acid–base titrations
1911	Donnan [10]	Potential between two aqueous phases separated by a membrane of limited permeability to ions
1923, 1924	Horvitz [15], Hughes [16]	Sensitivity of glass electrodes to other ions
1929, 1930	Guggenheim [11,12]	Theory of the phase-boundary potential
1935, 1936	Teorell [13], Meyer and Sievers [14]	Membrane potential as a sum of the phase-boundary and diffusion potential
1937	Kolthoff and Sanders [19]	Silver halide disk electrodes
1937	Nicolsky [17]	Potential response in the presence of two kinds of ions
1943	Sollner [22]	Potentiometric response of ion-exchanger membranes to cations and anions
1957, 1958	Stow et al. [23], Severinghouse and Bradley [24]	First gas-sensitive electrodes
1963	Katz and Rechnitz [42]	First potentiometric biosensor
1966, 1967	Stefanac and Simon [25,26]	Potentiometric response of membranes based on lipophilic antibiotics
1967	Ross [27], Bloch et al. [28]	Liquid membranes with charged ionophores
1967	Bloch et al. [28]	PVC-based solvent polymeric membranes
1967	Pedersen [29,30]	Crown ethers as the starting point of host–guest chemistry
1986	Oesch et al. [32]	ISEs for clinical use
1970	Cornwall et al. [36]	Intracellular measurements with liquid ion-exchanger microelectrodes
1975	Thomas et al. [38]	Intracellular measurements with ionophore-based microelectrodes
1976	Koryta et al. [40]	Electrochemical investigations of the ITIES
1979	Solsky and Rechnitz [41]	Antibody-selective membrane electrodes

On the other hand, Donnan in his contribution to "chemical physiology," described the potential between two aqueous phases separated by a membrane that is impermeable to a specific ion [10]. Later, this principle of "Donnan exclusion" became the basis for explaining the potentiometric response of liquid membranes. A full theoretical account of the phase-boundary potential was provided in 1929 and 1930 by Guggenheim, from The Royal Agricultural College in Copenhagen, who in two papers "On the conception of electrical potential difference between two phases" introduced the concept of electro-chemical potential (see Section 1.2) [11,12]. Teorell [13] as well as Meyer and Sievers [14] approximated the potential difference between two phases sepa-rated by a membrane, "the membrane potential," as the sum of three separate contributions, i.e., the two potential differences at the phase boundaries and the transmembrane potential difference (see also [5]). In spite of this early development of the fundamental theory, there were considerable debates about the working mechanisms of ISEs until the 1990s (see below).

Horvitz [15] and Hughes [16] realized that the glass electrode also responds to other ions, and Nicolsky extended the Nernst equation to describe the elec-trode response in the presence of two ions [17]. Later, the Nicolsky equation was extended considering many ions, which may have different charges. However, this so-called Nicolsky–Eisenman equation, which, in fact, is due to neither Nicolsky nor Eisenman, is inconsistent and invalid for ions of different charges. Although the potentiometric selectivity coefficient was first defined by Nicolsky in his equation, it has a more fundamental meaning in that it only depends on thermody-namic and weighing parameters. This also holds for ions of different charges, for which the Nicolsky–Eisenman equation does not correctly describe the response function (see Section 1.2.3).

Already in 1935, in a monograph on the practice of potentiometric analysis, Hiltner collected a great number of practical applications of electrodes of the first and second kind [18]. Two years later, Kolthoff and Sanders introduced solid-state electrodes based on silver halide disks for analyzing halide solutions [19]. Such membranes became commercially available in the 1960s, first on the basis of the work of Pungor in Hungary [20] and later introduced by Ross and Frant at Orion Research in the US [21]. The potentiometric application of ion-exchange membranes goes back to the 1940s. For example, Sollner reported the potentiometric detection of Li^+, Na^+, K^+, NH_4^+, and Mg^{2+} with collodion (nitrocellulose) membranes [22]. After impregnating these membranes with the polycation protamine, they responded to a series of anions [22].

The first gas sensor utilizing a CO_2 permeable membrane, which allowed CO_2 to diffuse from the sample into an aqueous layer held in a cellophane film on the surface of a conventional pH glass electrode, was described in 1957 [23,24].

There is only a limited number of ions that can be detected by solid-state ISEs, and the selectivity of ion-exchange membranes is mainly determined by the free energy of hydration of the ions. Therefore, the real versatility and strength of potentiometric sensors were only brought about by the invention of liquid membrane electrodes whose selectivity depends on the strength of the complexes formed between the analyte ions and the lipophilic complexing agent (ionophore or ion carrier) incorporated into the membrane. In 1966–1967, several groups started this research area simultaneously. Stefanac and Simon introduced ISEs based on complex-forming lipophilic antibiotics [25,26], while Ross [27] and Bloch et al. [28] described the first liquid-membrane ISEs with charged ionophores. In their very first paper, Bloch et al. also initiated the use of highly plasticized polyvinyl chloride (PVC) membranes [28], which still are the most common forms of liquid or solvent polymeric membranes. Incidentally, in the same year, Pedersen published his first papers on crown ethers, thus marking the starting point of host–guest chemistry, which became important for the design of ionophores [29,30].

A very fruitful research period followed, with contributions by numerous important research groups. Within a few years, ISEs were described for many ions, more than 50 by 1990 [31]. In clinical analysis, very soon polymeric-membrane-based ISEs replaced the earlier technology of flame photometry [32]. Today, in practically all clinical laboratories worldwide, a series of cations and anions are routinely determined by potentiometry. Blood gas analysis, which covers them as well as a number of other analytes, is mainly based on ISEs and other electrochemical sensors [33]. Today, it is an estimated worldwide market of US $6 billion. The development of a series of potentiometric microelectrodes also falls into the same period of time. Early pH [34] as well as Na^+- and K^+-selective glass microelectrodes [35] with diameters of a few tens of micron were followed, in the 1970s, by cation- or anion-exchanger-filled microcapillaries of diameters <1 μm [36,37]. Soon after that, ionophore-based microelectrodes were described for a number of alkali and alkaline earth metal ions [38,39]. Electrochemistry on liquid–liquid interfaces (ITIES) was also initiated in the 1970s [40], the same time as potentiometric bioanalysis with liquid membranes, an important emerging field today [41]. Actually, glass electrodes have already been used for bioanalysis since 1963 [42].

In the early 1990s, following this extremely rich research work, the interest in the field of ISEs declined and it seemed that no major developments were to be expected. This view has radically changed since the mid-1990s. The present chapter covers a selection of the most important developments. They include the deduction of the response function valid in the presence of several ions (mixed solutions), which led to equations replacing the semi-empirical and inconsistent Nicolsky–Eisenman equation in cases where ions of different charges are present. The understanding of the potential response in the presence of zero-current ion fluxes allowed dramatic improvements in the selectivity behavior and lower detection limits (LDLs) of ISEs. Ion fluxes can also be beneficial and may be used for different improvements and novel measuring techniques. A further topic of the present chapter is the measurement of ionophore–ion complex formation constants, which are decisive for the selectivity behavior of ISEs. Finally, recent developments of solid-contact electrodes will be reviewed. Although these are the subjects of ongoing research, it is very probable that they will be widely applied in the next generation of ISEs.

Other currently important research topics such as pulsed amperometric and coulometric techniques, the combination of potentiometry with stripping analysis, and potentiometric biosensors are not covered here.

1.2 RESPONSE CHARACTERISTICS OF ION-SELECTIVE ELECTRODES

1.2.1 DYNAMIC RESPONSE MODELS

ISEs may be mechanistically understood by considering the relationship between potential changes and zero-current ion distribution properties of the aqueous sample and polymeric membrane. This may be accomplished with static (considering equilibria) or dynamic models. The most comprehensive dynamic model is based on the Nernst–Planck equation describing the fluxes of charged species in the presence of concentration and potential gradients. It is written here for a one-dimensional system with mass transport in the x direction:

$$J_i = -\frac{D_i}{RT} c_i \frac{\delta \tilde{\mu}_i}{\delta x} \tag{1.1}$$

where J_i is the flux for the ion i; D_i, its diffusion coefficient; c_i, its concentration; and $\tilde{\mu}_i$, the electrochemical potential, which is a function of the standard chemical potential, μ_i^o, the activity of i, a_i, and the electrical potential, ϕ:

$$\tilde{\mu}_i = \mu_i^o + RT \ln a_i + z_i F \phi \tag{1.2}$$

The standard chemical potential is a direct function of the Gibb's energy of solvation (see next). Combining the Nernst–Planck equation with the Poisson equation and the law of mass conservation, one can derive an implicit description of the potential difference between two aqueous solutions separated by a membrane [43,44]. Numerical solutions of this equation have been reported, and the results related to the steady-state response of ISEs hardly differ from those obtained on the basis of the simpler phase-boundary potential model discussed further. In a more recent paper [45], it was claimed that the response of an ISE in a mixed solution of a trivalent and a monovalent ion can best be described on the basis of the Nernst–Planck diffusion model. However, the erroneous response calculated with the help of the phase-boundary model was based on a typographical error in an equation [46].

Undoubtedly, the Nernst–Planck approach is based on fewer and less restrictive model assumptions than the phase-boundary potential model [47]. However, the price to be paid for this more general approach is not only that an intuitive interpretation of the experimental observations is difficult but also that parameters such as the individual mobilities of all relevant ions and the rate constants for the ion transfer through the phase boundary must be known. Since such parameters are usually not accessible, simulations are based on assumed values. Therefore, so far, concerning the steady-state response of ISEs, no case has been presented proving any advantage whatsoever of the dynamic model. Moreover, it is important to realize that under the simplifying assumptions of equal mobilities for all ions in the membrane phase and of fast interfacial ion transfer, often suggested in response to the lack of experimental mobilities and ion transfer constants, the dynamic model leads to exactly the same predictions as the phase-boundary model. Evidently, there is no advantage of using a more complex model if experimental limitations lead to simplifying assumptions that reduce the complex model to the level of the simpler and more flexible one. Note, however, that for describing time-dependent phenomena, which are not covered here in detail, the Nernst–Planck–Poisson model and its simplifications are valuable approaches [48–52].

One interpretation pertaining to dynamic models strongly emphasizes the fluxes of individual kinds of ions across an ISE membrane. According to this approach, response characteristics of ISEs are related to transference numbers of the ions through the bulk of the membrane phase, while for the phase-boundary potential model, ion activities in the boundary layer of the membrane are decisive. At first glance, it seems astonishing that two so different models can equally fit the experimental data. However, assuming equal ionic mobilities in the membrane phase, the transference numbers (t_i) are directly related to the concentrations of the ions in the membrane near the phase boundary $(t_i = z_i c_i / R_T$, with R_T the concentration of a monovalent ion exchanger in the membrane), which are the relevant parameters of the phase-boundary potential. The intuitively deceptive consequence of the focus on transference is the overemphasized importance attributed to the movement of ions through the membrane. In the same way as the response of pH glass electrodes is usually governed by a constant H_3O^+ activity in the hydrated surface layer of the glass and is not influenced by the transfer of Na^+ ions through the glass, which is required to measure the electromotive force (*emf*) [53,54], the phase-boundary ion activity in solvent polymeric membranes usually governs their response irrespective of the conditions inside the membrane.

1.2.2 PHASE-BOUNDARY POTENTIAL MODEL

As early as 1929, Guggenheim set forth a simple equation relating the distribution of ions across the interface between two immiscible solutions and the associated potential change, which is valid under the assumption of electrochemical equilibrium across this interface [11,12]. At equilibrium, the electrochemical potentials for any ion, i, must be equal on either side of the interface. Therefore, rewriting Equation 1.2 for the aqueous (*aq*) and membrane (*m*) phase and after assuming $\tilde{\mu}_i(aq) = \tilde{\mu}_i(m)$, the resulting potential change across the interface is expressed as:

$$E_{PB} = \frac{RT}{z_i F} \ln k_i + \frac{RT}{z_i F} \ln \frac{a_i(aq)}{a_i(m)} \qquad (1.3)$$

where $E_{PB} = \phi(m) - \phi(aq)$ is the phase-boundary potential; the corresponding standard chemical potentials, μ_i°, are included in k_i:

$$\varepsilon_i^\circ = \frac{RT}{z_i F} \ln k_i = \frac{\mu_i^\circ(aq) - \mu_i^\circ(m)}{z_i F} \tag{1.4}$$

As indicated, the term is, in fact, the standard potential, ε_i°, which is constant for a given ion but varies from one ion to another.

Although the phase-boundary potential model has been known since the 1930s, dynamic models were more popular in the early days of polymeric membrane electrodes. This is perhaps due to the strong influence of the theoretical work of Eisenman's group, which grew out of the treatment of biological membranes for which no equilibria and no electroneutrality condition were required [55–57]. Since the response functions and the selectivities show correlations both with dynamic and equilibrium parameters, controversies about the governing mechanisms prevailed over many decades [58,59]. The group of Pungor especially had always advocated the phase-boundary model and collected experimental evidence to support it [60,61] (see also [62,63]). Eventually, the predictive power resulting from the discussion of equilibria at the phase boundary and, among others, the direct observation of concentrations in the membrane by optical means [64–66], led to a swift development of the phase-boundary potential model, by far surpassing the output of the dynamic models (see also [67] and examples below).

Teorell [13] and Meyer and Sievers [14] understood the potential difference between two phases separated by a membrane, the so-called membrane potential, as the sum of three separate contributions, i.e., the two potential differences on the phase boundaries and the transmembrane potential difference (see also [68]). Buck, later, called this the segmented potential model [69]. Assuming fast local equilibria across the phase boundary, the phase-boundary potential can be described, according to Guggenheim, by a simple relationship between the standard free energies and activities of the ion in the two phases and the electric potential across the phase boundary (cf. Equation 1.3). Later, Hung offered a comprehensive treatment for different kinds of ions [70,71], which was well explained by van den Berg [72] and extended by Kakiuchi [73] for phases of limited volumes. The phase-boundary potential model has recently been coined in this manner to distinguish it from the more general segmented potential model. It does not represent an entirely different approach to the prediction of ISE responses, but a special and highly relevant case of the segmented potential model. In particular, the assumptions underlying the phase-boundary potential model allow us to ignore ionic mobilities, which were central to the prevailing theory of ISEs in the 1970s and 1980s, but are not readily available experimentally. By the end of the 1980s, it began to become clear that ionic mobilities in ISE membranes give rise to subtle effects that, typically, may be ignored without significant loss of accuracy. Consequently, the membrane potential, E_M, can be understood as the sum of the two individual phase-boundary potentials. If the same ion, i, is potential-determining at both interfaces, E_M may be written as:

$$E_M = \frac{RT}{z_i F} \ln \frac{a_i(aq)}{a_i(m)} \frac{a_i(m)'}{a_i(aq)'} \qquad (1.5)$$

where activities labeled with (') denote those at the inner phase boundary of the membrane.

The phase-boundary potential arises from a charge separation of cations and anions across the interface due to their different standard free energies in the two phases. The simplicity and usefulness of the phase-boundary model lie in the fact that the actual distribution profile of ions within the charge separation layer between the two bulk phases is not important; the quantitative description of the *emf* only requires the knowledge of the relative k_i values and the ionic activities in both phases. The former can be estimated by measurements on ion-exchanger membranes (see below). The corresponding selectivity series is named after Hofmeister, who did not study membranes or free energies but investigated the diuretic and laxative effects of salts and found some correlations for cations and anions [74]. If a membrane, in addition, contains an ionophore, its *emf* response characteristics can be predicted on the basis of complexation equilibria. The preference of a strongly complexed ion is a consequence of its stabilization in the membrane, but not of an enhanced extraction from the sample into the bulk of the membrane. The total ionic content of the membrane does not change because of the presence of an electrically neutral ionophore. Since phase-boundary equilibria are normally fast on the time scale of potentiometric experiments, the phase-boundary potential model is still useful if the ion activities near the phase boundaries are not equal to those in the bulk phases, whose importance is evident in Sections 1.4 and 1.5.

The powerful phase-boundary potential model is valid if two important assumptions hold. In the first assumption, the diffusion potential is considered to be negligible. Diffusion potentials will, of course, be irrelevant if diffusion processes are negligible or if only one kind of ion is transported across the membrane. In cases when at least two kinds of ions of different mobilities are transported, a diffusion potential arises and affects the ion fluxes so that the electroneutrality condition is fulfilled. Therefore, the diffusion potential in a permselective membrane is negligible whenever only one kind of ion of a given charge sign is present in the membrane. Fortunately, experimental evidence shows that the diffusion potential is small in other cases as well (on the order of millivolts) [75,76]. The second assumption, i.e., that the organic phase boundary contacting the sample is in chemical equilibrium with the aqueous sample solution, holds whenever the phase transfer reaction is much faster than the relevant diffusion processes in both the aqueous and organic layers. While exceptions to this assumption are certainly conceivable, the literature, to the best of our knowledge, does not contain any references to a potentiometric sensor for which experimental observations indicate this second assumption to be invalid.

For simplicity, ion activities in the membrane are often substituted by concentrations. This substitution is usually reasonable since a constant concentration of ion exchanger defines the ionic strength in ion-exchanger membranes. However, it is not an underlying assumption of the phase-boundary model. Indeed, cases have been reported in which ion pairing has been accounted for [77–81].

The essence of the phase-boundary potential model is that Equation 1.3, as simple as it may look, straightforwardly describes the *emf* response of ISE membranes of diverse compositions exposed to any type of aqueous sample. The challenge in the application of the phase-boundary model to a particular membrane and type of sample lies in finding and quantifying the parameters that determine the concentration of the uncomplexed primary ion on the membrane side of the phase boundary and the activity of the primary ion on the sample side of this phase boundary. The latter is not identical with the ion activity in the bulk of the sample if ion fluxes are relevant. Consequently, a practical limitation of the phase-boundary potential approach lies in the investigator's ability to identify the relevant chemical species as well as qualitatively and quantitatively describe the chemical equilibria at the sample/membrane interface and, if required, the driving forces that transport chemical species toward and from the phase boundary. However, as the following examples will show, recent years have brought along a tremendous increase in the complexity of systems that can be rationalized quantitatively.

The complexity of the local chemical equilibrium at the sample/membrane interface to be considered depends not only on the number and kinds of involved chemical species in the membrane and sample solution but also on the range of phenomena that one attempts to account for. Let us consider some of the factors that may be necessary to fully rationalize the response of an ISE. The free energies of solvation of the uncomplexed primary and interfering ions, I and J, in the membrane are affected by its matrix and plasticizer and are quantified by k_I and k_J. In membranes containing an ionophore, the primary ions occur mostly in their complexed form and only in very small amounts in their free form. The parameters that describe the complexation equilibria are the formation constants and stoichiometric coefficients of the complexes between ionophore and primary or interfering ions. Moreover, the total charge of free and complexed primary and interfering ions in the bulk of the membrane must be counterbalanced by an equal amount of opposite charge of the lipophilic ion-exchanger (and the ionophore if it is electrically charged). Consequently, the charge and concentration of the ionophore and ion exchanger are also important parameters affecting the response of an electrode. To account for the upper detection limit, the distribution of co-ions of opposite charge between the sample and the membrane needs to be considered. This is achieved using the partition equilibria of the relevant electrolyte but, for a more accurate description, may also require ion pair formation constants in the membrane phase. The description of the lower detection limit, on the other hand, involves the distribution of the salts of primary and interfering ions and ion exchanger between the membrane and sample phases and/or the co- or counter-transport of primary, interfering, and co-ions across the membrane. To rationalize the selectivity loss of a membrane in long-term applications, the distribution of an electrically neutral ionophore between the membrane phase and a continuously renewed sample can be accounted for by the lipophilicity of the ionophore. Alternatively, to describe the loss of primary ion and charged ionophore into the sample requires, besides other parameters, the partition constant for the primary ion salt of that ionophore.

As the application of the phase-boundary potential to the theoretical understanding of electrochemical ion sensors will accompany us throughout this chapter, let us consider first the origin of the Nernstian response slope. Ion-selective membranes are generally formulated so that the resulting electrode response follows the Nernst equation, which is written as:

$$E_i = K_i + \frac{s}{z_i} \log a_i(aq) \tag{1.6}$$

where $s = RT/F$. The observed *emf*, E_i, is the sum of all individual potential changes in the galvanic cell. If they are all constant, with the exception of the phase-boundary potential, E_{PB}, at the sample/membrane interface, they can be included into a single constant potential term, K_{cell}, so that E_i can be expressed (cf. Equation 1.3) as:

$$E_i = K_{cell} + E_{PB} = K_{cell} + \frac{RT}{z_i F} \ln \frac{k_i a_i(aq)}{a_i(m)} \tag{1.7}$$

Clearly, Equation 1.7 reduces to the Nernst equation, Equation 1.6, if $a_i(m)$ is independent of the sample composition. This is illustrated by a general case in which we allow for the presence of an ion exchanger of total concentration R_T and charge z_R and an ionophore, L, of charge z_L. A single sample ion, i, interacts with the membrane. For simplicity, coulombic interactions between ion exchanger and any counterion are neglected in this treatment. The ionophore may form complexes, $[iL_n]^{z_i}$, in the membrane that are described with overall complex formation constants:

$$\beta_n = a_{iL_n}/\left(a_L^n a_i(m)\right) \tag{1.8}$$

Note that the phase label (m) is omitted for species confined to the membrane. The charge balance in the membrane is written as:

$$z_R R_T + z_L c_L + \sum_n (nz_L + z_i) c_{iL_n} + z_i c_i(m) = 0 \tag{1.9}$$

After inserting complex formation constants and activity coefficients into Equation 1.9, we obtain:

$$z_R R_T + \frac{z_L a_L}{\gamma_L} + \frac{z_i a_i (m)}{\gamma_i (m)} \left(1 + \sum_n \frac{nz_L + z_i}{z_i} \frac{\gamma_i (m)}{\gamma_{iL_n} (m)} \beta_n a_L^n \right) = 0 \qquad (1.10)$$

For the common situation of electrically neutral ionophores, $z_L = 0$, so that Equation 1.10 simplifies to:

$$z_R R_T + \frac{z_i a_i (m)}{\gamma_i (m)} \left(1 + \sum_n \frac{\gamma_i (m)}{\gamma_{iL_n} (m)} \beta_n a_L^n \right) = 0 \qquad (1.11)$$

In the simple case of an ionophore-free ion-exchanger membrane in the presence of monovalent ions i only, Equation 1.11 further simplifies to:

$$R_T = \frac{a_i (m)}{\gamma_i (m)} \qquad (1.12)$$

The Nernst equation, Equation 1.6, is obtained by solving Equation 1.10 for $1/a_i(m)$:

$$\frac{1}{a_i (m)} = -\frac{z_i}{\gamma_i (m) \left(z_R R_T + \frac{z_L a_L}{\gamma_L} \right)} \left(1 + \sum_n \frac{nz_L + z_i}{z_i} \frac{\gamma_i (m)}{\gamma_{iL_n} (m)} \beta_n a_L^n \right) \qquad (1.13)$$

and inserting Equation 1.13 into Equation 1.3 to give:

$$E_{PB} = E_i^{\circ} + \frac{s}{z_i} \log a_i (aq) \qquad (1.14)$$

with

$$E_i^{\circ} = \frac{s}{z_i} \log \left(\frac{-z_i k_i}{\gamma_i (m)} \frac{1 + \sum_n \frac{nz_L + z_i}{z_i} \frac{\gamma_i (m)}{\gamma_{iL_n} (m)} \beta_n a_L^n}{z_R R_T + \frac{z_L a_L}{\gamma_L}} \right) \qquad (1.15)$$

For an ionophore-free ISE membrane and a monovalent ion exchanger ($z_R = -z_i/|z_i|$), Equation 1.15 reduces to:

$$E_i^\circ = \frac{s}{z_i} \log\left(\frac{k_i z_i}{\gamma_i\,(m)\,R_T} \right) \tag{1.16}$$

In all these cases, a Nernstian response behavior is expected if the ion activities in the membrane phase are independent of the sample composition. This is accomplished by suppressing ion-exchange and coextraction processes with co- and counterions in the sample. The ion-exchange selectivity is particularly important with ISEs, and the selectivity coefficient discussed further next is a direct measure of this characteristic.

1.2.3 SELECTIVITY COEFFICIENT

Historically, the potentiometric selectivity coefficient was first introduced by Nicolsky in his equation describing the *emf* response in the presence of two monovalent ions [17]. Later, this equation was empirically extended to ions of different charges (often called the Nicolsky–Eisenman equation). Later on, it was realized that this inconsistent equation is not valid if the sample simultaneously contains ions of different charges, and it was erroneously assumed that the selectivity coefficient is not an appropriate measure in this case [82]. We show here that the selectivity coefficient has a fundamental meaning also for ions of different charges and that its definition is independent of the Nicolsky equation.

For simple cases of ions of the same charge, the selectivity coefficient, K_{ij}^{pot}, may be intuitively understood as a weighting factor for the interfering ion. In other cases, the relationship is more complex, but the usefulness of the selectivity coefficient derives from the fact that it can be directly related to the respective E_i° values described above (Equation 1.15) [83,84]. It is, therefore, a constant characteristic for a given electrode and may constitute the basis for the chemical optimization of sensor selectivities.

The selectivity coefficient is derived from the separately measured Nernstian responses to the primary ion, I:

$$E_I = K_{cell} + E_I^\circ + \frac{s}{z_I} \log a_I\,(aq) \tag{1.17}$$

and to the interfering ion, J, as follows:

$$E_J = K_{cell} + E_J^\circ + \frac{s}{z_J} \log a_J\,(aq) \tag{1.18}$$

The response according to Equation 1.18 can, alternatively, be formulated using E_I° and the selectivity coefficient, K_{IJ}^{pot}, as follows:

$$E_J = K_{cell} + E_I^\circ + \frac{s}{z_J} \log K_{IJ}^{pot} a_J\,(aq)^{z_J/z_I} \tag{1.19}$$

Subtracting Equation 1.18 from Equation 1.19 gives the following expression for the selectivity coefficient:

$$\log K_{IJ}^{pot} = \frac{z_I}{s}\left(E_J^\circ - E_I^\circ\right)$$

(1.20)

For the most general case covered here, Equation 1.15 for E_I° and E_J° is inserted into Equation 1.20 to give:

$$K_{IJ}^{pot} = \left(\frac{-z_J k_J}{\gamma_J(m)}\frac{1+\sum\limits_{n=n_J}\frac{nz_L + z_J}{z_J}\frac{\gamma_J(m)}{\gamma_{JL_n}(m)}\beta_n a_{L,J}^n}{z_R R_T + \frac{z_L a_{L,J}}{\gamma_L}}\right)^{z_J/z_I}$$

$$\times \frac{-\gamma_I(m)}{z_I k_I}\frac{z_R R_T + \frac{z_L a_{L,I}}{\gamma_L}}{1+\sum\limits_{n=n_I}\frac{nz_L + z_I}{z_I}\frac{\gamma_I(m)}{\gamma_{IL_n}(m)}\beta_n a_{L,I}^n}$$

(1.21)

If, for simplicity, one assumes that all activity coefficients in Equation 1.21 are unity, the general expression for K_{IJ}^{pot} reduces to:

$$K_{IJ}^{pot} = \left(-z_J k_J \frac{1+\sum\limits_{n=n_J}\frac{nz_L + z_J}{z_J}\beta_n c_{L,J}^n}{z_R R_T + z_L c_{L,J}}\right)^{z_J/z_I}\frac{-1}{z_I k_I}\frac{z_R R_T + z_L c_{L,I}}{1+\sum\limits_{n=n_I}\frac{nz_L + z_I}{z_I}\beta_n c_{L,I}^n}$$

(1.22)

These relationships may be effectively used to predict optimal membrane concentrations for any given complex stoichiometries and complex formation constants. For the practically important case of an electrically neutral ionophore ($z_L = 0$) and a monovalent ion exchanger, Equation 1.22 simplifies to:

$$K_{IJ}^{pot} = \left(-z_J k_J \frac{1+\sum\limits_{n=n_J}\beta_n c_{L,J}^n}{R_T}\right)^{z_J/z_I}\frac{-1}{z_I k_I}\frac{R_T}{1+\sum\limits_{n=n_I}\beta_n c_{L,I}^n}$$

(1.23)

The concentration of the uncomplexed ionophore for the measurement of any ion i in the absence of the other ion (shown in Equation (1.23) as $c_{L,I}$ and $c_{L,J}$) can be expressed by the following implicit relationship (where L_T denotes the total concentration of the ionophore):

$$R_T \sum_{n=n_i} n\beta_n c_{L,i}^n = z_i(L_T - c_{L,i})\left\{1 + \sum_{n=n_i} \beta_n c_{L,i}^n\right\} \tag{1.24}$$

If, for simplicity, one assumes single stoichiometries, n_I and n_J, for each primary and interfering ion complex and disregards the influence of uncomplexed ions on the charge balance, Equation 1.24 can easily be solved for $c_{L,i}$. For I, this is written as:

$$c_{L,I} = L_T - n_I R_T / z_I \tag{1.25}$$

which, together with the analogous expression for J, is inserted into Equation 1.23 to give:

$$K_{IJ}^{pot} = \left(-z_J k_J \frac{1 + \beta_{n_J}\left(L_T - n_J R_T / z_J\right)^{n_J}}{R_T}\right)^{z_J/z_I} \frac{-1}{z_I k_I} \frac{R_T}{1 + \beta_{n_I}\left(L_T - n_I R_T / z_I\right)^{n_I}} \tag{1.26}$$

The change of the logarithmic selectivity coefficient may be written in simplified form if all complex formation constants are sufficiently large so that uncomplexed primary and interfering ions can be neglected. It may be used to estimate the selectivity modifying influence of variations in membrane composition as follows:

$$\Delta \log K_{IJ}^{pot} = \log\left(\left(-\frac{\left(L_T - n_J R_T / z_J\right)^{n_J}}{R_T}\right)^{z_J/z_I} \frac{-R_T}{\left(L_T - n_I R_T / z_I\right)^{n_I}}\right) \tag{1.27}$$

In Figure 1.1, Equation 1.27 is illustrated for a typical case with divalent ions I and J and of different complex stoichiometries. The results show how the selectivity coefficient depends on the ion-exchanger concentration and that it can be used to optimize membrane compositions. In the case shown, the selectivity optimum is found at ca. 77 mol% ion exchanger relative to the ionophore, which agrees with earlier predictions for a Mg^{2+}-selective electrode with respect to Ca^{2+} as its main interfering ion [85]. Note that when the relative concentration of the free ionophore drastically changes as a function of the sample composition, ISEs may exhibit potential drifts and are not generally recommended for practical use.

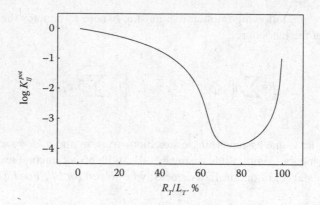

FIGURE 1.1   Dependence of the selectivity coefficient, $\log K_{IJ}^{pot}$, on the total molar concentration of the ionophore, L_T, relative to that of the ion exchanger, R_T, for the case of divalent primary (I) and interfering (J) ions forming metal–ligand complexes of stoichiometry 1:2 (I) and 1:3 (J). The ion J is more strongly discriminated if the concentration of the ion exchanger is so high that the amount of ligand in the membrane is not sufficient to form a complex of optimal stoichiometry (see Equation 1.27).

For the simple case of an uncharged ionophore forming 1:1 complexes and monovalent primary and interfering ions, Equation 1.26 transforms to:

$$K_{IJ}^{pot} = \frac{k_J}{k_I} \frac{1+\beta_{JL}c_{L,J}}{1+\beta_{IJ}c_{L,I}} \tag{1.28}$$

This relationship is further simplified into the well-established relationship when the primary and interfering ions are predominantly complexed by the ionophore:

$$K_{IJ}^{pot} = \frac{k_J}{k_I} \frac{\beta_{JL}}{\beta_{IJ}} \tag{1.29}$$

Conversely, in the absence of an ionophore (or large molar excess of ion exchanger over the ionophore), the selectivity coefficient becomes a function of the free energies of transfer only:

$$K_{IJ}^{pot} = \frac{k_J}{k_I} \tag{1.30}$$

Intermediate cases may be obtained from Equation 1.28. In the above example, the concentration of the uncomplexed ionophore for any i is related to the total concentrations of the ionophore, L_T, and ion exchanger, R_T, by considering the charge balance and complex formation equations:

$$L_T = c_{L,i} + \frac{R_T \beta_{iL} c_{L,i}}{1 + \beta_{iL} c_{L,i}} \tag{1.31}$$

which is solved for $c_{L,i}$ to give:

$$c_{L,i} = \frac{1}{2\beta_{iL}} \left(\beta_{iL}(L_T - R_T) - 1 + \sqrt{4\beta_{iL}L_T + (\beta_{iL}(L_T - R_T) - 1)^2} \right) \tag{1.32}$$

For each i, Equation 1.32 is inserted into Equation 1.28 to give the selectivity coefficient as a function of membrane composition, complex formation constants, and free energies of transfer for the involved ions:

$$K_{IJ}^{pot} = \frac{k_J}{k_I} \frac{1 + \frac{1}{2}\left(\beta_{JL}(L_T - R_T) - 1 + \sqrt{4\beta_{JL}L_T + (\beta_{JL}(L_T - R_T) - 1)^2} \right)}{1 + \frac{1}{2}\left(\beta_{IL}(L_T - R_T) - 1 + \sqrt{4\beta_{IL}L_T + (\beta_{IL}(L_T - R_T) - 1)^2} \right)} \tag{1.33}$$

For selected values of complex formation constants, Figure 1.2 illustrates how the selectivity coefficient depends on the ion-exchanger concentration. Note that the selectivity behavior approaches that of a membrane based on an ion exchanger

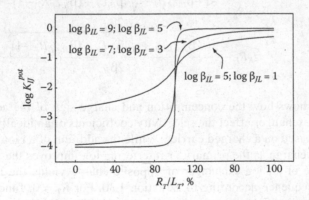

FIGURE 1.2 Dependence of the selectivity coefficient, $\log K_{IJ}^{pot}$, on the total molar concentration of the ionophore, L_T, relative to that of the ion exchanger, R_T, for two monovalent ions I and J forming complexes of different strength (ratio of complex formation constants, $\beta_{IL}/\beta_{JL} = 10^4$ in all cases). At an excess of ion exchanger (>100 mol%), the selectivity is mainly determined by the standard free energy of transfer of the two ions (their difference is assumed to be 0). The transition from an ISE membrane whose selectivity is dictated by complexation ($\log K_{IJ}^{pot} = -4$) to that of a simple ion-exchanger membrane ($\log K_{IJ}^{pot} = 0$) is well-defined only with ligands forming sufficiently strong complexes. For weaker complexes, the selectivity of complex formation in the sensor membrane cannot be fully exploited.

alone if its molar concentration exceeds that of the ionophore. The transition becomes less pronounced for ionophores that form weaker complexes with primary and interfering ions.

For an ionophore that is electrically charged in its uncomplexed form, the selectivity coefficient may be simplified in a similar way from Equation 1.22. Assuming again that all charged species are monovalent ($z_L = -z_I = -z_J = -1$) and assuming a 1:1 complex between ionophore and primary or interfering ion, one obtains:

$$K_{IJ}^{pot} = \frac{k_J}{k_I} \frac{z_R R_T - c_{L,I}}{z_R R_T - c_{L,J}} \tag{1.34}$$

As earlier, the concentration of the uncomplexed ionophore for any ion i is related to the total concentrations of ionophore, L_T, and ion exchanger, R_T, when charge balance and complex formation equations are considered:

$$\beta_{iL} c_{L,i}^2 + (1 - z_R R_T \beta_{iL}) c_{L,i} - L_T = 0 \tag{1.35}$$

Solving this equation for $c_{L,i}$ and inserting the result for each ion i into Equation 1.34 gives [86]:

$$K_{IJ}^{pot} = \frac{k_J}{k_I} \frac{z_R R_T - \dfrac{-1 + \beta_{IL} z_R R_T + \sqrt{4\beta_{IL} L_T + (\beta_{IL} z_R R_T - 1)^2}}{2\beta_{IL}}}{z_R R_T - \dfrac{-1 + \beta_{JL} z_R R_T + \sqrt{4\beta_{JL} L_T + (\beta_{JL} z_R R_T - 1)^2}}{2\beta_{JL}}} \tag{1.36}$$

Figure 1.3 shows how the concentration and charge sign of the added monovalent ion exchanger affect the selectivity coefficients of an ideally behaving membrane based on a charged carrier. While the addition of an ion exchanger of the same charge as the primary or interfering ion improves the selectivity, the presence of an ion exchanger of opposite charge yields the Hofmeister selectivity sequence according to Equation 1.30. For $R_T = 0$, functional ISE membranes are obtained since the ionophore itself exhibits ion-exchanger properties, but the resulting selectivity is predicted to be less than optimal [86,87].

1.2.4 STEADY-STATE RESPONSE FUNCTION (NICOLSKY–BAKKER EQUATIONS)

In this section, the ISE response to samples containing a mixture of ions is derived in the most general form [88]. It is based on equilibrium considerations, and transmembrane ion fluxes are not taken into account here (cf. Section 1.3.4).

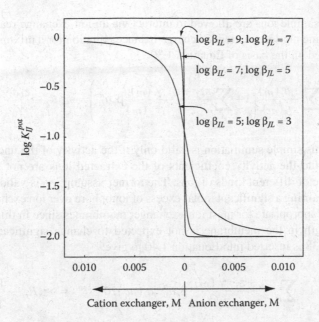

FIGURE 1.3 Influence of the concentration and charge sign of the added monovalent ion exchanger on the selectivity coefficient, $\log K_{IJ}^{pot}$, of a charged-carrier-based anion-selective membrane (ratio of complex formation constants, $\beta_{IL}/\beta_{JL} = 10^2$ in all cases).

The derivations are both a generalization and simplification of earlier treatments by Bakker et al. [83] and Bakker, Nägele, and Pretsch [89].

The phase-boundary potential for any ion i is formulated as follows:

$$E_{PB} = \frac{RT}{z_i F} \ln k_i \frac{a_i(aq)}{a_i(m)} = \frac{1}{z_i} s \log k_i \frac{a_i(aq)}{a_i(m)} \qquad (1.37)$$

Solved for the membrane activity of i gives:

$$a_i(m) = a_i(aq) k_i 10^{-z_i E_{PB}/s} \qquad (1.38)$$

The charge balance equation for one single ion i interacting with the membrane (in the absence of any interfering ion) is rewritten from Equation 1.9:

$$z_R R_T + \frac{z_L a_L}{\gamma_L} + \frac{z_i a_i(m)}{\gamma_i(m)} \left(1 + \sum_n \frac{n z_L + z_i}{z_i} \frac{\gamma_i(m)}{\gamma_{iL_n}(m)} \beta_n a_L^n \right) = 0 \qquad (1.39)$$

If multiple sample ions are allowed to interact via an ion-exchange reaction with the membrane phase, an extended charge balance equation for a mixture of ions i is formulated on the basis of Equation 1.39:

$$\sum_i \frac{z_i a_i(m)}{\gamma_i(m)} \left(1 + \sum_n \frac{nz_L + z_i}{z_i} \frac{\gamma_i(m)}{\gamma_{iL_n}(m)} \beta_n a_L^n \right) = -z_R R_T - \frac{z_L a_L}{\gamma_L} \tag{1.40}$$

Note that this simple summation is valid only if the activity of the uncomplexed ionophore and the activity coefficients of the extracted ions are not altered by the presence of different kinds of ions. The former assumption is valid for membranes containing a significant molar excess of ionophore over ion exchanger, and the latter is appropriate for most ion-exchanger membranes since in this case, the ionic strength in the membrane is not expected to change significantly. Now, Equation 1.38 is inserted into Equation 1.40 to give:

$$\sum_i \frac{z_i k_i}{\gamma_i(m)} \left(1 + \sum_n \frac{nz_L + z_j}{z_j} \frac{\gamma_i(m)}{\gamma_{iL_n}(m)} \beta_n a_L^n \right) a_i(aq) 10^{-z_i E_{PB}/s} = -z_R R_T - \frac{z_L a_L}{\gamma_L} \tag{1.41}$$

The following relationship is obtained from Equation 1.15:

$$\frac{z_i k_i}{\gamma_i(m)} \left(1 + \sum_n \frac{nz_L + z_i}{z_i} \frac{\gamma_i(m)}{\gamma_{iL_n}(m)} \beta_n a_L^n \right) 10^{-z_i E_i^o/s} = -z_R R_T - \frac{z_L a_L}{\gamma_L} \tag{1.42}$$

and inserted into Equation 1.41 to yield the following compact expression:

$$10^{-z_i E_{PB}/s} \sum_i a_i(aq) \, 10^{z_i E_i^o/s} = 1 \tag{1.43}$$

Summing up for all ions of interest may be performed by grouping the ions i according to their valency to obtain separate summations for monovalent, divalent, trivalent ions, and so on, as follows:

$$10^{-E_{PB}/s} \sum_{i(z_i=1)} 10^{E_i^o/s} a_i(aq) + 10^{-2E_{PB}/s} \sum_{i(z_i=2)} 10^{2E_i^o/s} a_i(aq)$$

$$+ 10^{-3E_{PB}/s} \sum_{i(z_i=3)} 10^{3E_i^o/s} a_i(aq) + \cdots = 1 \tag{1.44}$$

Note that inserting Equation 1.20, rewritten as:

$$10^{z_J E_J^o/s} = K_{IJ}^{pot} 10^{z_I E_I^o/s} \tag{1.45}$$

gives the same expression but now as a function of the selectivity coefficient:

$$10^{\left(E_I^o - E_{PB}\right)/s} \sum_{i(z_i=1)} K_{I,i}^{pot} a_i(aq) + 10^{2\left(E_I^o - E_{PB}\right)/s} \sum_{i(z_i=2)} K_{I,i}^{pot} a_i(aq)$$

$$+ 10^{3\left(E_I^o - E_{PB}\right)/s} \sum_{i(z_i=3)} K_{I,i}^{pot} a_i(aq) + \cdots = 1 \tag{1.46}$$

If the primary ion I is not included in the summation, which would then involve interfering ions J only, Equation 1.46 is alternatively written as:

$$10^{z_I\left(E_I^o - E_{PB}\right)/s} a_I(aq) + 10^{2\left(E_I^o - E_{PB}\right)/s} \sum_{J(z_J=1)} K_{IJ}^{pot} a_J(aq)$$

$$+ 10^{2\left(E_I^o - E_{PB}\right)/s} \sum_{J(z_J=2)} K_{IJ}^{pot} a_J(aq) + \cdots = 1 \tag{1.47}$$

Within mathematical constraints, Equations 1.45 or 1.46 can be solved for the phase-boundary potential, E_{PB}, for any desired combination of ion charges to describe the response of the potentiometric sensor to samples of any mixture of ions in the absence of zero-current ion fluxes. For any monovalent and divalent ions, e.g., the following mixed-ion response results from Equation 1.46:

$$E_{PB} = E_I^o + s \log \left[\frac{\dfrac{1}{2} \sum_{i(z_i=1)} K_{I,i}^{pot1/z_I} a_i(aq) +}{\sqrt{\left(\dfrac{1}{2} \sum_{i(z_i=1)} K_{I,i}^{pot1/z_I} a_i(aq)\right)^2 + \sum_{i(z_i=2)} K_{I,i}^{pot2/z_I} a_i(aq)}} \right] \tag{1.48}$$

Note again that the primary ion is included in the summations, in which case the selectivity coefficient $K_{II}^{pot} = 1$. The response to any combination of monovalent, divalent, and trivalent ions is obtained in complete analogy:

$$E_{PB} = E_I^o + s \log \left[\frac{1}{12} \left(4\Sigma_1 + \frac{2 \times 2^{1/3}\left(1+\sqrt{-3}\right)\left\{\Sigma_1^2 + 3\Sigma_2\right\}}{\left(-2\Sigma_1^3 - 9\Sigma_1\Sigma_2 - 27\Sigma_3 + 3\sqrt{3}\sqrt{-\Sigma_1^2\Sigma_2^2 - 4\Sigma_2^3 + 4\Sigma_1^3\Sigma_3 + 18\Sigma_1\Sigma_2\Sigma_3 + 27\Sigma_3^2}\right)^{1/3}} + 2^{2/3}\left(1-\sqrt{-3}\right)\left(-2\Sigma_1^3 - 9\Sigma_1\Sigma_2 - 27\Sigma_3 + 3\sqrt{3}\sqrt{-\Sigma_1^2\Sigma_2^2 - 4\Sigma_2^3 + 4\Sigma_1^3\Sigma_3 + 18\Sigma_1\Sigma_2\Sigma_3 + 27\Sigma_3^2}\right)^{1/3} \right) \right] \tag{1.49}$$

where

$$\Sigma_1 = \sum_{i(z_i=1)} K_{I,i}^{pot1/z_I} a_i (aq)$$

$$\Sigma_2 = \sum_{i(z_i=2)} K_{I,i}^{pot2/z_I} a_i (aq) \qquad (1.50)$$

$$\Sigma_3 = \sum_{i(z_i=3)} K_{I,i}^{pot3/z_I} a_i (aq)$$

Equations 1.48 and 1.49 are self-consistent relationships that are derived directly from ion-exchange equilibrium considerations and are, therefore, more accurate as compared with the empirical so-called Nicolsky–Eisenman equation known as:

$$E_{PB} = E_I^{\circ} + \frac{s}{z} \log \left(a_I (aq) + \sum_J K_{IJ}^{pot} a_J (aq)^{z_I/z_J} \right) \qquad (1.51)$$

and which has been shown to give significantly different predicted potentials depending on which ion is treated as the primary ion [83]. In Figure 1.4, calculated response curves for a monovalent primary ion and interfering ions of different charges are given according to Equation 1.49, and the results are compared with those from the less accurate Equation 1.51. If all ions have the same valency z, however, the following simple relationship is obtained from Equation 1.47:

$$E_{PB} = E_I^{\circ} + \frac{s}{z} \log \left(a_I (aq) + \sum_J K_{IJ}^{pot} a_J (aq) \right) \qquad (1.52)$$

This is known as the Nicolsky equation and has a thermodynamic meaning, in contrast to Equation 1.51.

1.3 IMPROVING THE LOWER DETECTION LIMIT AND SELECTIVITY BEHAVIOR

1.3.1 INTRODUCTION

One of the most spectacular developments during the last decade was the improvement of the LDL of a series of ISEs by up to six orders of magnitude. Simultaneously, many selectivity coefficients turned out to be much better than that determined earlier, and improvements by up to 10 orders of magnitude have

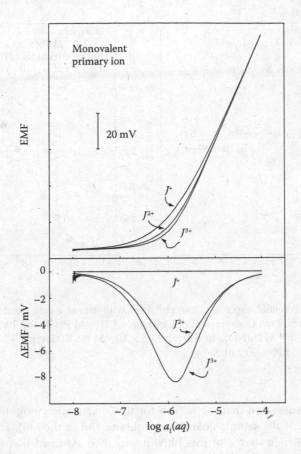

FIGURE 1.4 *Top*: Potential response curves of a membrane selective to a monovalent primary cation, I^+, in the presence of mono-, di-, and trivalent interfering cations, J, according to Equation 1.49. *Bottom*: Difference between the responses predicted by the exact model (Equation 1.49) and the empirical Nicolsky–Eisenman equation (Equation 1.51).

been reported. Surprisingly, these results were achieved by simply eliminating the biases caused by the conventional design of ISEs rather than by synthesizing new ionophores or introducing novel membrane chemistry. In the meantime, the underlying mechanism is well explored, and it has become a kind of routine task to design ISEs having optimal performance.

Several previous observations indicated that the LDL of ISEs was not fully understood. In many early cases, for example, contrary to theoretical expectations, ISEs did not show a response to interfering ions. On the other hand, it was not clear why LDLs and selectivity coefficients were much better when the primary ion activities in the samples were adjusted with ion buffers [90,91]. The observation that optical sensors based on the same ionophores as their potentiometric counterparts had often much better detection limits [92] gave rise to a systematic investigation of the factors defining the LDL of

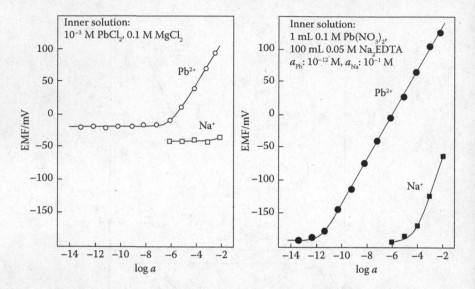

FIGURE 1.5 Potential response of two Pb^{2+}-ISEs with the same membrane but different inner electrolytes. *Left*: Conventional (1:1 mixture of 10^{-3} M $PbCl_2$ and 10^{-1} M $MgCl_2$). *Right*: 1 mL of 10^{-1} M $Pb(NO_3)_2$ in 100 mL of 5×10^{-2} M Na_2EDTA, pH 4.34 (calculated activities: 10^{-12} M Pb^{2+}, 10^{-1} M Na^+).

ISEs. It was suspected that the reason for this discrepancy might be that the compositions of the sample near the membrane and in the bulk are different [92]. As a possible source of this bias, it was then assumed that the activity of the primary ion near the membrane was enhanced due to its flux from the inner solution [93]. Indeed, some improvements of the detection limit of a K^+-ISE were observed with a lower primary ion concentration of the inner solution, which usually was kept rather high according to the traditional recipe. However, the improvement was limited because the beneficial effect of further dilution declined [93]. A spectacular breakthrough was achieved when the primary ion activity of the inner solution of a Pb^{2+}-ISE was buffered to a very low level by using a rather concentrated Na_2EDTA solution (Figure 1.5) [94]. From these observations, it was concluded that ion fluxes across the membranes increased the primary ion activity in the sample layer adjacent to the membrane. This was proven by scanning electrochemical microscopic investigations (Figure 1.6) [95].

Zero-current primary ion fluxes into the sample may be of different mechanistic origins [96]. For the investigated systems, insufficient lipophilicity and chemical instability of relevant membrane components may often be ruled out [97,98], although they may turn out to be relevant for ISE membranes with further improved detection limits. The main cause of ion fluxes in conventional

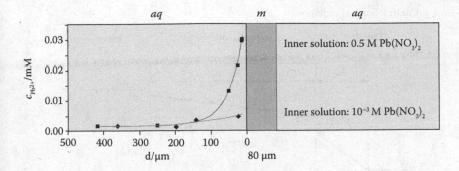

FIGURE 1.6 Scanning electrochemical microscopic investigation of concentration profiles near a Pb^{2+}-selective liquid membrane (m; thickness, 80 μm). With a 0.5 M inner solution, the coextraction of $Pb(NO_3)_2$ generates a significant transmembrane concentration gradient leading to an increased Pb^{2+} concentration in the aqueous sample in the proximity of the membrane.

systems, which had a rather high concentration of primary ions in the inner solution, proved to be the coextraction of primary ions together with their counterions into the membrane [93]. The thus generated concentration gradient across the membrane induced the fluxes of primary ions and their counterions from the inner solution to the sample (Figure 1.7, top). However, zero-current ion fluxes also occur if the inner solution is dilute enough to avoid significant coextraction [93]. Their origin is the exchange of a small amount of primary ions by interfering ones at the sample side of the membrane. When for ions of the same charge ion fluxes are negligible, 50% of primary ions are replaced by interfering ions at the detection limit (see Equation 1.62 below). Since, however, the exchange of <1% of primary ions by interfering ones already generates a significant concentration gradient in the membrane, ionic interferences become relevant even with very weak interferences, i.e., at much lower concentrations of the interfering ions. In such cases, the zero-current ion flux of primary ions into the sample is accompanied by a corresponding counterflux of interfering ions into the inner solution (Figure 1.7, bottom). It must be noted that because of the rather high ionic concentration in the sensing membranes of the order of millimole kilogram^{-1}, such fluxes induced by weak interferences may also be relevant in solid-contact electrodes, which have no inner solution.

1.3.2 LOWER DETECTION LIMIT

According to the 1976 IUPAC definition [99,100], the LDL is given as the ion activity at the cross section of the two extrapolated linear segments of the calibration curve. If one assumes that these linear segments correspond, respectively,

Coextraction and codiffusion

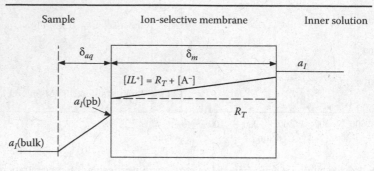

Ion-exchange and counterdiffusion

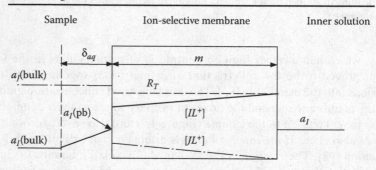

FIGURE 1.7 Two mechanisms of inducing transmembrane concentration gradients in an ISE membrane. *Top*: A rather high concentration of the primary ion (I^+) in the inner solution induces its coextraction together with the counterion (A^-) into the membrane (see also Figure 1.6). *Bottom*: The exchange of a small portion of I^+ by the interfering ion (J^+) reduces its concentration in the membrane on the sample side. The thicknesses of the unstirred Nernstian diffusion layer and the membrane are denoted by δ_m and δ_{aq}, respectively.

to the Nernstian response to I (Equation 1.17) and to the background ion, J (Equation 1.19), the two equations can be subtracted to give:

$$E_I^o + \frac{s}{z_I} \log a_I = E_I^o + \frac{s}{z_J} \log K_{IJ}^{pot\, z_J/z_I}\, a_J \tag{1.53}$$

which is further simplified and solved for the primary ion activity to give the so-called static detection limit:

$$a_I\left(aq, LDL, static\right) = K_{IJ}^{pot} a_J\left(aq\right)^{z_I/z_J} \tag{1.54}$$

Note that the LDL directly depends on the concentration of the interfering ion and the corresponding selectivity coefficient. As shown below, for dilute solutions with $a_I <$ ca. 10^{-6} M, this is only a limiting value that, in general, cannot be reached owing to ion fluxes. In all cases, better membrane selectivities lead to a better LDL. How the LDL depends on the selectivity coefficients and other parameters is discussed next (Section 1.3.3).

Importantly, however, the definition by Equation 1.54 differs from that for analytical methods in general (also by IUPAC), which is based on the standard deviation of the noise multiplied by a factor [101]. Commonly, the value of 3 is used as the multiplier, but based on a careful consideration of type I and type II errors (false positives and false negatives), a factor of 3.29 would be correct (see current IUPAC recommendations) [101]. At the cross section of the two extrapolated linear segments of the calibration curve, the *emf* for monovalent primary ions is higher by 17.8 mV than at zero analyte concentration. This is much higher than the standard deviation of typically 10–50 µV of the noise. Therefore, the detection limit according to the general definition is lower by about two orders of magnitude [102]. This significant difference should be considered if detection limits of various analytical methods are to be compared.

1.3.3 RESPONSE FUNCTION IN THE PRESENCE OF ION FLUXES

Owing to the fast equilibrium at the membrane/sample interface, the response function of an ISE can be described with the phase-boundary potential model even in the presence of ion fluxes [103–107]. From the slightly different approaches, we summarize here the description based on the potentiometric selectivity coefficients. For simplicity, the discussion is limited to ions of the same charge, but solutions are also available for ions of different charges [107].

The phase-boundary potential, E_{PB}, was given above for any monovalent or divalent primary and interfering ions by Equation 1.48, which describes the ISE response as a function of the activity of the interfering ion and the corresponding selectivity coefficient. An alternative description by the phase-boundary potential model makes use of the activity of the primary ion in both phases:

$$E_{PB} = \frac{s}{z_I} \log \frac{k_I a_I (aq)}{a_I (m)} \tag{1.37}$$

After inserting the complex formation constant, Equation 1.37 is rewritten as:

$$E_{PB} = \frac{s}{z_I} \log \frac{k_I a_L^n \beta_{IL_n} a_I (aq)}{a_{IL_n}} \tag{1.55}$$

A combination of Equations 1.48 and 1.55 relates the activity of the primary ion in the membrane to the activities of primary and interfering ions in the sample:

$$\frac{s}{z_I}\log\frac{k_I a_L^n \beta_{IL_n} a_I(aq)}{a_{IL_n}} = E_I^o + \frac{s}{z_I}\log\left[\frac{\frac{1}{2}\sum_{J1} K_{I,J1}^{pot\ 1/z_I} a_{J1}(aq)}{+\sqrt{\left(\frac{1}{2}\sum_{J1} K_{I,J1}^{pot\ 1/z_I} a_{J1}(aq)\right)^2 + \sum_{J2} K_{I,J2}^{pot\ 2/z_I} a_{J2}(aq)}}\right] \quad (1.56)$$

In the absence of interfering ions, Equation 1.48 reduces to the Nernst equation:

$$E_{PB} = E_I^o + \frac{s}{z_I}\log a_I(aq) \quad (1.57)$$

In most cases of practical relevance, one may assume that the primary ion in the membrane exists predominantly in its complexed form, IL_n, and that the activity of the uncomplexed ionophore is constant. In the absence of interference from other ions, therefore, a_{IL_n} in Equation (1.55) can be approximated by the ion-exchanger concentration, R_T, as follows:

$$E_{PB} = \frac{s}{z_I}\log\frac{k_I a_L^n \beta_{IL_n} a_I(aq)}{\gamma_{IL_n} R_T} \quad (1.58)$$

For this special case, the combination of the Equations 1.57 and 1.58 results in:

$$\frac{s}{z_I}\log\frac{k_I a_L^n \beta_{IL_n} a_I(aq)}{\gamma_{IL_n} R_T} = E_I^o + \frac{s}{z_I}\log a_I(aq) \quad (1.59)$$

which is simplified by solving it for E_I^o and inserting the result into Equation 1.56 to obtain:

$$\frac{R_T a_I(aq)}{c_{IL_n}} = \frac{1}{2}\sum_{J1} K_{I,J1}^{pot\ 1/z_I} a_{J1}(aq)$$

$$+ \sqrt{\left(\frac{1}{2}\sum_{J1} K_{I,J1}^{pot\ 1/z_I} a_{J1}(aq)\right)^2 + \sum_{J2} K_{I,J2}^{pot\ 2/z_I} a_{J2}(aq)} \quad (1.60)$$

Equation 1.60 assumes that the activity coefficient for the complexed primary ion remains unchanged during the ion-exchange process with the interfering ion. It allows us to express the primary ion concentration in the membrane as a function of measurable parameters.

If the sample contains only one type of interfering ion of the same charge as the primary ion, Equation 1.60 simplifies to:

$$\frac{R_T}{c_{IL_n}} = \frac{a_I(aq) + K_{IJ}^{pot} a_J(aq)}{a_I(aq)} \tag{1.61}$$

This equation can be used to calculate the fraction of the primary ion replaced by the interfering one in a given sample. For example, according to the 1976 IUPAC definition, the LDL, as the ion activity at the cross section of the two extrapolated linear segments of the calibration curve (see Section 1.3.2) [99], for ions of the same charge ideally corresponds to:

$$a_I(aq, LDL, static) = K_{IJ}^{pot} a_J(aq)^{z_I/z_J} \tag{1.54}$$

Inserting this relationship into Equation 1.61 gives the following concentration of the primary ion complex in the membrane at the static detection limit:

$$c_{IL_n}(LDL) = \frac{R_T}{2} \tag{1.62}$$

In practice, however, Equation 1.62 describes the thermodynamic limit of the LDL, and the concentration of the primary ion in the membrane is in practical situations often higher by several orders of magnitude. If the back side of the membrane is in contact with an aqueous inner solution, the LDL at steady state can be described by considering linear concentration gradients across the respective diffusion layers. Under these circumstances, the ratio of the concentration gradients in the unstirred Nernstian diffusion layer and in the membrane is constant (cf. Figure 1.7):

$$\frac{a_I(aq) - a_{I,bulk}(aq)}{c'_{IL_n} - c_{IL_n}} = q \tag{1.63}$$

Here, Equation 1.63 is written with ion activities in the aqueous phase, which is valid if the activity coefficient, γ_I, remains constant within the Nernst diffusion layer. The proportionality factor, q, depends on the diffusion coefficients and the thicknesses of the diffusion layers in both phases and includes the activity coefficient for the primary ion in the aqueous phase:

$$q = \frac{\gamma_I(aq) D_{org} \delta_{aq}}{D_{aq} \delta_{org}} \tag{1.64}$$

The concentration of the primary ion at the sample side of the membrane is obtained from Equation 1.63 as:

$$c_{IL_n} = c'_{IL_n} - \frac{a_I(aq) - a_{I,bulk}(aq)}{q} \tag{1.65}$$

It can be eliminated by combining this equation with Equation 1.61:

$$c'_{IL_n} - \frac{a_I(aq) - a_{I,bulk}(aq)}{q} = \frac{R_T a_I(aq)}{a_I(aq) + K_{IJ}^{pot} a_J(aq)} \tag{1.66}$$

Equation 1.66 is formulated here for ions of the same charge but can be obtained in complete analogy for ions of any charge by combining Equations 1.44 and 1.65. It describes an implicit function of the activity of the primary ion at the membrane surface, $a_I(aq)$, on the activity of these ions in the bulk of the sample, $a_{I,bulk}(aq)$. The concentration of the primary ions at the inner side of the membrane, $c'_I(org)$, can be obtained from Equation 1.60. To get the ISE response function, $a_I(aq)$ is calculated from Equation 1.66 and inserted into the Nicolsky equation, Equation 1.52:

$$E_I = E_I^o + \frac{s}{z_I} \log \left(\frac{1}{2} \left(\begin{array}{c} a_{I,bulk}(aq) - K_{IJ}^{pot} a_J(aq) + q\left(c'_{IL_n} - R_T\right) \\ \overline{+ \sqrt{\begin{array}{c} 4K_{IJ}^{pot} a_J(aq)\left\{a_{I,bulk}(aq) + qc'_{IL_n}\right\} \\ + \left\{a_{I,bulk}(aq) - K_{IJ}^{pot} a_J(aq) + q\left(c'_{IL_n} - R_T\right)\right\}^2 \end{array}}} \end{array} \right) + K_{IJ}^{pot} a_J(aq) \right) \tag{1.67}$$

From Equation 1.67 with $a_{I,bulk}(aq) = 0$, the detection limit is found as:

$$a_I(aq,LDL) = \frac{1}{2}\left(\begin{array}{c} -K_{IJ}^{pot} a_J(aq) + q\left(c'_{IL_n} - R_T\right) \\ \overline{+ \sqrt{4qc'_{IL_n} K_{IJ}^{pot} a_J(aq) + \left\{K_{IJ}^{pot} a_J(aq) + q\left(R_T - c'_{IL_n}\right)\right\}^2}} \end{array} \right)$$
$$+ K_{IJ}^{pot} a_J(aq) \tag{1.68}$$

Obviously, the kinetic detection limit depends strongly on the composition of the inner solution. Three hypothetical cases will be considered here. In the first one, significant electrolyte coextraction at the inner membrane side is allowed, resulting in $c'_{IL_n} = 2R_T$. For this situation, the LDL is found as:

$$a_I\left(aq,\text{LDL}\right) = \frac{1}{2}\left(\frac{-K_{IJ}^{pot}a_J\left(aq\right)+qR_T}{+\sqrt{8qR_T K_{IJ}^{pot}a_J\left(aq\right)+\left\{K_{IJ}^{pot}a_J\left(aq\right)+qR_T\right\}^2}}\right) + K_{IJ}^{pot}a_J\left(aq\right)$$

(1.69)

If ion fluxes are predominant, $qR_T \gg K_{IJ}^{pot}a_J(aq)$ so that Equation 1.69 simplifies to:

$$a_I\left(aq,\text{LDL}\right) = qR_T$$

(1.70)

Inserting typical values for the diffusion coefficients in the aqueous and membrane phases (10^{-5} and 10^{-8} cm^2 s^{-1}, respectively) as well as for the thicknesses of the membrane (200 µm) and aqueous Nernst diffusion layer (ca. 100 µm), and assuming an ion-exchanger concentration of 10 mM, the LDL is predicted as ca. $10^{-3} \times 0.5 \times 0.01$ M = 5 µM.

In the second case, the inner solution is chosen so as to bring about a significant exchange of the primary ion by an interfering one at the inner membrane side, inducing a strong inward ion flux. For the most extreme case, this is approximated with $c'_{IL_n} = 0$, and the LDL is derived from Equation 1.68 as:

$$a_I\left(aq,\text{LDL},static\right) = K_{IJ}^{pot}a_J\left(aq\right)^{z_I/z_J}$$

(1.54)

While this is the most desired thermodynamic LDL, it should be noted that the ISE exhibits no sensitivity in this activity range owing to a precipitous potential change at a much higher concentration. For this reason, a practically relevant LDL has been proposed as the maximum deviation from the Nernstian response [108], which in this case is much closer to the one calculated according to Equation 1.70.

A robust LDL is found in the third case, in which ion-exchange or electrolyte coextraction equilibria are either negligibly small or perfectly compensated, and $R_T = c'_{IL_n}$ so that Equation 1.68 can be simplified to:

$$a_I\left(aq,\text{LDL}\right) = \frac{1}{2}\left(-K_{IJ}^{pot}a_J\left(aq\right)+\sqrt{4qR_T K_{IJ}^{pot}a_J\left(aq\right)+\left\{K_{IJ}^{pot}a_J\left(aq\right)\right\}^2}\right)$$

(1.71)

If ion fluxes are not relevant, $qR_T \ll K_{IJ}^{pot}a_J\left(aq\right)$, and the thermodynamic LDL is obtained (see Equation 1.54). Conversely, if ion fluxes are dominant, $qR_T \gg K_{IJ}^{pot}a_J\left(aq\right)$ so that Equation 1.71 yields:

$$a_I(aq,\text{LDL}) = \left(\frac{qR_T}{z_I}\sum K_{IJ}^{pot}c_J\right)^{1/2}$$

(1.72)

For any combination of mono- and divalent ions, the more general Equation 1.73 must be applied. It can be seen that the power term is 1/3 if the primary ion is divalent and the dominating interfering ion monovalent, whereas in the opposite case, it is 2/3.

$$\log c_I (aq,\text{LDL}) = \left(\frac{1}{2} + \frac{1}{6}(z_J - z_I) \right) \log \left(\left(\frac{qR_T}{z_I} \right)^{z_I/z_J} \sum K_{IJ}^{pot} c_J^{\; z_I/z_J} \right) \quad (1.73)$$

The relevance of Equations 1.72 and 1.73 is discussed further in more detail.

1.3.4 OPTIMIZING THE INNER SOLUTION

The bias, i.e., the difference between the concentrations of the sample ions in the bulk and near the surface of the membrane is given by the second term of Equation 1.74, which is obtained by combining Equations 1.64 and 1.65:

$$c_I (aq) = c_{I,bulk} (aq) + \frac{D_{org} \delta_{aq}}{D_{aq} \delta_{org}} \left(c'_{IL_n} - c_{IL_n} \right) \quad (1.74)$$

More than ten different methods have been suggested for minimizing this bias [109,110]. One consists in reducing the transmembrane concentration gradient, $c'_{IL_n} - c_{IL_n}$. Because c_{IL_n} depends on the sample composition, the ideal situation of zero concentration difference (see Figure 1.7) is only reached for special combinations of primary and interfering ion concentrations. In all other cases, depending on the sign of $(c'_{IL_n} - c_{IL_n})$, either an inward or an outward flux of the primary ion is induced. While the outward flux deteriorates the LDL, the inward flux induces a super-Nernstian response (see Figure 1.8). In practice, c_{IL_n} is first estimated for a typical sample composition with Equation 1.60 and then, an appropriate composition of the inner solution is calculated so that $c_{IL_n} = c'_{IL_n}$. Depending on the typical sample composition, the optimal inner solution may strongly vary (see Figure 1.9).

Strictly speaking, the outlined model is only valid when steady state is established over the whole membrane, which for a typical PVC membrane would require several hours after each sample change [111]. This is the reason why signals are strongly drifting in the region of strong super-Nernstian response. The increased sensitivity (i.e., the slope of the response curve, cf. Figure 1.8) is, therefore, of no practical utility. For small deviations from the ideal response, the remaining concentration gradients only induce small *emf* drifts so that slightly super-Nernstian responses are acceptable. In analogy to the definition of the LDL, a deviation of $RT/z_I F \times \ln 2$ from the extrapolated linear Nernstian segment of the calibration curve (i.e., $17.8/z_I$ mV at 25°C) was proposed as a practical limit for a tolerable super-Nernstian response (see Figure 1.8) [103].

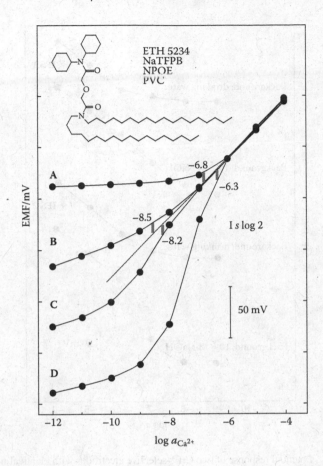

FIGURE 1.8 Potential responses of four Ca^{2+}-selective electrodes with identical membranes but different inner electrolytes: A, 10^{-2} M $CaCl_2$ (conventional); B, 10^{-3} M $CaCl_2$, 5×10^{-2} M Na_2EDTA, pH 5.39; C, 10^{-3} M $CaCl_2$, 5×10^{-2} M Na_2EDTA, pH 6.91; D, 10^{-3} M $CaCl_2$, 5×10^{-2} M Na_2EDTA, pH 8.60. Indicated LDLs according to the extended definition.

In the absence of relevant ion fluxes, the LDL is a simple function of the activity of the interfering ion and the corresponding selectivity coefficient (see Equation 1.54 above). At the LDL, a significant portion of the primary ions is replaced by interfering ions in the membrane (50% for ions of the same charge). If ion fluxes are relevant, the replacement of a much smaller portion of the primary ions is already relevant. Therefore, the required selectivity must be much better than in the absence of significant ion fluxes, i.e., when the detection limit is <ca. 10^{-6} M (for conventional PVC membranes). The lower the detection limit, the smaller must be the ion fluxes, i.e., the portion of the replaced primary ions. Therefore, the LDL in the presence of ion fluxes (dynamic LDL) requires a stronger discrimination of interfering ions and is a more complicated function of the selectivity coefficient. General equations have been derived above

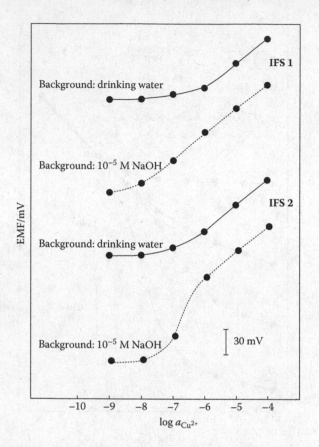

FIGURE 1.9 Potential response of two Cu^{2+}-selective electrodes with identical membranes but two different inner solutions (IFS) consisting of 10^{-4} M $Cu(NO_3)_2$ with 1.9×10^{-4} M Et_4NNO_3 (IFS1) and with 3.5×10^{-4} M Et_4NNO_3 (IFS2). Based on the selectivity coefficients, 28.6% and 46.0% of Cu^{2+} are replaced by Et_4N^+ with IFS1 and IFS2, respectively, on the inner membrane side. The response is optimal with IFS1 if the sample background is 10^{-5} M NaOH (dashed lines), and with IFS2 with a cationic background representative of drinking water (1.3×10^{-3} M Ca^{2+}, 3.1×10^{-4} M Mg^{2+}, 2.2×10^{-4} M Na^+, and 3.3×10^{-5} M K^+; solid line).

for those cases where coextraction at the inner membrane side is negligible, see Equation 1.72 [107]. For ISE membranes, in which neither ion exchange nor coextraction is relevant at their inner side, one obtains for primary and interfering ions of the same charge:

$$\log a_I\left(aq, \text{LDL}\right) = \frac{1}{2}\log\left(\frac{qR_T}{z_I}\sum K_{IL}^{pot} a_J\left(aq\right)\right) \qquad (1.75)$$

This equation shows that, due to ion fluxes, an improvement of the relevant selectivity coefficient by a factor of n decreases the LDL by only $\sqrt{n}$, rather than n as is the case in the absence of ion fluxes. For example, for a typical PVC membrane with $q\,R_T \cdot 10^{-6}$ M, the LDL of a monovalent primary ion is 10^{-7} M when the concentration of a dominant interfering ion, J^+, is 10^{-3} M and $K_{IJ}^{pot} = 10^{-5}$. The selectivity coefficient would have to be improved to 10^{-9} in order to reach an LDL of 10^{-9} M.

The above equations and Equation 1.73 for any monovalent and divalent ions were derived for membranes in which neither ion exchange nor coextraction is relevant at their inner side. The LDL can be improved if the primary ion concentration at the inner membrane side is somewhat reduced by ion exchange with the inner solution so that a just acceptable super-Nernstian response is induced. For ions of the same charge (Equation 1.75), an improvement by 0.38 logarithmic units can be achieved [107]. This relationship is illustrated in Figure 1.10 for an Ag^+-ISE [107]. Calibration curves were taken with a background of 10^{-5} M $LiNO_3$. Based on the relevant selectivity coefficient of $\log K_{AgLi}^{pot} = -7.7$, the static LDL would be $10^{-12.7}$ M. According to Equation 1.75, with $q\,R_T \cdot 10^{-6}$ M, the dynamic detection limit should be ca. $10^{-9.35}$ M, which is in good agreement with the observed calibration curve (Figure 1.10, top). At lower sample concentrations, the calculated replacement of just 0.02% of Ag^+ by Li^+ is sufficient to induce an ion flux that determines the LDL. With an inner solution that brings about a replacement of 1% of Ag^+ by an interfering ion on the inner membrane side, a significant super-Nernstian response is observed along with the estimated improvement of the LDL by ca. 0.4 logarithmic units (Figure 1.10, bottom). The theoretical curves are in rather good agreement with the observed responses. This example shows that small changes in the composition of the inner and sample solutions can strongly affect the responses below nanomolar concentrations.

In another example, the LDL of $10^{-7.4}$ and $10^{-8.7}$ M at pH 6.0 and pH 3.0, respectively, were correctly predicted for I^--ISE with $\log K_{IOH}^{pot} = -1.65$ and $q\,R_T \cdot 7 \times 10^{-6}$ M [98]. The detection limits without ion fluxes would have been $10^{-9.65}$ and $10^{-12.6}$ M. Note that, in contrast to ISEs, the LDL of optode membranes is not influenced by ion fluxes (see next). Indeed, microsphere optodes of analogous membrane composition exhibit the expected LDL of $10^{-11.5}$ M at pH 3.5 [112].

As a further example, the response of a Cd^{2+}-ISE is shown in Figure 1.11 for two different pH values of the sample [106]. In this case, H^+ is the dominating interfering ion with $\log K_{CdH}^{pot} = -3.0$. In perfect agreement with the observed values, LDLs of $10^{-7.4}$ and $10^{-9.8}$ M are predicted with Equation 1.73 if an optimal exchange of primary ions by interfering ones (replacement of 0.05% of Cd^{2+} by Na^+) is assumed at the inner membrane side. Here, EDTA was used as an ion buffer to stabilize the composition of the inner solution. It must be noted that in this case, the composition of the inner solution for achieving the above replacement was not correctly predicted with the equation derived in analogy to Equation 1.60. Instead of the expected 1.9×10^{-5} M free Cd^{2+}, the experimentally optimized inner solution only contained 1.6×10^{-11} M free Cd^{2+}. Such a strong discrepancy, which was also observed in other cases when using EDTA [113], but not in the above examples of Ag^+- and I^--ISEs with an EDTA-free

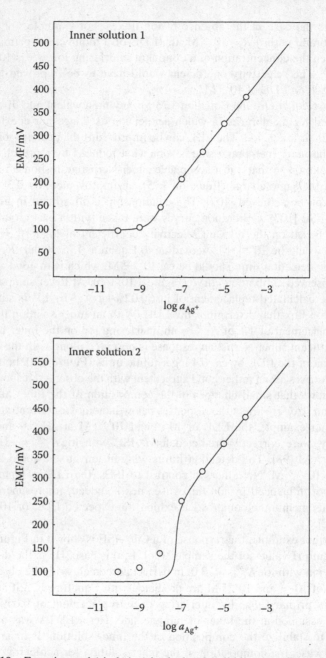

FIGURE 1.10 Experimental (circles) and calculated (solid lines) *emf* responses of Ag^+-selective electrodes with identical membranes but different Li^+-containing inner solutions: 1 (top) gives an optimal LDL, the calculated exchange of Ag^+ by Li^+ on the inner membrane side being only 0.02%, whereas 2 (bottom) induces the exchange of a larger portion (calculated, 1%) of silver ions on the inner membrane side.

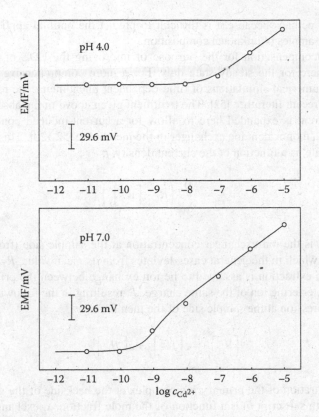

FIGURE 1.11 Experimental (circles) and calculated response (solid lines) of a Cd^{2+}-selective electrode in 10^{-4} M $NaNO_3$ at pH 4.0 (top) and 7.0 (bottom). Here, H^+ is the dominant interfering ion (log $K_{CdH}^{pot} = -3.0$).

inner solution, seems to be due to the favored extraction of EDTA complexes into lipophilic phases [114].

1.3.5 CURRENT COMPENSATION

As an alternative approach to the chemical optimization steps described above, zero-current ion fluxes can also be compensated for by applying an appropriate external current, causing the migration of ions in the opposite direction [115–119]. This has, indeed, yielded spectacular improvements in the LDL for a Ca^{2+}-ISE ($10^{-8.9}$ M) [116] and a Pb^{2+}-ISE ($10^{-11.5}$ M) [115] in transient experiments involving sequential dilutions. A theoretical treatment of steady-state responses in the presence of external currents was based on the phase-boundary potential model and was in good qualitative agreement with the experimental findings [117]. However, despite the enormous promise of this method, the approach has not found broader application, perhaps because it is difficult to predict the optimal applied current required in samples of unknown composition.

Current compensation for the purpose of improving the LDL of an ISE is described here for the steady state only. For a more comprehensive treatment including numerical simulations of time-dependent phenomena, the reader may consult the recent literature [52]. The treatment given above in the absence of an applied current is expanded here to allow for a current-induced concentration gradient of a monovalent ion exchanger in the membrane [52,120]. This gradient, at steady state, is a function of the current density, j:

$$\frac{j}{F} = \frac{2D_m}{\delta_m}\left(R_T - \left[R^-\right]_f\right) \tag{1.76}$$

where $[R^-]_f$ is the ion-exchanger concentration at the sample side (front) of the membrane, which in the general case, deviates from its mean value, R_T. The only mode of ion extraction is assumed to be ion exchange between the primary ion, I^z, and an interfering ion of the same charge, J^z, resulting in the following charge balance expression at the sample side of the membrane:

$$\left[R^-\right]_f = z\left[IL^z\right]_f + z\left[JL^z\right]_f \tag{1.77}$$

The concentration of the primary ion complex at the backside of the membrane (marked with subscript b) is a function of the mole fraction, x, exchanged at the inner side:

$$\left[IL^z\right]_b = x\frac{2R_T - \left[R^-\right]_f}{z} \tag{1.78}$$

The value of x can be calculated from the composition of the contacting inner solution and the selectivity of the membrane in analogy to previous work [120]. The relationship between the concentration gradients of the primary ion in the membrane and the aqueous diffusion layer of the contacting sample is expressed as above in Equation 1.63:

$$q = \frac{a_{I,bulk} - a_I}{\left[IL^z\right]_f - \left[IL^z\right]_b} \tag{1.79}$$

The term $[R^-]_f$ is obtained by solving Equation 1.76; inserting it into Equations 1.61 and 1.78 yields two expressions that are combined with Equation 1.79 to eliminate $[IL^z]_b$ and $[IL^z]_f$. This gives an implicit equation describing a_I as a function of $a_{I,bulk}$:

$$q = \frac{2FD_m\left(a_I - a_{I,bulk}\right)\left(a_I + a_J K_{IJ}^{pot}\right)}{d_m xza_J K_{IJ}^{pot}\left(j + 2F\dfrac{D_m}{d_m}R_T\right) + a_I\left(j + xzj + 2F\dfrac{D_m}{d_m}R_T\left(xz - 1\right)\right)} \qquad (1.80)$$

Equation 1.80 is solved with appropriate software for a_I and inserted into Equation 1.7 to describe the phase-boundary potential change. Changes in the primary ion complex concentration, $[IL^z]_f$, are often sufficiently small to have no appreciable effect on the observed potential (Equation 1.7), but are obtained from Equation 1.79 if needed.

Figure 1.12 shows calculated steady-state chronopotentiometric calibration curves with applied cathodic currents of varying amplitude [52]. The predicted behavior as a function of the applied current mimics the situation achieved by varying the composition of the inner solution [103]. Cathodic currents, for example, will result in an imposed inward cation flux, which might be used to yield optimal detection limits in complete analogy to the chemical optimization of the inner solution. Recently, evaluation of the so-called stir effect has been proposed

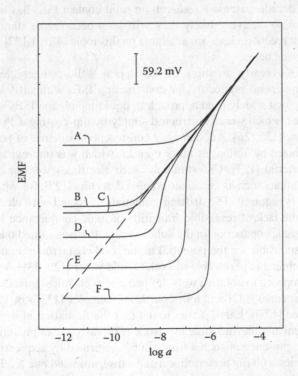

FIGURE 1.12 Calculated steady-state *emf* response of a Ag⁺-selective membrane under galvanostatic polarization assuming a 1% level of coextraction at the inner membrane side. Applied currents: A: 0, B: –6.32, C: –6.38, D: –6.96, E: –12.64, and F: –66.80 nA; membrane area, 0.79 cm².

to find the optimal current, which would allow the approach to be used for measuring samples of unknown composition. Note, however, that the steady-state theory suggests that galvanostatic optimization may not lead to lower detection limits than chemical optimization steps [52].

1.4 SOLID CONTACT ION-SELECTIVE ELECTRODES

For decades, conventional polymeric membrane electrodes have been constructed mainly with liquid inner contact consisting of a rather concentrated solution of the primary ion. In the meantime, it is known that such a solution may lead to leaching of ions into the sample, thus, adversely affecting the LDL and the selectivity behavior of the ISE (see Section 1.3). Nevertheless, liquid contacts are still generally used even with miniaturized sensors, such as for clinical analysis. Often, a hydrogel is applied to stabilize the miniaturized liquid contact, requiring considerable time to rehydrate the inner layer when first using the ISE. In spite of this drawback and the more complicated construction, it is still widely used, obviously because ISEs with solid inner contact have had serious drawbacks.

In the last decade, extensive research on solid-contact ISEs has led to important progresses so that very likely they will soon become the standard design. Since excellent recent reviews are available on this topic [47,121,122], only a brief summary is given here.

The numerous early attempts of developing solid-contact electrodes did not result in a general procedure for constructing ISEs with sufficient stability [123,124]. The first widely used approach to liquid-membrane ISEs were coated-wire electrodes, which were constructed simply by dip-coating a Pt wire with a PVC membrane [125,126]. Another easy construction made use of porous graphite, hydrophobized by Teflon, as inner contact, which was impregnated with the membrane material [127]. Unfortunately, such electrodes were only useful for special applications such as chromatographic detectors [128], for which no long-term stability is required. The drifting potentials of coated-wire electrodes were attributed to the lack of reversible transition of ionic conductance in the membrane to electronic conduction in the solid contact. It was assumed that an oxygen half-cell is responsible for the potential at the solid contact/membrane interface, since the membrane is permeable to both O_2 and water [129–131]. Various inner layers were proposed to obtain a well-defined electrochemical interface including poly(vinyl ferrocene) [132] and a lipophilic Ag complex [133] (for various other approaches, see [123]). Later, it was found that the formation of a thin aqueous film between membrane and inner electrode [134] also causes potential instabilities [135]. The presence of such a film can be confirmed by sequential measurements of samples with the preferred ion and a discriminated one (cf. Figure 1.13). This "water layer test" shows typical drifts, faster when changing to a solution of the interfering ion and slower when going back again to that of the primary ion, which can be attributed to changes in the composition of the aqueous layer due to the transport of the respective ions through the sensing membrane [135]. For

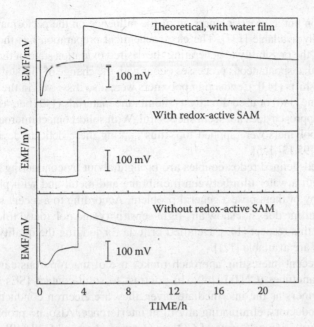

FIGURE 1.13 Potential response of K$^+$-selective solid-contact membrane electrodes, conditioned in 0.1 M KCl, from $t = 0$ for 3.5 h in contact with 0.1 M NaCl, $t = 4$ h again with 0.1 M KCl. Top: Theoretical response assuming the presence of a thin water film between membrane and inner solid contact. Center: Experimental response for a K$^+$-ISE having a lipophilic, redox-active, self-assembled monolayer (SAM) between membrane and inner contact. Bottom: Same membrane directly placed on Au as solid contact. This "water layer test" shows typical potential drifts, which can be attributed to changes in the composition of the aqueous layer between membrane and solid contact owing to transport of the sample ions through the sensing membrane.

membranes with much lower diffusion coefficients such as polyacrylates, similar experiments with CO_2 are useful [136]. Direct evidence for the presence of a 10 ± 1 nm thick water film between a PVC-based ISE membrane and a silicon wafer was provided by De Marco with neutron reflectometry [137]. Apparently, with polyacrylate-based membranes, water pockets but no contiguous water layer are formed [138].

One way of obtaining stable potentials is the generation of self-assembled, redox-active, lipophilic monolayers on the surface of the inner electrode [139–141]. A simpler method made use of conducting polymers. Since their first application as inner contact of a polymeric membrane ISE in 1992 [142], more than 100 papers were published on this topic (for recent reviews, see [47,122]). Numerous polymers have been investigated, mainly in the group of Ivaska, including polypyrroles, poly-thiophenes, the hydrophilic poly(3,4-ethylenedioxythiophene), polyanilines, and others. The solid-contact layer can be generated by drop-casting or by *in situ* elec-trochemical polymerization applying both cyclic voltammetry and galvanostatic electropolymerization. It seems that the different groups had their preferred method of preparation but a direct comparison of the influence on the performance of the ISEs is hardly available [137]. The electrochemical preparation has the apparent advantage of the possibility of generating the desired oxidation state of the polymer, but it seems that spontaneous processes occur with the change of this state and concomitant E^0 shifts [143]. Conducting polymers were also dissolved in the ion-selective membrane [144] or used directly as membrane materials [145] but, as a general recipe, this approach does not seem successful. With solid-contact microelectrodes, conducting polymers were applied too, thus making these delicate devices much more robust [95,131,146].

While well-defined redox couples are brought about by conducting polymers, problems with a water film between membrane and metal and with photosensitivity of many of them pose a general problem. According to a recent systematic study, polyaniline does not show any light sensitivity and poly(octyl thiophene) is the worst in this respect [147]. Detailed criteria for judging the quality of solid-contact ISEs are available [121].

A more recent interesting approach makes use of macroporous carbon [148] or carbon nanotubes (CNTs) [149]. In contrast to solid-contact ISEs with conducting polymers as the intermediate layer, they are electron conductors rather than semiconductors, eliminating any light interference. Also, no problems with water film formation or effects of CO_2 or O_2 were observed [148,149]. Based on studies with electrochemical impedance spectroscopy (EIS) and cyclic voltammetry (CV), it was concluded that CNTs can be described schematically as an asymmetric capacitor where one side is formed by electronic charges (electrons/holes) in the single-walled CNT wall and the other side is formed by ions in the ion-selective membrane [150]. Initially, the preparation of CNT-based contacts was hampered by the limited solubility and low dispersibility in both aqueous and organic solvents. Very recently, Qin et al. showed that multiwall carbon nanotubes (MWCNTs) can be dispersed with the aid of poly(ethylene-co-acrylic acid) in different ISE membrane matrices including plasticized PVC, plasticizer-free methacrylate copolymer, and PVC-ionic liquid [151]. Thus, a general approach is available to prepare all solid-state ion-selective sensors by a simple one-step drop casting method.

The performance of solid-contact ISEs in terms of LDL and selectivity behavior came into the focus of interest only recently [152–154]. At first sight, the absence of an inner solution as a source of leaking of primary ions may be expected to be an advantage. However, having a rather high ionic concentration of typically ca. 10 mM, the membrane itself may be an important source of contamination of the sample. Moreover, it is not possible to design an optimized inner solution that brings about the same degree of ion exchange as the typical target sample. Additionally, the conditioning time also increases considerably.

So far, in terms of LDL and selectivity behavior, solid-contact ISEs have been proved at least to be equivalent to their liquid-contact counterparts [152–154]. The use of membrane material of low diffusivity, such as polyacrylates and polymethacrylates [155], seems to be especially advantageous. Probably, a reduction of the membrane thickness, which, today, typically is in the 100 μm range, will help to improve the performance. It is expected that in the coming years, the solid-contact in ISEs will replace the currently used liquid-contact as the typical design.

1.5 MAKING USE OF ION FLUXES

Zero-current counterdiffusion and codiffusion processes are undesired in view of reaching optimal LDLs, but may give rise to useful applications that place ISEs closer to the realm of dynamic electrochemistry. These concepts are of quite recent origin and may drastically expand the utility of membrane electrodes. Some examples given here may inspire the field to look beyond the classical direct potentiometry for a number of very promising applications. They include the determination of unbiased selectivity coefficients, reliable polyion sensors, better endpoints in potentiometric titrations, the measurement of total ion concentrations, high amplitude sensing, backside-calibration potentiometry, and constant-current as well as controlled-potential coulometry.

As discussed in Section 1.3, ion fluxes may be forced upon an ISE membrane by chemical or electrochemical means. The chemical adjustment of such fluxes under zero-current conditions has been described above in sufficient detail, but the electrochemical treatment merits some discussion. In the case of ion conductors such as ISE membranes, an applied current, j, leads to a defined total net flux of ions, J_T, across the membrane, which is the sum of all individual ion fluxes, J_k:

$$j = zFA \sum_k J_k = zFAJ_T \qquad (1.81)$$

The galvanostatic control of ion fluxes can be done on ion sensing membranes of conventional composition, for example, to counteract ion fluxes of chemical origin. Alternatively, such a galvanostatic flux adjustment can also be used to control the ion extraction process in its entirety, with ion-selective membranes that ideally possess no ion-exchanger properties. Figure 1.14 shows the measurement sequence of such a pulstrode proposed recently [156]. In an initial galvanostatic pulse, ions are extracted from the sample into the ion-selective membrane. The potential may, in principle, be measured at a specified time during this initial pulse and used as sensor output. The undesired potential drop associated with the solution resistance is avoided by switching to open-circuit potential measurement during a second pulse. The resulting potentials are a direct measure of the ion activities at either side of the sample/membrane phase boundary and are transient because of the continuously changing concentration profiles in this dynamic experiment. For this reason, the sequence ends with a potential pulse to expel the hydrophilic ions back into the aqueous solutions and to prepare the membrane for the next pulse sequence.

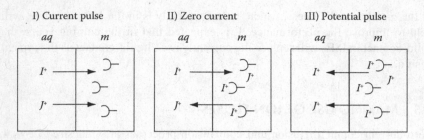

FIGURE 1.14 Measurement sequence with a pulstrode. I) By the initial galvanostatic pulse, primary (I^+) and interfering (J^+) ions are extracted from the aqueous sample (aq) into the ion-selective membrane (m). II) The open-circuit potential during the second pulse is applied to eliminate the potential drop associated with the solution resistance (so-called iR drop). III) Hydrophilic ions are expelled from the membrane back into the aqueous solution in the regeneration step, in which the membrane is held at the baseline potential.

The principles of such pulstrodes exhibit an important resemblance to those of ion transfer voltammetry [157–162]. However, there are some clear differences. Polymeric membrane materials are essentially the same as those of their potentiometric counterparts, making them relatively robust devices for practical use. By contrast, ion transfer voltammetry, for the most part, still works with simple organic solvents such as 1,2-dichloroethane, which are a reminiscence of the early days in ISE research. However, using more practical, rugged polymeric materials brings about significant differences in the functioning of such sensors since the diffusion coefficients in the membrane phase are smaller by orders of magnitude. As a consequence, the rate-limiting step in the transfer process of ions is often given by their diffusion in the membrane, and not in the aqueous phase as in classical ion transfer voltammetry [117,163–165]. This different response mechanism often gives different sensor characteristics for the two classes of sensors [156].

1.5.1 SELECTIVITY MEASUREMENTS

For many decades, ISE membranes were characterized after conditioning in a solution of the analyte ion by recording calibration curves either in separate ion solutions (so-called separate solution method) or for the analyte ion in solutions containing interfering ions as a background (so-called fixed interference method) [99]. As explained in Section 1.3, an outward ion flux of analyte ions may result in an incomplete ion exchange with interfering ions of interest if the membrane selectivity is high. Instead, at low analyte concentrations, the electrode is sensitive to the concentration of expelled analyte ions from the membrane. The bias occurring in such selectivity determinations is highly significant and can reach 10 orders of magnitude. It can be eliminated by utilizing ISE membranes that exhibit an inward ion flux. This flux does not influence the sensor response above a critical concentration (typically $>10^{-5}$ M), but eliminates the bias because it siphons off small amounts of analyte ions in the direction of the inner solution. This procedure has been successfully applied for the characterization of a number of ISE membranes, e.g., for detecting Pb^{2+} and Ca^{2+} [166]. Galvanostatically treated ion-selective membranes can serve exactly the same purpose [167]. Figure 1.15 shows separate calibration curves for a PVC–NPOE membrane containing only the inert lipophilic salt ETH 500, responding to sample anions of decreasing lipophilicity, from ClO_4^- to Cl^-. The potential responses upon imposing anodic currents show near-Nernstian electrode slopes. Prior exposure to ClO_4^- solutions has no influence on measurements in samples containing less lipophilic ions, and the resulting selectivity coefficients are not biased.

1.5.2 POLYION SENSORS

Potentiometric polyion sensors are the earliest examples of useful sensors that rely on strong ion fluxes. Pioneered by the groups of Meyerhoff and Yang for the detection of the polyanionic anticoagulant, heparin, and its polycationic

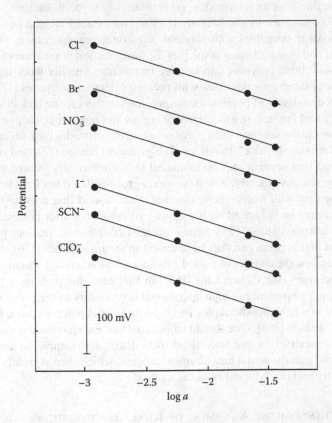

FIGURE 1.15 Pulsed chronopotentiometric calibration curves for a PVC–NPOE membrane containing only the inert lipophilic salt, ETH 500, measured in separate solutions of anions following the order of decreasing lipophilicity (from ClO_4^- to Cl^-).

antidote, protamine, this class of sensors initially had rather unlikely prospects of success [168,169]. Such polyions are rather hydrophilic, bear a very high total charge (up to −70), and must be reliably measured in blood, which is a complex sample containing a very large excess of other electrolytes. However, the high polyion charge would render the observed sensitivity of the measurements analytically useless since large values of z reduce the Nernstian slope. This dilemma was solved for heparin sensors, quite serendipitously, with membranes that contained the chloride form of the optimized anion exchanger. It turned out that the extraction of such polyions by ion exchange into properly formulated hydrophobic matrices, such as plasticized PVC, is thermodynamically favored in the presence of physiological NaCl concentrations. This resulted in a strong inward flux of any extracted polyion heparin, leading to significantly super-Nernstian response slopes owing to the analyte being depleted from the aqueous diffusion layer near the membrane [170].

Much progress has been made since the early understanding of this response mechanism. Selective and accurate heparin assays are now preferably performed with protamine-sensing electrodes via protamine–heparin titration endpoint analyses. Recent work by the Meyerhoff group introduced elegant potentiometric immunoassay principles with polyions–substrate conjugates using such electrodes [171] and explored assay principles for the detection of proteases [172], i.e., enzymes that cleave polyions into smaller fragments. Another work introduced the concept of dendrimer detection with polyion-selective electrodes [173].

Two key drawbacks of potentiometric polyion sensors are the lack of chemical reversibility and the inability to control or trigger the polyion extraction process. Voltammetric polyion sensors and polyion-selective pulstrodes have recently been introduced to address these limitations. The groups of Samec [174] and Amemiya [175] showed that heparin may be extracted voltammetrically into properly formulated ion-selective membranes. Amemiya's group utilized ion transfer voltammetry principles with micropipette electrodes and found that polyion detection sensitivities may be enhanced with stripping voltammetry after prior accumulation and adsorption steps [175]. Samec utilized coulometric readout principles and showed that heparin can thus be assessed in serum samples [176]. As one of its first examples, the concept of pulsed galvanostatic ion sensing was also applied to detect polyions (see Figure 1.16) [165]. In this case, the prolonged potentiostatic stripping step ensures a high operational reversibility of the polyion sensor, and heparin in whole blood samples was determined via potentiometric heparin–protamine titration [165]. One should emphasize that the electrochemical control of the polyion extraction step also allows us to distinguish sample incubation and mixing steps from the actual measurement process, which should greatly improve the reliability and reproducibility of the method.

1.5.3 POTENTIOMETRIC ASSESSMENT OF TOTAL CONCENTRATIONS

Potentiometry with ISEs is rather unique because it may be considered as a nonperturbing technique that provides information on ion activities. For this reason,

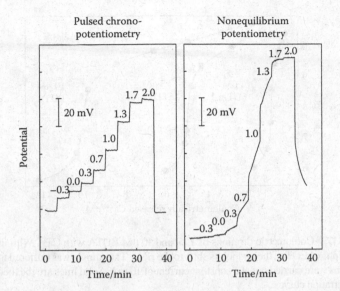

Pulsed chrono-
potentiometry

Nonequilibrium
potentiometry

FIGURE 1.16 Calibration curves for protamine in 0.1 M NaCl obtained by pulsed galvanostatic control (left) and with a conventional protamine ISE (right).

it may help answering questions about sample speciation and chemical driving forces associated with electrolytes. More often than not, however, for the same sample, information on the total concentration is also desired. This is frequently obtained by titrimetric techniques using potentiometric detectors, e.g., to determine the total acidity in addition to the pH of the analyte solution. Since extremely accurate volumetric titrations are labor-intensive and require consumables and standard solutions, new concepts have been explored in recent years to make titrations as simple and convenient as direct pH measurements.

Without involving ionophore-based sensing membranes, de Rooij and coworkers utilized Pt electrodes to coulometrically deliver a pulse of H_3O^+ in a localized sample environment, coupled with potentiometric pH measurements at a specified distance [177]. The observed diffusion time of the pH perturbation was found to be a function of the buffer capacity of the sample. Drawbacks of this approach are the relatively poor selectivity of the H_3O^+ generation and measurement processes used, the dependence of the signal on the diffusion coefficients in the sample, which can vary, and the required engineering to accurately machine and space all electrochemical elements in a confined volume.

One may easily envision the accurate delivery of ionic reagents from an ion-selective membrane by current control. If spontaneous zero-current ion fluxes are negligible, an imposed current pulse will deliver, with a very high degree of selectivity, a defined quantity of reagents into the sample without any need for titrimetric standards or calibration procedures. This concept has, indeed, just recently been demonstrated with a number of polymeric ion-selective membranes for the purpose of accurately delivering reagents [178]. Figure 1.17 demonstrates

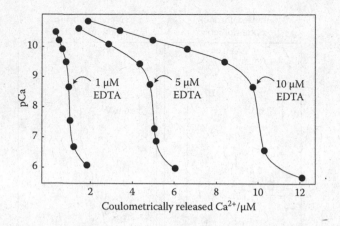

FIGURE 1.17 Coulometric titrations of 1, 5, and 10 μM EDTA with Ca^{2+}. With increasing concentrations of EDTA, the endpoints shift to the right. The sample was buffered to pH 10.0 and titrations were carried out at a constant current of 10 mA. Solid lines are the theoretically predicted titration curves.

the coulometric delivery of Ca^{2+} for the titration of EDTA. Here, one ion-selective membrane was electrochemically controlled to deliver Ca^{2+}, and another one of similar composition was used as the detector electrode. The simple relationship between the applied number of coulombs and the expelled amount of ions (Faraday's law) coupled with the high ion selectivity of such membranes makes this a very promising technique.

Ion-selective membranes can also be used to perform localized titrations with a built-in detection step, without the need for separate electrodes to deliver or remove reagents and detect the endpoint, as discussed above. In one example, a Ca^{2+}-selective membrane devised to exhibit an inward ion flux was shown to give information on total rather than free ion concentrations in the sample [179]. In the example of EDTA-buffered Ca^{2+} solutions, the dissociation of the Ca^{2+}–EDTA complex helped to maintain a sufficiently high flux of Ca^{2+} in the direction of the membrane, thus reducing its depletion near the membrane surface. With increasing Ca^{2+} buffer capacity of the sample, this caused response functions to deviate less from Nernstian slopes. In this early example, however, ion fluxes were difficult to adjust and the resulting electrode response as a function of the total ion concentration was difficult to quantify [179]. Later, pulsed galvanostatically controlled Ca^{2+}-selective electrodes were used for the same purpose [180]. Here, the ion fluxes were adjusted electrochemically, which should improve accuracy and repeatability over a classical potentiometric readout. More recently, such pulstrodes have been used to more accurately monitor the total ion concentrations, by plotting the differential potential as a function of the applied current density or at a fixed current as a function of pulse time [181]. In the latter example, shown in Figure 1.18 for assessing the total acidity with a pH electrode, chemically selective, localized titrations are possible within a few seconds, without the need for

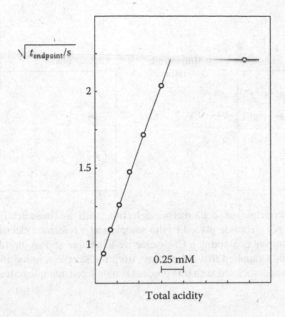

FIGURE 1.18 Determination of total acidity by chronopotentiometric flash titrations at polymer membrane ISEs: Observed linear relationship of the square root of titration end point as a function of total acidity.

microfluidics or complex electronics. Note that similar concepts have also been reported with metal electrodes [182], although the underlying chemical selectivity is certainly inferior to that of ion-selective membranes.

Controlled-current coulometry was described above to accurately deliver reactants from ion-selective membranes. Recently, controlled-potential coulometry was introduced by the group of Kihara as a very promising calibration-free analysis method in the field of ion sensing [183,184]. The original setup is illustrated in Figure 1.19. A suitable potential applied between a Ag/AgCl electrode placed in the sample and a reference electrode in contact with an organic solvent containing a Ca^{2+}-selective ionophore and an inert lipophilic salt drives Ca^{2+} from the sample solution into the organic solvent. The current associated with the process is monitored and integrated over the entire analysis time. This coulometric analysis is possible if most Ca^{2+} are selectively transferred and non-Faradaic processes (charging currents) are kept small. Kihara's group demonstrated the selective transfer of K^+, Ca^{2+}, and Mg^{2+} by this approach [183]. To make this promising technique even more practical for routine use, the system will need to be drastically miniaturized to reduce analysis times and sample volumes, the direct contact of Ag/AgCl with the sample should be avoided and, perhaps most importantly, the solvent used, 1,2-dichloroethane, must be replaced by a more robust and inert sensing material. Coulometry might be the key solution to many of the stability problems that plague the field of ion sensing in nontraditional applications today.

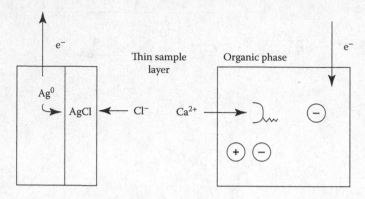

FIGURE 1.19 Principle of coulometric detection with an ion-selective membrane. Between a Ag/AgCl electrode placed in the sample and a reference electrode in contact with an organic solvent containing a Ca^{2+}-selective ionophore and an inert lipophilic salt, a suitable potential is applied that drives Ca^{2+} from the sample solution into the organic solvent. The current associated with this process is monitored and integrated over the time of analysis.

1.5.4 STEPTRODES: SENSORS WITH ENHANCED SENSITIVITY

Conventionally, ISEs are measured against a reference electrode placed in the same sample, and the potentiometric response ideally follows the Nernst equation. While this assembly gives thermodynamically well-defined readouts, it has numerous drawbacks. Reference electrodes have not changed much in the past hundred or so years and normally require a bridge electrolyte placed between the sample and, typically, a Ag/AgCl electrode. The associated liquid junction is prone to clogging in real-world samples, which can happen even during prolonged periods of time in dilute solutions (AgCl has a higher solubility in concentrated chloride solutions because of the formation of higher chloride complexes). Besides, the electrolyte reservoir makes reference electrodes the stumbling block in miniaturization. Moreover, the Nernst equation itself often gives too small potential responses for a given concentration change. A 10% ion activity increase, for instance, gives a mere 1.2 mV change in the *emf* of a Ca^{2+}-selective electrode. This places enormous demands on temperature control and stabilities of reference and indicator electrodes and requires frequent recalibration procedures. All these demands have been met in routine clinical analysis [33,185], but numerous analyses have to be done outside of such a controlled laboratory environment, where instabilities in potential readings make reliable measurements very difficult. Note, for example, that *in vivo* chemical sensing probes are normally used to observe large swings of concentrations that are purposely induced in short experiment times [186]. In these cases, the sensor signal is compared with that before the perturbation step. Prolonged reliable monitoring of small concentration changes is often not possible.

Recent work has attempted to address the above-mentioned limitations. If the ISEs used exhibit an ion flux from the sample to the membrane, localized analyte depletion may occur near the membrane surface, giving a much larger potential response than that predicted on the basis of the Nernst equation. If such an ISE is measured against a similarly configured electrode exhibiting a slightly different inward flux, a very large bell-shaped concentration-dependent electrochemical response is observed. Figure 1.20 illustrates this concept with two Cu^{2+}-selective electrodes measured against each other [187]. The observed potential swing reaches >100 mV for a change in the logarithmic concentration of 0.5, translating to a sensitivity increase by a factor of 7 relative to that predicted by the Nernst equation. In addition, no conventional reference electrode is necessary. Note that such sensor responses are no longer a simple function of the analyte ion activity in the sample but also depend on the kinetics of mass transport and, if applicable, of dissociation.

Potentiometric sensors are widely used as endpoint indicators in titrimetric analysis. Indeed, the sometimes extremely wide concentration range in which they respond to ion activity changes makes them highly suitable for this task, especially if they possess sufficient selectivity. The expected potential change at the endpoint is ideally calculated by the Nernst equation and given by the ion

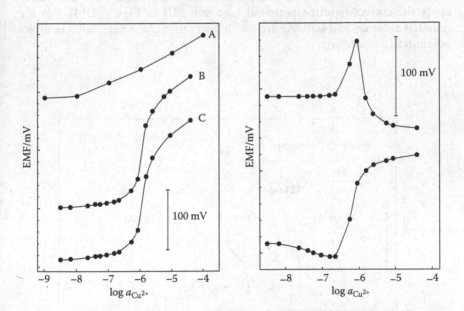

FIGURE 1.20 Concept of steptrodes. Left: Response of three Cu^{2+}-selective electrodes based on a conventional membrane (A) and two membranes with ca. fivefold and 10-fold concentrations of lipophilic sites and ionophore (B and C, respectively). The inner solutions of ISEs B and C had a buffered low activity of Cu^{2+} and a high one of K^+. Right: Differences in the responses of ISEs B and C (top) and of A and C (bottom).

activity change. If ISE membranes that exhibit super-Nernstian response slopes are used, the expected endpoint may be much larger. This was demonstrated with Pb^{2+}-selective membranes with an inward ion flux [188]. In a titration of 10 μM EDTA with Pb^{2+}, the potential change at the endpoint was found to be as much as three times larger than calculated on the basis of the Nernst equation. Note that with more dilute samples, the inward flux may give erroneous endpoints since some of the added reagents are locally consumed by the membrane. The magnitude of this bias depends on the transmembrane ion flux and is adjustable.

More recently, the concept of pulstrodes discussed above was also applied to high-amplitude chemical sensing. Since the ion flux can be adjusted instrumentally, the differential potential is obtained by comparing subsequent pulses of different amplitudes on the same sensor [189]. Alternatively, it was shown that one can sample the potentials at different pulse times and utilize their differences as the sensor output. Since current and pulse times can be varied, the resulting bell-shaped potential response is shifted over a wide concentration range. Figure 1.21 illustrates the measurement of Ca^{2+} at concentration levels mimicking those of 10-fold diluted blood samples, with sensitivities that are about 20 times higher than traditionally given in potentiometry [189]. Note that some potential control is still required to accurately apply the correct baseline potential (see pulse III in Figure 1.14), but the required accuracy and stability are much lower than those demanded in direct potentiometric sensing.

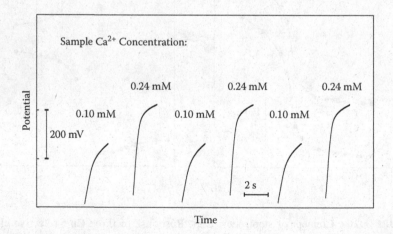

FIGURE 1.21   Reproducibility of the pulstrode response in Ca^{2+} solutions with 1 mM NaCl as background. Each current pulse sequence at –10 μA and at zero current is followed by a 60-s baseline potential pulse, (for clarity, not shown here). The sensitivity achieved is about 20 times higher than by conventional potentiometry.

1.5.5 Response Behavior of Membranes with Fast Ion Fluxes: Backside Calibration

Among the many new sensing concepts involving ISEs introduced in the past few years, backside calibration potentiometry is rather unique and potentially very promising [190]. Rather than relying on the magnitude of the observed potential (or potential difference for the concepts just discussed), one may choose a completely different readout principle. In backside calibration potentiometry, as the name implies, the sample side of the sensor membrane is never altered for calibration purposes, and the reference electrodes, in some cases, are unimportant as well. The concept, illustrated in Figure 1.22, utilizes relatively thin supported ISE membranes across which steady-state concentration gradients build up in a matter of seconds. In this experiment, the composition of the inner solution is altered until the concentration gradient across the membrane is reduced to zero (comparable to the zeroing of a Wheatstone bridge). This is accomplished by changing the stirring rate in the two aqueous solutions and monitoring the potential: once the ion concentration gradient disappears, the influence of stirring on the *emf* disappears too. What makes this technique unique is that the magnitude of the potential is not important. Of course, the technique alone does not allow us to determine single ion activities, which is thermodynamically impossible. Instead, it has been shown that the activity ratio of the analyte ion to its dominant interferent is determined because the concentration gradients are dictated by ion-exchange equilibrium processes. If the activity of the interferent is known, the analyte ion activity can be determined. Figure 1.23 shows how this concept was used to assess Pb^{2+} activities in samples buffered at pH 4.0, with H_3O^+ being the dominant interferent [120,190]. Indeed, the Pb^{2+} activities found by this technique in environmental water samples corresponded well with independently measured values.

1.6 IONOPHORE COMPLEX FORMATION CONSTANTS

1.6.1 Introduction

For proper functioning, polymer membrane ISEs and their voltammetric and optical counterparts rely on a selective extraction step. The underlying sensing phase usually shows a selectivity sequence in line with the lipophilicity of the ions, i.e., more lipophilic ions are preferred over less lipophilic ones. Selective complexation is required to achieve a selectivity sequence that differs from this so-called Hofmeister series. Formation constants for ion-ionophore complexes, β_{iLn}, determined in sensing phases are at the heart of thoroughly understanding these sensors. As shown above, they dictate the most important sensor characteristics, including its selectivity, upper and LDLs, and, partly, its response time.

A large number of highly selective ionophores have been reported over the last four decades [31,108], and many of the best ones are also commercially available. Since the methods of determining complex formation constants are more recent, for many ionophores only selectivity data are available. Early selectivity

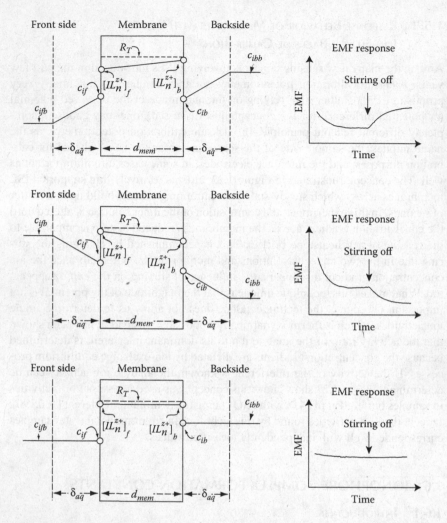

FIGURE 1.22 Schematic representation of backside-calibration potentiometry. *Top* and *center*, *left panels*: The different concentrations of the primary ion, c_i, in the aqueous phases at both sides of the membrane ($c_{ifb} \neq c_{if}$ and $c_{ibb} \neq c_{ib}$) induce a concentration polarization of the complex, $[IL_n^{z+}]$, in the membrane phase; *right panels*: The stirring effect on the *emf* response depends on the sign of the concentration gradient in the membrane. *Bottom*: A symmetrical situation of concentrations in the membrane causes the stirring effect to disappear.

coefficients were often biased by ion fluxes so that, in reality, they are often much better than the published values [191]. A recent summary of complex formation constants of relevant ionophores is given in Ref. [192].

An important trend is the covalent binding of ionophores to the polymer backbone. Initially, the main motivation was the increase in lifetime of the ISEs [193,194], but more recently the reduction of ion fluxes and thus the beneficial

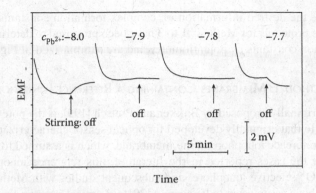

FIGURE 1.23 Application of backside-calibration potentiometry. Time curves of the *emf* response for determining Pb²⁺ activities at a constant level of the dominant interfering ion, H_3O^+ (pH adjusted to 4.0) in a sample of Warsaw tap water. Labels show the logarithmic Pb²⁺ concentrations in the calibration solution on the front side of the Pb²⁺-selective Celgard membrane. Arrows mark the time when stirring was switched off.

influence of immobilization on the LDL was demonstrated [195–197]. The same effect can be achieved if the ionophore is attached to nanoparticles instead of the polymer backbone [198].

ISEs are, in principle, ideal candidates to assess complex formation constants in the membrane phase. Consider again Equation 1.3 for the phase-boundary potential:

$$E_{PB} = \frac{RT}{z_i F} \ln k_i + \frac{RT}{z_i F} \ln \frac{a_i(aq)}{a_i(m)} \qquad (1.3)$$

In established applications, ISEs are utilized to trace changes in the ion activity of the aqueous sample for the purpose of determining complex formation constants, e.g., in chemical titrations. Equation 1.3 makes clear that ISEs can provide the same type of information about the sensing phase, i.e., as changes in $a_i(m)$. So, why have ISEs not traditionally been used for this purpose? With ISE membranes, the overall membrane potential, E_M, is also a function of the phase-boundary potential at the inner solution. For equal compositions of sample and inner solutions, this potential may be described as:

$$E_M = \frac{RT}{z_i F} \ln \frac{a_i(m)'}{a_i(m)} \qquad (1.82)$$

Usually, ISEs are built symmetrically, hence Equation 1.82 predicts a potential change of zero. Consequently, experiments must be designed to either decouple the two activities in a defined fashion or to change the activity at one sample side so as to give the desired information on complex formation constants. The following three protocols for Methods 1 to 3 have been proposed to determine complex formation constants by potentiometry and are summarized in Figure 1.24.

1.6.2 METHOD 1: MEMBRANES CONTAINING A REFERENCE IONOPHORE

Method 1, originally proposed by Bakker and Pretsch [199], is the potentiometric counterpart to that originally developed for optical sensor characterization [200]. It utilizes a reference ionophore in the membrane, which is assumed to bind only one ion. For the cases reported in the literature, this reference ionophore was always a H_3O^+-selective ionophore, and subsequent studies with Method 3 confirmed the above assumption to be correct [201]. In optical sensors, the reference

Method 1: Reference ionophore

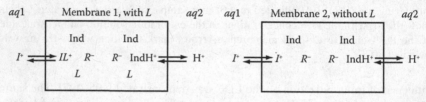

Method 2: Reference ion

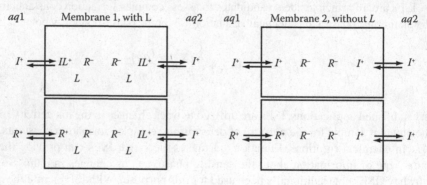

Method 3: Sandwich membrane

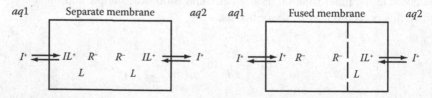

FIGURE 1.24 Protocols of the three different potentiometric methods for determining complex formation constants in the membrane phase (L: ionophore, Ind: H^+-selective ionophore; cf. text).

ionophore also acts as the optical reporter molecule (chromoionophore) to follow the ion-exchange equilibrium between H_3O^+ and a competing ion of interest into a membrane that also contains a lipophilic ion exchanger. If the experiment is repeated under the same conditions but with an added ionophore selective for this competing ion in the optical sensing film, the extraction properties are drastically shifted to lower activities of the competing ion. This information is used to determine the complex formation constant of the ionophore, assuming that ion pair formation between complexed or uncomplexed ions with the lipophilic ion exchanger is of similar strength.

The analogous technique with ISEs evaluates the LDL of the pH response functions of similarly devised membranes. If an ionophore for the competing ion is present, the LDL must be higher, and this information is used to determine the complex formation constant. The experiment may be performed with an inner solution whose composition keeps the inner membrane side responsive to pH in both experiments (i.e., equal values of $a_i(m)'$ in Equation 1.82). The membrane potentials below the LDL, where the outer side of the membrane only responds to the activity of the competing ion, are then effectively decoupled from the chemical processes occurring at the inner membrane side. This method has been mainly explored to characterize the complex formation of alkali and alkaline earth metal ions [199]. A key drawback of this method, in some cases, may be the requirement of having to change the sample pH over a wide range, which is not practicable due to the limited solubility of some ions.

1.6.3 METHOD 2: USING A REFERENCE ION IN THE SAMPLE

The group of Pretsch introduced an attractive method to estimate apparent complex formation constants [202]. It is based on the determination of selectivity coefficients as with Method 1, but avoids the use of a reference ionophore. Instead, a lipophilic and bulky reference ion in the sample, such as Et_4N^+, is chosen, which is considered not to interact with the ionophore. Selectivity coefficients are determined by recording separate calibration curves for all ions of interest, including the reference ion, using membranes with and without the ionophore. The change in the selectivity coefficients with respect to the reference ion for the two membranes is a direct measure of the complex formation constant.

The strength of Method 2 lies in its experimental simplicity and in the fact that off-the-shelf ISEs can be used for the determination. The assumption that the reference ion does not interact with the ionophore can, in principle, be verified by applying Method 3 described next, which, if necessary, may be used for correction purposes. This method is especially attractive to characterize ionophores for transition metal ions, since the pH of the sample can be kept constant. The protocol also yields unbiased selectivity coefficients as part of the same experiment. It has been recently used in screening efforts to identify promising ionophores for the detection of Pb^{2+} [202] and Ag^+ [203].

Figure 1.25 shows the separate calibration curves obtained for PVC–NPOE membranes with two different ionophores, 1,3-alt-5,11,17,23-tetra-*tert*-butyl-25,27-di-*n*-

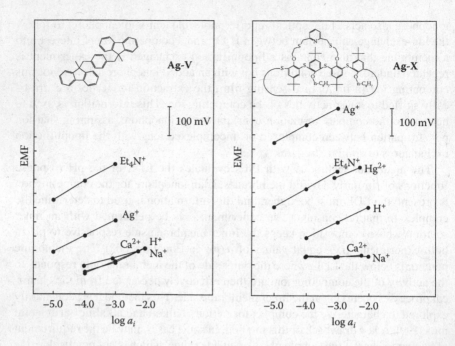

FIGURE 1.25 Potentiometric response behavior of two Ag^+-selective ionophores, Ag-V and Ag-X, used for determining their complex formation constants with the reference ion method (see Figure 1.24).

octyloxy-26,28-[pyridine-2,6-bis(methylthioethoxy)]thiacalix[4]arene (Ag-X) and 1,4-bis[(9-methyl-9*H*-fluoren-9-yl)methyl]benzene (Ag-V), using Et_4N^+ as the reference ion [203]. Note that with Ag-X, Et_4N^+ gives much lower potential readings relative to Ag^+, suggesting a larger complex formation constant. The selectivity coefficient for the analogous ionophore-free membrane was determined as $\log K_{AgNH_4}^{pot} = 5.06 \pm 0.01$ ($n = 3$). For the two examples given in Figure 1.25, the complex formation constants were determined as $\log \beta_{AgL} = 3.89 \pm 0.07$ (for Ag-V) and 11.00 ± 0.03 (for Ag-X), assuming a 1:1 complex stoichiometry. Aromatic π-bond forming ionophores such as Ag-V appear to interact too weakly with Ag^+, compared with thioether-containing calixarenes, to be useful in ISE membranes [113].

1.6.4   METHOD 3: THE SEGMENTED SANDWICH MEMBRANE METHOD

As early as 1984, researchers from St. Petersburg, Russia, proposed an interesting method to decouple the two phase-boundary potentials of an ISE membrane (cf. Equation 1.82) in order to gain information on complex formation constants, complex stoichiometries, and ionophore diffusion coefficients in the membrane phase [204,205]. Their work was much ahead of its time; the publications, written in Russian, were not noticed until some 15 years later, when the method was again described by the original authors [206] and enthusiastically reevaluated and endorsed by Bakker [207]. It uses a concentration-polarized membrane to decouple the two phase boundaries from each other. This is accomplished by fusing together two differently formulated membrane segments, each of known composition. The membrane potential is determined before and after assembling the sandwich membrane. Complex formation constants can then be determined by using two similarly configured membrane segments, only one of which contains the ionophore under study. The resulting activity ratio gives a nonzero membrane potential (see Equation 1.82), and is used to determine the complex formation constant of the ionophore. Method 3 assumes that the polarized interface between the two membrane segments does not disturb the measurement, and experimental care must be taken to avoid a water film to form between the two fused segments. The method also assumes that membrane diffusion potentials are irrelevant. This was confirmed for valinomycin by showing that the addition of excess inert lipophilic salt, which suppresses diffusion potentials, gives the same results [207]. As with the other two methods described above, Method 3 yields apparent complex formation constants if information on ion pair formation and relevant activity coefficients is not known.

Figure 1.26 illustrates the large potential change observed upon forming a membrane with only one side containing the ionophore valinomycin, but both sides containing a lipophilic ion exchanger in PVC–DOS [207]. Note that the potential of the fused membrane remains stable for about 20 min, after which time the diffusion of ionophore across the sandwich membrane starts to become

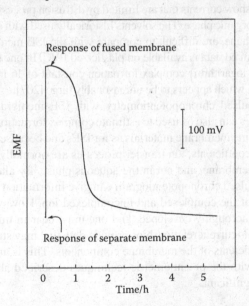

FIGURE 1.26 Potential change observed upon fusing two membranes, one with and the other without valinomycin as ionophore. Both sides contain a lipophilic ion exchanger in PVC–DOS. Owing to the diffusion of valinomycin into the ionophore-free part of the double membrane, the potential starts drifting back to the original value after about 20 min.

noticeable. The complex formation constants must be determined within this lag time, which depends on the membrane thickness and diffusion coefficients and requires some experimental practice.

The method is not limited to determining complex formation constants. It gives information on diffusion coefficients by monitoring the relaxation time of concentration-polarized membranes. It can also be used to determine the extraction capability of a dissociated electrolyte into membrane material by taking measurements on sandwich membranes of which one side contains a cation exchanger and the other one an anion exchanger [208]. In addition, it was found useful to characterize ion pair formation constants in the membrane phase [208–210] as well as to study H_3O^+- and anion-selective ionophores and electrically charged ionophores [211].

1.6.5 CYCLIC VOLTAMMETRY AND RELATED METHODS

In some cases, ion transfer voltammetry at the ITIES has also been used to estimate complex formation constants [212]. The cell is usually set up to ensure that all potentials are constant with the exception of that at the ITIES. The presence of an ionophore in the organic phase facilitates the transfer of the ion of interest, and an ion transfer wave at smaller potentials is observed. Calculating the difference in the Gibbs energies of ion transfer for membranes with and without ionophore allows us to estimate complex formation constants. The underlying theory is complicated by the fact that most such ion transfer waves show currents that are limited by diffusion processes in the aqueous sample, not in the organic phase. The solvents historically used with such experiments, e.g., 1,2-dichloroethane, are difficult to compare with the ISE membranes discussed above, and very limited data is available on plasticized PVC. In one such example, the authors reported a logarithmic complex formation constant of 16 for valinomycin in PVC–NPOE [213], which appears to be unreasonably large [207].

Conceivably, pulsed chronopotentiometry with ISE membranes lacking ion-exchange properties can also be used to estimate complex formation constants [163]. They utilize the same membrane materials as for ISEs and because of the associated smaller diffusion coefficients, ion transfer processes are normally limited by mass transport in the membrane, and not in the aqueous phase. By altering the applied current, a single pulsed chronopotentiogram can give information on the extraction thermodynamics of the complexed and uncomplexed ion. However, the approach has not yet been thoroughly developed, and one must bear in mind that methods analyzing potential–current relationships provide data that may strongly depend on the diffusion coefficients of the membrane components. This is an important limitation compared with the potentiometric techniques described above, where such influences are insignificant.

1.7 OUTLOOK

After early developments of the glass and solid-state electrodes, potentiometry exhibited an impressively rapid development following the introduction of liquid membrane electrodes by the end of the 1960s. Within about two decades, ISEs for more than 50 ions were described, and potentiometry had succeeded in replacing flame photometry in clinical analysis. Microelectrodes and, in today's terminology, nanoelectrodes were applied mainly in physiological studies, and a number of companies offered ISEs for various ions. Potentiometry seemed to reach its peak, and quite a few scientists lost their interest in this field of research because they did not see much prospect for further important achievements. By describing a series of important new developments during the last decades, this chapter demonstrates that they were wrong.

Interestingly, potentiometry has largely developed outside the mainstream of electrochemistry. In fact, from the beginning, the research mostly was not motivated by electrochemical research. Cremer worked at the Physiological Institute in Munich and the first part of the title of his seminal paper on glass electrodes is "About the cause of the electromagnetic properties of tissue." The final part of the title of Donnan's paper was "A contribution to physical-chemical physiology." Guggenheim worked at the Royal Agricultural College in Copenhagen, Teorell at the Rockefeller Institute of Medical Research in New York, and Eisenman in the Department of Physiology, University of California Medical Center. Also, most key contributors to the developments in the 1970s and 1980s, such as Simon and Rechnitz, were originally from other fields, and true electrochemists, such as Buck and Koryta were the exception. More recently, the number of electrochemists contributing to the field has been increasing.

Several developments reviewed in this chapter are close to being finalized. This holds for the theoretical description of the response function of ISEs, including cases with relevant ion fluxes. It is foreseeable that electrodes with solid contact and a membrane other than of PVC will replace the current ones with liquid contact and PVC membrane. Exciting new developments are expected in connection with miniaturization, including nanotechnology and bioapplications of potentiometry, and especially in the area where potentiometry comes closer to traditional electrochemistry, including amperometric, galvanostatic, and coulometric methods.

ACKNOWLEDGMENTS

The authors are grateful to the National Institutes of Health (R01-EB002189) and the Australian Research Council (DP0987891) for financial support and to Dr. D. Wegmann for careful reading of the manuscript.

REFERENCES

1. Nernst, W., Die elektromotorische Wirksamkeit der Jonen, *Z Phys Chem*, Vol. 4, (1889): p. 129.
2. Nernst, W., Zur kenntnis der in Lösung befindlichen Körper. I. Theorie der Diffusion, *Z Phys Chem*, Vol. 2, (1888): p. 613.
3. Nernst, W., and R. H. Riesenfeld, Ueber elektrolytische Erscheinungen an der Grenzfläche zweier Lösungsmittel, *Ann Physik*, Vol. 8, (1902): p. 600.

4. Riesenfeld, R. H., Concentrationsketten mit nicht mischbaren Lösungsmitteln, *Ann Physik*, Vol. 8, (1902): p. 616.
5. Cremer, M., Über die Ursache der elektromotorischen Eigenschaften der Gewebe, zugleich ein Beitrag zur Lehre von den polyphasischen Elektrolytketten, *Z Biol*, Vol. 47, (1906): p. 562.
6. Helmholtz, H. L. F. v., *Vorträge und Reden von Hermann von Helmholtz*, Vieweg, Braunschweig, 1896.
7. Meyer, G., Über die electromotorischen Kräfte zwischen Glas und Amalgamen, *Ann Phys*, Vol. 276, (1890): p. 244.
8. Haber, F., and E. Klemensiewicz, Über elektrische Phasengrenzkräfte, *Z Phys Chem*, Vol. 67, (1909): p. 385.
9. "Wir haben uns danach vorgestellt, dass die Oberflächenschichten gut benetzten Glases eine Wasserphase darstellen könnten, in welcher die Konzentration der Wasserstoffionen und der Hydroxylionen keine merkliche oder wenigstens keine erhebliche Veränderung erfährt, wenn man die benetzende äussere Flüssigkeit einmal sauer und einmal alkalisch wählt."
10. Donnan, F. G., Theorie der Membrangleichgewichte und Membranpotentiale bei Vorhandensein von nicht dialysierenden Elektrolyten. Ein Beitrag zur physikalischchemischen Physiologie, *Z Elektrochem*, Vol. 17, (1911): p. 572.
11. Guggenheim, E. A., The conception of electrical potential difference between two phases and the individual activities of ions, *J Phys Chem*, Vol. 33, (1929): p. 842.
12. Guggenheim, E. A., On the conception of electrical potential difference between two phases. II, *J Phys Chem*, Vol. 34, (1930): p. 1540.
13. Teorell, T., An attempt to formulate a qualitative theory of membrane permeability, *Proc Soc Exp Biol Med*, Vol. 33, (1935): p. 282.
14. Meyer, K. H., and J.-F. Sievers, La perméabilité des membranes. I. Théorie de la perméabilité ionique, *Helv Chim Acta*, Vol. 19, (1936): p. 649.
15. Horvitz, K., Der Ionenaustausch am Dielektrikum. I. Die Elektrodenfunktion der Gläser, *Z Physik*, Vol. 15, (1923): p. 369.
16. Hughes, W. S., The potential difference between glass and electrolytes in contact with the glass, *J Am Chem Soc*, Vol. 44, (1924): p. 2860.
17. Nicolsky, B. P., Theory of glass electrodes (in Russian), *Acta Physicochim USSR*, Vol. 7, (1937): p. 597.
18. Hiltner, W., *Ausführung Potentiometrischer Analysen*, Julius Springer, Berlin, 1935.
19. Kolthoff, I. M., and H. L. Sanders, Electric potentials at crystal surfaces, and at silver halide surfaces in particular, *J Am Chem Soc*, Vol. 59, (1937): p. 416.
20. Buck, R. P., and E. Lindner, Tracing the history of selective ion sensors, *Anal Chem*, Vol. 73, (2001): p. 88A.
21. Frant, M. S., Where did ion selective electrodes come from? *J Chem Educ*, Vol. 74, (1997): p. 159.
22. Sollner, K., A new potentiometric method to determine cations and anions with collodion and protamine-collodion membranes, *J Am Chem Soc*, Vol. 65, (1943): p. 2260.
23. Stow, R. W., R. F. Baer, and B. F. Randall, Rapid measurement of the tension of carbon dioxide in blood, *Arch Phys Med*, Vol. 38, (1957): p. 646.
24. Severinghaus, J. W., and A. F. Bradley, Electrodes for blood pO_2 and pCO_2 determination, *J Appl Physiol*, Vol. 13, (1958): p. 515.
25. Stefanac, Z., and W. Simon, *In-vitro*-Verhalten von Makrotetroliden in Membranen als Grundlage für hochselektive kationenspezifische Elektrodensysteme, *Chimia*, Vol. 20, (1966): p. 436.
26. Stefanac, Z., and W. Simon, Ion specific electrochemical behavior of macrotetrolides in membranes, *Microchem J*, Vol. 12, (1967): p. 125.

27. Ross, J. W., Calcium-selective electrode with liquid ion exchanger, *Science*, Vol. 156, (1967): p. 1378.
28. Bloch, R., A. Shatkay, and H. A. Saroff, Fabrication and evaluation of membranes as specific electrodes for calcium ions, *Biophys J*, Vol. 7, (1967): p. 865.
29. Pedersen, C. J., Cyclic polyethers and their complexes with metal salts, *J Am Chem Soc*, Vol. 89, (1967): p. 2495.
30. Pedersen, C. J., Cyclic polyethers and their complexes with metal salts, *J Am Chem Soc*, Vol. 89, (1967): p. 7017.
31. Bühlmann, P., E. Pretsch, and E. Bakker, Carrier-based ion-selective electrodes and bulk optodes. 2. Ionophores for potentiometric and optical sensors, *Chem Rev*, Vol. 98, (1998): p. 1593.
32. Oesch, U., D. Ammann, and W. Simon, Ion-selective membrane electrodes for clinical use, *Clin Chem*, Vol. 32, (1986): p. 1448.
33. Meyerhoff, M. E., *In vivo* blood-gas and electrolyte sensors: Progress and challenges, *Trends Anal Chem*, Vol. 12, (1993): p. 257.
34. Caldwell, P. C., Studies on the internal pH of large muscle and nerve fibres, *J Physiol*, Vol. 142, (1968): p. 22.
35. Hinke, J. A. M., Glass micro-electrodes for measuring intracellular activities of sodium and potassium, *Nature*, Vol. 184, (1959): p. 1257.
36. Cornwall, M. C., D. F. Peterson, D. L. Kunze, J. L. Walker, and A. M. Brown, Intracellular potassium and chloride activities measured with liquid ion exchanger, *Brain Res*, Vol. 23, (1970): p. 433.
37. Khuri, R. N., J. J. Hajjar, and S. K. Agulian, Measurement of intracellular potassium with liquid ion-exchanger microelectrodes, *J Appl Physiol*, Vol. 32, (1972): p. 419.
38. Thomas, R. C., W. Simon, and M. Oehme, Lithium accumulation in snail neurons measured by a new Li^+-sensitive microelectrode, *Nature*, Vol. 258, (1975): p. 754.
39. Ammann, D., *Ion-Selective Microelectrodes*, Springer-Verlag, Berlin, 1986.
40. Koryta, J., P. Vanysek, and M. Brezina, Electrolysis with an electrolyte dropping electrode, *J Electroanal Chem*, Vol. 67, (1976): p. 263.
41. Solsky, R. L., and G. A. Rechnitz, Antibody-selective membrane electrodes, *Science*, Vol. 204, (1979): p. 1308.
42. Katz, S. A., and G. A. Rechnitz, Direct potentiometric determination of urea after urease hydrolysis, *Z Anal Chem*, Vol. 196, (1963): p. 248.
43. Brumleve, T. R., and R. P. Buck, Numerical solution of the Nernst–Planck and Poisson equation system with applications to membrane electrochemistry and solid state physics, *J Electroanal Chem*, Vol. 90, (1978): p. 1.
44. Sokalski, T., and A. Lewenstam, Application of Nernst–Planck and Poisson equations for interpretation of liquid-junction and membrane potentials in real-time and space domains, *Electrochem Commun*, Vol. 3, (2001): p. 107.
45. Sokalski, T., P. Lingenfelter, and A. Lewenstam, Numerical solution of the coupled Nernst–Planck and poisson equations for liquid junction and ion-selective membrane potentials, *J Phys Chem B*, Vol. 107, (2003): p. 2443.
46. Bakker, E., P. Bühlmann, and E. Pretsch, The phase-boundary potential model, *Talanta*, Vol. 63, (2004): p. 3.
47. Bobacka, J., A. Ivaska, and A. Lewenstam, Potentiometric ion sensors, *Chem Rev*, Vol. 108, (2008): p. 329.
48. Lingenfelter, P., I. Bedlechowicz-Sliwakowska, T. Sokalski, M. Maj-Zurawska, and A. Lewenstam, Time-dependent phenomena in the potential response of ion-selective electrodes treated by the Nernst–Planck–Poisson model. 1. Intramembrane processes and selectivity, *Anal Chem*, Vol. 78, (2006): p. 6783.

49. Sokalski, T., W. Kucza, M. Danielewski, and A. Lewenstam, Time-dependent phenomena in the potential response of ion-selective electrodes treated by the Nernst–Planck–Poisson model. Part 2: Transmembrane processes and detection limit, *Anal Chem*, Vol. 81, (2009): p. 5016.

50. Morf, W. E., E. Pretsch, and N. F. de Rooij, Computer simulation of ion-selective membrane electrodes and related systems by finite-difference procedures, *J Electroanal Chem*, Vol. 602, (2007): p. 43.

51. Morf, W. E., E. Pretsch, and N. F. de Rooij, Memory effects of ion-selective electrodes: Theory and computer simulation of the time-dependent potential response to multiple sample changes, *J Electroanal Chem*, Vol. 633, (2009): p. 137.

52. Höfler, L., I. Bedlechowitz, T. Vigassy, R. E. Gyurcsányi, E. Bakker, and E. Pretsch, Limitations of current polarization for lowering the detection limit of potentiometric polymeric membrane sensors, *Anal Chem*, Vol. 81, (2009): p. 3592.

53. Baucke, F. G. K., Glass electrodes: Why and how they function, *Ber Bunsen-Ges Phys Chem*, Vol. 100, (1996): p. 1466.

54. Baucke, F. G. K., Functioning of glass electrodes. A discussion of interfacial equilibria, *Phys Chem Glasses*, Vol. 42, (2001): p. 220.

55. Sandblom, J., G. Eisenman, and J. L. Walker, Electrical phenomena associated with the transport of ions and ion pairs in liquid ion-exchange membranes. I. Zero current properties, *J Phys Chem*, Vol. 71, (1967): p. 3862.

56. Sandblom, J., G. Eisenman, and J. L. Walker, Electrical phenomena associated with the transport of ions and ion pairs in liquid ion-exchange membranes. II. Nonzero current properties, *J Phys Chem*, Vol. 71, (1967): p. 3871.

57. Ciani, S. M., G. Eisenman, and G. Szabo, A theory for the effects of neutral carriers such as the macrotetralide actin antibiotics on the electric properties of bilayer membranes, *J Membr Biol*, Vol. 1, (1969): p. 1.

58. Morf, W. E., and W. Simon, Cation-response mechanism of neutral carrier based ion-selective electrode membranes, *Helv Chim Acta*, Vol. 69, (1986): p. 1120.

59. Pungor, E., Working mechanism of ion-selective electrodes, *Pure Appl Chem*, Vol. 64, (1992): p. 503.

60. Horvai, G., V. Horváth, A. Farkas, and E. Pungor, Studies on neutral carrier ion-selective electrodes. 3. Radiotracer measurements, *Magy Kém Foly*, Vol. 94, (1988): p. 367.

61. Tóth, K., E. Lindner, E. Pungor, E. Zippel, and R. Kellner, FTIR-ATR-spectroscopic analysis of bis-crown ether based PVC-membranes. III. *In situ* spectroscopic and electrochemical methods to study surface and bulk membrane processes, *Fresenius' Z Anal Chem*, Vol. 331, (1988): p. 448.

62. Umezawa, K., X. M. Lin, S. Nishizawa, M. Sugawara, and Y. Umezawa, Cation permselectivity in the phase boundary of ionophore-incorporated solvent polymeric membranes as studied by Fourier transform infrared attenuated total reflection spectrometry, *Anal Chim Acta*, Vol. 282, (1993): p. 247.

63. Tohda, K., Y. Umezawa, S. Yoshiyagawa, S. Hashimoto, and M. Kawasaki, Cation permselectivity at phase boundary of ionophore-incorporated solvent polymeric membranes as studied by optical second harmonic generation, *Anal Chem*, Vol. 67, (1995): p. 570.

64. Schneider, B., T. Zwickl, B. Federer, E. Pretsch, and E. Lindner, Spectropotentiometry: A new method for *in situ* imaging of concentration profiles in ion-selective membranes with simultaneous recording of potential-time transients, *Anal Chem*, Vol. 68, (1996): p. 4342.

65. Lindner, E., T. Zwickl, E. Bakker, B. T. T. Lan, K. Tóth, and E. Pretsch, Spectroscopic *in situ* imaging of acid coextraction processes in solvent polymeric ion-selective electrode and optode membranes, *Anal Chem*, Vol. 70, (1998): p. 1176.

66. Lindner, E., R. E. Gyurcsanyi, and B. D. Pendley, A glance into the bulk of solvent polymeric pH membranes, *Pure Appl Chem*, Vol. 73, (2001): p. 17.

67. Bakker, F., M. Nägele, U. Schaller, and E. Pretsch, Applicability of the phase boundary potential model to the mechanistic understanding of solvent polymeric membrane-based ion-selective electrodes, *Electroanalysis*, Vol. 7, (1995): p. 817.

68. Morf, W. E., *The Principles of Ion-Selective Electrodes and of Membrane Transport*, Elsevier, New York, 1981.

69. Buck, R. P., Theory and Principles of membrane electrodes, in: *Ion-Selective Electrodes in Analytical Chemistry*, Vol. 1, H. Freiser (ed.), Plenum Press, New York (1978): p. 1.

70. Hung, L. Q., Electrochemical properties of the interface between two immiscible electrolyte solutions: Part I. Equilibrium situation and galvani potential difference, *J Electroanal Chem*, Vol. 115, (1980): p. 159.

71. Hung, L. Q., Electrochemical properties of the interface between two immiscible electrolyte solutions: Part III. The general case of the galvani potential difference at the interface and of the distribution of an arbitrary number of components interacting in both phases, *J Electroanal Chem*, Vol. 149, (1983): p. 1.

72. van den Berg, A., P. D. van der Wal, M. Skowronska-Ptasinska, E. J. R. Sudhölter, P. Bergveld, and D. N. Reinhoudt, Membrane potentials of membranes with fixed ionic sites, *J Electroanal Chem*, Vol. 284, (1990): p. 1.

73. Kakiuchi, T., Limiting behavior in equilibrium partitioning of ionic components in liquid-liquid two-phase systems, *Anal Chem*, Vol. 68, (1996): p. 3658.

74. Hofmeister, F., Zur Lehre von der Wirkung der Salze, *Arch exp Pathol Pharmakol*, Vol. 24, (1888): p. 247.

75. Kakiuchi, T., and M. Senda, The liquid-junction potential at the contact of two immiscible electrolyte solutions in the absence of supporting electrolytes. Reference electrodes reversible to alkylammonium ions and tetraphenylborate ion in nitrobenzene, *Bull Chem Soc Jpn*, Vol. 60, (1987): p. 3099.

76. Mikhelson, K. N., A. Lewenstam, and S. E. Didina, Contribution of the diffusion potential to the membrane potential and the ion-selective electrode response, *Electroanalysis*, Vol. 11, (1999): p. 793.

77. Egorov, V. V., and Y. F. Lushchik, H+-selective electrodes based on neutral carriers: Specific features in behaviour and quantitative description of the electrode response, *Talanta*, Vol. 37, (1990): p. 461.

78. Rosatzin, T., E. Bakker, K. Suzuki, and W. Simon, Lipophilic and immobilized anionic additives in solvent polymeric membranes of cation-selective chemical sensors, *Anal Chim Acta*, Vol. 280, (1993): p. 197.

79. Cobben, P. L. H. M., R. J. M. Egberink, J. G. Bomer, P. Bergveld, and D. N. Reinhoudt, Chemically modified field effect transistors: the effect of ion-pair association on the membrane potentials, *J Electroanal Chem*, Vol. 368, (1994): p. 193.

80. Yoshida, Y., M. Matsui, K. Maeda, and S. Kihara, Physichochemical understanding of the selectivity at an ion selective electrode of the liquid membrane type and relation between the selectivity and the distribution ratios in the ion-pair extraction, *Anal Chim Acta*, Vol. 374, (1998): p. 269.

81. Bühlmann, P., S. Amemiya, S. Nishizawa, K. P. Xiao, and Y. Umezawa, Hydrogen-bonding ionophores for inorganic anions and nucleotides and their application in chemical sensors, *J Inclusion Phenom Mol Recognit Chem*, Vol. 32, (1998): p. 151.

82. Umezawa, Y., K. Umezawa, and H. Sato, Selectivity coefficients for ion-selective electrodes: Recommended methods for reporting values (Technical Report), *Pure Appl Chem*, Vol. 67, (1995): p. 507.

83. Bakker, E., R. K. Meruva, E. Pretsch, and M. E. Meyerhoff, Selectivity of polymer membrane-based ion-selective electrodes: Self-consistent model describing the potentiometric response in mixed ion solutions of different charge, *Anal Chem*, Vol. 66, (1994): p. 3021.

84. Bakker, E., E. Pretsch, and P. Bühlmann, Selectivity of potentiometric ion sensors, *Anal Chem*, Vol. 72, (2000): p. 1127.

85. Eugster, R., P. M. Gehrig, W. E. Morf, U. E. Spichiger, and W. Simon, Selectivity-modifying influence of anionic sites in neutral-carrier-based membrane electrodes, *Anal Chem*, Vol. 63, (1991): p. 2285.

86. Schaller, U., E. Bakker, U. E. Spichiger, and E. Pretsch, Ionic additives for ion-selective electrodes based on electrically charged carriers, *Anal Chem*, Vol. 66, (1994): p. 391.

87. Amemiya, S., P. Bühlmann, Y. Umezawa, B. Rusterholz, and E. Pretsch, Cationic or anionic sites? Selectivity optimization of ion-selective electrodes based on charged ionophores, *Anal Chem*, Vol. 72, (2000): p. 1618.

88. Bakker, E., Generalized selectivity description for polymeric ion-selective electrodes based on the phase boundary potential model, *J Electroanal Chem*, Vol. 639, (2010): p. 1.

89. Nägele, M., E. Bakker, and E. Pretsch, General description of the simultaneous response of potentiometric ionophore-based sensors to ions of different charge, *Anal Chem*, Vol. 71, (1999): p. 1041.

90. Schefer, U., D. Ammann, E. Pretsch, U. Oesch, and W. Simon, Neutral carrier based Ca^{2+}-selective electrode with detection limit in the sub-nanomolar range, *Anal Chem*, Vol. 58, (1986): p. 2282.

91. Sokalski, T., M. Maj-Zurawska, and A. Hulanicki, Determination of true selectivity coefficients of neutral carrier calcium selective electrode, *Mikrochim Acta*, Vol. 114, (1991): p. 285.

92. Bakker, E., M. Willer, and E. Pretsch, Detection limit of ion-selective bulk optodes and corresponding electrodes, *Anal Chim Acta*, Vol. 282, (1993): p. 265.

93. Mathison, S., and E. Bakker, Effect of transmembrane electrolyte diffusion on the detection limit of carrier-based potentiometric ion sensors, *Anal Chem*, Vol. 70, (1998): p. 303.

94. Sokalski, T., A. Ceresa, T. Zwickl, and E. Pretsch, Large improvement of the lower detection limit of ion-selective polymer membrane electrodes, *J Am Chem Soc*, Vol. 119, (1997): p. 11347.

95. Gyurcsányi, R. E., A.-S. Nybäck, K. Tóth, G. Nagy, and A. Ivaska, Novel polypyrrole based all-solid-state potassium-selective microelectrodes, *Analyst*, Vol. 123, (1998): p. 1339.

96. Mi, Y., S. Mathison, R. Goines, A. Logue, and E. Bakker, Detection limit of polymeric membrane potentiometric ion sensors: How can we go down to trace levels?, *Anal Chim Acta*, Vol. 397, (1999): p. 103.

97. Telting-Diaz, M., and E. Bakker, Effect of lipophilic ion-exchanger leaching on the detection limit of carrier-based ion-selective electrodes, *Anal Chem*, Vol. 73, (2001): p. 5582.

98. Malon, A., A. Radu, W. Qin, Y. Qin, A. Ceresa, M. Maj-Zurawska, E. Bakker, and E. Pretsch, Improving the detection limit of anion-selective electrodes: An iodide-selective membrane with a nanomolar detection limit, *Anal Chem*, Vol. 75, (2003): p. 3865.

99. Guilbault, G. G., R. A. Durst, M. S. Frant, H. Freiser, E. H. Hansen, T. S. Light, E. Pungor, et al., Recommendations for nomenclature of ion-selective electrodes, *Pure Appl Chem*, Vol. 48, (1976): p. 127.

100. Buck, R. P., and E. Lindner, Recommendations for nomenclature of ion-selective electrodes, *Pure Appl Chem*, Vol. 66, (1994): p. 2527.
101. Currie, L. A., Nomenclature in evaluation of analytical methods including detection and quantification capabilities (IUPAC Recommendations 1995), *Pure Appl Chem*, Vol. 67, (1995): p. 1699.
102. Bakker, E., and E. Pretsch, Potentiometric sensors for trace-level analysis, *Trends Anal Chem*, Vol. 24, (2005): p. 199.
103. Sokalski, T., T. Zwickl, E. Bakker, and E. Pretsch, Lowering the detection limit of solvent polymeric ion-selective electrodes. 1. Modeling the influence of steady-state ion fluxes, *Anal Chem*, Vol. 71, (1999): p. 1204.
104. Zwickl, T., T. Sokalski, and E. Pretsch, Steady-state model calculations predicting the influence of key parameters on the lower detection limit and ruggedness of solvent polymeric membrane ion-selective electrodes, *Electroanalysis*, Vol. 11, (1999): p. 673.
105. Morf, W. E., M. Badertscher, T. Zwickl, N. F. de Rooij, and E. Pretsch, Effects of ion transport on the potential response of ionophore-based membrane electrodes: A theoretical approach, *J Phys Chem B*, Vol. 103, (1999): p. 11346.
106. Ion, A. C., E. Bakker, and E. Pretsch, Potentiometric Cd^{2+}-selective electrode with a detection limit in the low ppt range (with Erratum), *Anal Chim Acta*, Vol. 440, (2001): p. 71; ibid., Vol. 452, (2002): p. 329.
107. Ceresa, A., A. Radu, S. Peper, E. Bakker, and E. Pretsch, Rational design of potentiometric trace level ion sensors. A Ag^+-selective electrode with a 100 ppt detection limit, *Anal Chem*, Vol. 74, (2002): p. 4027.
108. Bakker, E., P. Bühlmann, and E. Pretsch, Polymer membrane ion-selective electrodes— What are the limits?, *Electroanalysis*, Vol. 11, (1999): p. 915.
109. Ceresa, A., T. Sokalski, and E. Pretsch, Influence of key parameters on the lower detection limit and response function of solvent polymeric membrane ion-selective electrodes, *J Electroanal Chem*, Vol. 501, (2001): p. 70.
110. Szigeti, Z., T. Vigassy, E. Bakker, and E. Pretsch, Approaches to improving the lower detection limit of polymeric membrane ion-selective electrodes, *Electroanalysis*, Vol. 18, (2006): p. 1254.
111. Bakker, E., Determination of unbiased selectivity coefficients of neutral carrier based cation-selective electrodes, *Anal Chem*, Vol. 69, (1997): p. 1061.
112. Wygladacz, K., and E. Bakker, Fluorescent microsphere fiber optic microsensor array for direct iodide detection at low picomolar concentrations, *Analyst*, Vol. 132, (2007): p. 268.
113. Szigeti, Z., I. Bitter, K. Tóth, C. Latkoczy, D. J. Fliegel, D. Günther, and E. Pretsch, A novel polymeric membrane electrode for the potentiometric analysis of Cu^{2+} in drinking water, *Anal Chim Acta*, Vol. 532, (2005): p. 129.
114. Juang, R.-S., and I.-P. Huang, Liquid-liquid extraction of copper(II)-EDTA chelated anions with microporous hollow fibers, *J Chem Technol Biotechnol*, Vol. 75, (2000): p. 610.
115. Pergel, E., R. E. Gyurcsányi, K. Tóth, and E. Lindner, Picomolar detection limits with current-polarized Pb^{2+} ion-selective membranes, *Anal Chem*, Vol. 73, (2001): p. 4249.
116. Lindner, E., R. E. Gyurcsányi, and R. P. Buck, Tailored transport through ion-selective membranes for improved detection limits and selectivity coefficients, *Electroanalysis*, Vol. 11, (1999): p. 695.
117. Morf, W. E., M. Badertscher, T. Zwickl, N. F. de Rooij, and E. Pretsch, Effects of controlled current on the response behavior of polymeric membrane ion-selective electrodes, *J Electroanal Chem*, Vol. 526, (2002): p. 19.

118. Bedlechowicz, I., T. Sokalski, A. Lewenstam, and M. Maj-Zurawska, Calcium ion-selective electrodes under galvanostatic current control, *Sens Actuators B*, Vol. 108, (2005): p. 836.

119. Peshkova, M. A., T. Sokalski, K. N. Mikhelson, and A. Lewenstam, Obtaining Nernstian response of a Ca^{2+}-selective electrode in a broad concentration range by tuned galvanostatic polarization, *Anal Chem*, Vol. 80, (2008): p. 9181.

120. Xu, Y., W. Ngeontae, E. Pretsch, and E. Bakker, Backside calibration chronopotentiometry: Using current to perform ion measurements by zeroing the transmembrane ion flux, *Anal Chem*, Vol. 80, (2008): p. 7516.

121. Lindner, E., and R. E. Gyurcsányi, Quality control criteria for solid-contact, solvent polymeric membrane ion-selective electrodes, *J Solid State Electrochem*, Vol. 13, (2009): p. 51.

122. Bobacka, J., Conducting polymer-based solid-state ion-selective electrodes, *Electroanalysis*, Vol. 18, (2006): p. 7.

123. Nicolsky, B. P., and E. A. Materova, Solid-contact in membrane ion-selective electrodes, *Ion-Sel Electrode Rev*, Vol. 7, (1985): p. 3.

124. Cattrall, R. W., and I. C. Hamilton, Coated-wire ion-selective electrodes, *Ion-Sel Electrode Rev*, Vol. 6, (1984): p. 125.

125. Cattrall, R. W., and H. Freiser, Coated wire ion-selective electrodes, *Anal Chem*, Vol. 43, (1971): p. 1905.

126. Cattrall, R. W., I. C. Hamilton, and P. J. Iles, Photocured polymers in ion-selective electrode membranes, *Anal Chim Acta*, Vol. 169, (1985): p. 403.

127. Ruzicka, J., C. G. Lamm, and J. C. Tjell, Selectrode – the universal ion-selective electrode: Part III. Concept, constructions and materials, *Anal Chim Acta*, Vol. 62, (1972): p. 15.

128. Schnierle, P., T. Kappes, and P. C. Hauser, Capillary electrophoresis determination of different classes of organic ions by potentiometric detection with coated-wire ion-selective electrodes, *Anal Chem*, Vol. 70, (1998): p. 3585.

129. Cattrall, R. W., D. W. Drew, and I. C. Hamilton, Some alkylphosphoric acid esters for use in coated-wire calcium-selective electrodes, *Anal Chim Acta*, Vol. 76, (1975): p. 269.

130. Hulanicki, A., and M. Trojanowicz, Calcium-selective electrodes with PVC membranes and solid internal contacts, *Anal Chim Acta*, Vol. 87, (1976): p. 411.

131. Sundfors, F., T. Lindfors, L. Höfler, R. Bereczki, and R. E. Gyurcsányi, FTIR-ATR Study of water uptake and diffusion through ion-selective membranes based on poly(acrylates) and silicon rubber, *Anal Chem*, Vol. 81, (2009): p. 5925.

132. Hauser, P. C., D. W. L. Chiang, and G. A. Wright, A potassium-ion selective electrode with valinomycin based poly(vinyl chloride) membrane and poly(vinyl ferrocene) solid contact, *Anal Chim Acta*, Vol. 302, (1995): p. 241.

133. Liu, D., R. K. Meruva, R. B. Brown, and M. E. Meyerhoff, Enhancing EMF stability of solid-state ion-selective sensors by incorporating lipophilic silver-ligand complexes within polymeric films, *Anal Chim Acta*, Vol. 321, (1996): p. 173.

134. Cha, G. S., D. Liu, M. E. Meyerhoff, H. C. Cantor, A. R. Midgley, H. D. Goldberg, and R. B. Brown, Electrochemical performance, biocompatibility, and adhesion properties of new polymer matrices for solid-state ion sensors, *Anal Chem*, Vol. 63, (1991): p. 1666.

135. Fibbioli, M., W. E. Morf, M. Badertscher, N. F. de Rooij, and E. Pretsch, Potential drifts of solid-contacted ion-selective electrodes due to zero-current ion fluxes through the sensor membrane, *Electroanalysis*, Vol. 12, (2000): p. 1286.

136. Grygolowicz-Pawlak, E., K. Plachecka, Z. Brzozka, and E. Malinowska, Further studies on the role of redox-active monolayer as intermediate phase of solid-state sensors, *Sens Actuators B*, Vol. 123, (2007): p. 480.

137. De Marco, R., J.-P. Veder, G. Clarke, A. Nelson, K. Prince, E. Pretsch, and E. Bakker, Evidence of a water layer in solid-contact polymeric ion sensors, *Phys Chem Chem Phys*, Vol. 10, (2008): p. 73.
138. Veder, J.-P., R. De Marco, G. Clarke, R. Chester, A. Nelson, K. Prince, E. Pretsch, and E. Bakker, Elimination of undesirable water layers in solid-contact polymeric ion-selective electrodes, *Anal Chem*, Vol. 80, (2008): p. 6731.
139. Fibbioli, M., K. Bandyopadhyay, S.-G. Liu, L. Echegoyen, O. Enger, F. Diederich, P. Bühlmann, and E. Pretsch, Redox-active self-assembled monolayers as novel solid contacts for ion-selective electrodes, *Chem Commun*, Vol. 5, (2000): p. 339.
140. Fibbioli, M., K. Bandyopadhyay, S.-G. Liu, L. Echegoyen, O. Enger, F. Diederich, D. Gingery, et al., Redox-active self-assembled monolayers for solid-contact polymeric membrane ion-selective electrodes, *Chem Mater*, Vol. 14, (2002): p. 1721.
141. Grygolowicz-Pawlak, E., K. Wygladacz, S. Sek, R. Bilewicz, Z. Brzozka, and E. Malinowska, Studies on ferrocene organothiol monolayer as an intermediate phase of potentiometric sensors with gold inner contact, *Sens Actuators B*, Vol. 111–112, (2005): p. 310.
142. Cadogan, A., Z. Gao, A. Lewenstam, A. Ivaska, and D. Diamond, All-solid-state sodium-selective electrode based on a calixarene ionophore in a poly(vinyl chloride) membrane with a polypyrrole solid contact, *Anal Chem*, Vol. 64, (1992): p. 2496.
143. Dumanska, J., and K. Maksymiuk, Studies on spontaneous charging/discharging processes of polypyrrole in aqueous electrolyte solutions, *Electroanalysis*, Vol. 13, (2001): p. 567.
144. Bobacka, J., T. Lindfors, M. McCarrick, A. Ivaska, and A. Lewenstam, Single-piece all-solid-state ion-selective electrode, *Anal Chem*, Vol. 67, (1995): p. 3819.
145. Vazquez, M., J. Bobacka, and A. Ivaska, Potentiometric sensors for Ag^+- based on poly(3-octylthiophene) (POT), *J Solid State Electrochem*, Vol. 9, (2005): p. 865.
146. Gyetvai, G., S. Sundblom, L. Nagy, A. Ivaska, and G. Nagy, Solid contact micropipette ion selective electrode for potentiometric SECM, *Electroanalysis*, Vol. 19, (2007): p. 1116.
147. Lindfors, T., Light sensitivity and potential stability of electrically conducting polymers commonly used in solid contact ion-selective electrodes, *J Solid State Electrochem*, Vol. 13, (2009): p. 77.
148. Lai, C., M. A. Fierke, A. Stein, and P. Bühlmann, Ion-selective electrodes with three-dimensionally ordered macroporous carbon as the solid contact, *Anal Chem*, Vol. 79, (2007): p. 4621.
149. Crespo, G. A., S. Macho, and F. X. Rius, Ion-selective electrodes using carbon nanotubes as ion-to-electron transducers, *Anal Chem*, Vol. 80, (2008): p. 1316.
150. Crespo, G. A., S. Macho, J. Bobacka, and F. X. Rius, Transduction mechanism of carbon nanotubes in solid-contact ion-selective electrodes, *Anal Chem*, Vol. 81, (2009): p. 676.
151. Zhu, J., Y. Qin, and Y. Zhang, Preparation of all solid-state potentiometric ion sensors with polymer-CNT composites, *Electrochem Commun*, Vol. 11, (2009): p. 1684.
152. Sutter, J., E. Lindner, R. E. Gyurcsányi, and E. Pretsch, A polypyrrole-based solid-contact Pb^{2+}-selective PVC membrane electrode with nanomolar detection limit, *Anal Bioanal Chem*, Vol. 380, (2004): p. 7.
153. Sutter, J., A. Radu, S. Peper, E. Bakker, and E. Pretsch, Solid-contact polymeric membrane electrodes with detection limits in the subnanomolar range, *Anal Chim Acta*, Vol. 523, (2004): p. 53.
154. Chumbimuni-Torres, K. Y., N. Rubinova, A. Radu, L. T. Kubota, and E. Bakker, Solid contact potentiometric sensors for trace level measurements, *Anal Chem*, Vol. 78, (2006): p. 1318.

155. Heng, L. Y., K. Tóth, and E. A. H. Hall, Ion-transport and diffusion coefficients of non-plasticized methacrylic-acrylic ion-selective membranes, *Talanta*, Vol. 63, (2004): p. 73.

156. Shvarev, A., and E. Bakker, Reversible electrochemical detection of nonelectroactive polyions, *J Am Chem Soc*, Vol. 125, (2003): p. 11192.

157. Shirai, O., S. Kihara, Y. Yoshida, and M. Matsui, Ion transfer through a liquid membrane or a bilayer lipid membrane in the presence of sufficient electrolytes, *J Electroanal Chem*, Vol. 389, (1995): p. 61.

158. Homolka, D., L. Q. Hung, A. Hofmanova, M. W. Khalil, J. Koryta, V. Marecek, Z. Samec, et al., Faradaic ion transfer across the interface of two immiscible electrolyte solutions: chronopotentiometry and cyclic voltammetry, *Anal Chem*, Vol. 52, (1980): p. 1606.

159. Girault, H. H., Electrochemistry at the interface between two immiscible electrolyte solutions, *Electrochim Acta*, Vol. 32, (1987): p. 383.

160. Tomaszewski, L., F. Reymond, P.-F. Brevet, and H. H. Girault, Facilitated ion transfer across oil-water interfaces. Part III. Algebraic development and calculation of cyclic voltammetry experiments for the formation of a neutral complex, *J Electroanal Chem*, Vol. 483, (2000): p. 135.

161. Yuan, Y., and S. Amemiya, Facilitated protamine transfer at polarized water/1,2-dichloroethane interfaces studied by cyclic voltammetry and chronoamperometry at micropipet electrodes, *Anal Chem*, Vol. 76, (2004): p. 6877.

162. Samec, Z., E. Samcova, and H. H. Girault, Ion amperometry at the interface between two immiscible electrolyte solutions in view of realizing the amperometric ion-selective electrode, *Talanta*, Vol. 63, (2004): p. 21.

163. Jadhav, S., and E. Bakker, Voltammetric and amperometric transduction for solvent polymeric membrane ion sensors, *Anal Chem*, Vol. 71, (1999): p. 3657.

164. Shvarev, A., and E. Bakker, Pulsed galvanostatic control of ionophore-based polymeric ion sensors, *Anal Chem*, Vol. 75, (2003): p. 4541.

165. Long, R., and E. Bakker, Spectral imaging and electrochemical study on the response mechanism of ionophore-based polymeric membrane amperometric pH sensors, *Electroanalysis*, Vol. 15, (2003): p. 1261.

166. Sokalski, T., A. Ceresa, M. Fibbioli, T. Zwickl, E. Bakker, and E. Pretsch, Lowering the detection limit of solvent polymeric ion-selective membrane electrodes. 2. Influence of composition of sample and internal electrolyte solution, *Anal Chem*, Vol. 71, (1999): p. 1210.

167. Gemene, K. L., A. Shvarev, and E. Bakker, Title: Selectivity enhancement of anion-responsive electrodes by pulsed chronopotentiometry, *Anal Chim Acta*, Vol. 583, (2007): p. 190.

168. Ma, S. C., V. C. Yang, and M. E. Meyerhoff, Heparin-responsive electrochemical sensor: a preliminary study, *Anal Chem*, Vol. 64, (1992): p. 694.

169. Meyerhoff, M. E., B. Fu, E. Bakker, J. H. Yun, and V. C. Yang, Polyion-sensitive membrane electrodes for biomedical analysis, *Anal Chem*, Vol. 68, (1996): p. 168A.

170. Fu, B., E. Bakker, J. H. Yun, V. C. Yang, and M. E. Meyerhoff, Response mechanism of polymer membrane-based potentiometric polyion sensors, *Anal Chem*, Vol. 66, (1994): p. 2250.

171. Dai, S., and M. E. Meyerhoff, Nonseparation binding/immunoassays using polycation-sensitive membrane electrode detection, *Electroanalysis*, Vol. 13, (2001): p. 276.

172. Yun, J. H., V. C. Yang, and M. E. Meyerhoff, Protamine-sensitive polymer membrane electrode: Characterization and bioanalytical applications, *Anal Biochem*, Vol. 224, (1995): p. 212.

173. Buchanan, S. A. N., L. P. Balogh, and M. E. Meyerhoff, Potentiometric response characteristics of polycation-sensitive membrane electrodes toward poly(amidoamine) and poly(propylenimine) dendrimers, *Anal Chem*, Vol. 76, (2004): p. 1474.

174. Samec, Z., A. Trojánek, J. Langmaier, and E. Samcová, Cyclic voltammetry of biopolymer heparin at PVC plasticized liquid membrane, *Electrochem Commun*, Vol. 5, (2003): p. 867.

175. Guo, J., and S. Amemiya, Voltammetric heparin-selective electrode based on thin liquid membrane with conducting polymer-modified solid support, *Anal Chem*, Vol. 78, (2006): p. 6893.

176. Langmaier, J., J. Olscaronák, E. Samcová, Z. Samec, and A. Trojánek, Amperometry of heparin polyion using a rotating disk electrode coated with a plasticized PVC membrane, *Electroanalysis*, Vol. 18, (2006): p. 115.

177. Guenat, O. T., W. E. Morf, B. H. van der Schoot, and N. F. de Rooij, Universal coulometric nanotitrators with potentiometric detection, *Anal Chim Acta*, Vol. 361, (1988): p. 261.

178. Bhakthavatsalam, V., A. Shvarev, and E. Bakker, Selective Coulometric release of ions from ion selective polymeric membranes for calibration-free titrations, *Analyst*, Vol. 131, (2006): p. 895.

179. Ceresa, A., E. Pretsch, and E. Bakker, Direct potentiometric determination of total concentrations, *Anal Chem*, Vol. 72, (2000): p. 2050.

180. Shvarev, A., and E. Bakker, Distinguishing free and total calcium with a single pulsed galvanostatic ion-selective electrode, *Talanta*, Vol. 63, (2004): p. 195.

181. Gemene, K. L., and E. Bakker, Direct sensing of total acidity by chronopotentiometric flash titrations at polymer membrane ion-selective electrodes, *Anal Chem*, Vol. 80, (2008): p. 3743.

182. Daniele, S., C. Bragato, and A. M. Baldo, A steady-state voltammetric procedure for the determination of hydrogen ions and total acid concentration in mixtures of a strong and a weak monoprotic acid, *Electrochim Acta*, Vol. 52, (2006): p. 54.

183. Yoshizumi, A., A. Uehara, M. Kasuno, Y. Kitatsuji, Z. Yoshida, and S. Kihara, Rapid and coulometric electrolysis for ion transfer at the aqueous-organic solution interface, *J Electroanal Chem*, Vol. 581, (2005): p. 275.

184. Kasuno, M., Y. Kakitani, Y. Shibafuji, T. Okugaki, K. Maeda, T. Matsushita, and S. Kihara, Rapid and precise coulometric determination of calcium based on electrolysis for ion transfer at the aqueous-organic solution interface, *Electroanalysis*, Vol. 21, (2009): p. 2022.

185. Bakker, E., D. Diamond, A. Lewenstam, and E. Pretsch, Ion sensors: current limits and new trends, *Anal Chim Acta*, Vol. 393, (1999): p. 11.

186. Lindner, E., V. V. Cosofret, S. Ufer, T. A. Johnson, R. B. Ash, H. T. Nagle, M. R. Neuman, and R. P. Buck, In vivo and in vitro testing of microelectronically fabricated planar sensors designed for applications in cardiology, *Fresenius' J Anal Chem*, Vol. 346, (1993): p. 584.

187. Vigassy, T., W. E. Morf, M. Badertscher, A. Ceresa, N. F. de Rooij, and E. Pretsch, Making use of ion fluxes through potentiometric sensor membranes: ISEs with step responses at critical ion activities, *Sens Actuators B*, Vol. 76, (2001): p. 477.

188. Peper, S., A. Ceresa, E. Bakker, and E. Pretsch, Improved detection limits and sensitivities of potentiometric titrations, *Anal Chem*, Vol. 73, (2001): p. 3768.

189. Makarychev-Mikhailov, S., A. Shvarev, and E. Bakker, Calcium pulstrodes with 10-fold enhanced sensitivity for measurements in the physiological concentration range, *Anal Chem*, Vol. 78, (2006): p. 2744.

190. Malon, A., E. Bakker, and E. Pretsch, Backside calibration potentiometry: Ion activity measurements with selective supported liquid membranes by calibrating from the inner side of the membrane, *Anal Chem*, Vol. 79, (2007): p. 632.

191. Bakker, E., and E. Pretsch, Potentiometry at trace levels, *Trends Anal Chem*, Vol. 20, (2001): p. 11.

192. Bakker, E., and E. Pretsch, Modern potentiometry, *Angew Chem, Int Ed Engl*, Vol. 46, (2007): p. 5660.

193. Daunert, S., and L. G. Bachas, Ion-selective electrodes using an ionophore covalently attached to carboxylated poly(vinyl chloride), *Anal Chem*, Vol. 62, (1990): p. 1428.

194. Bereczki, R., R. E. Gyurcsányi, B. Agai, and K. Tóth, Synthesis and characterization of covalently immobilized bis-crown ether based potassium ionophore, *Analyst*, Vol. 130, (2005): p. 63.

195. Püntener, M., M. Fibbioli, E. Bakker, and E. Pretsch, Response and diffusion behavior of mobile and covalently immobilized H^+-ionophores in polymer membrane ion-selective electrodes, *Electroanalysis*, Vol. 14, (2002): p. 1329.

196. Püntener, M., T. Vigassy, E. Baier, A. Ceresa, and E. Pretsch, Improving the lower detection limit of potentiometric sensors by covalent immobilization of the ionophore, *Anal Chim Acta*, Vol. 503, (2004): p. 187.

197. Qin, Y., S. Peper, A. Radu, A. Ceresa, and E. Bakker, Plasticizer-free polymer containing a covalently immobilized Ca^{2+}-selective ionophore for potentiometric and optical sensing, *Anal Chem*, Vol. 75, (2003): p. 3038.

198. Jágerszki, G., A. Grün, I. Bitter, K. Tóth, and R. E. Gyurcsányi, Ionophore-gold nanoparticle conjugates for Ag^+-selective sensors with nanomolar detection limit, *Chem Commun*, Vol. 46, (2010): p. 607.

199. Bakker, E., and E. Pretsch, Ion-selective electrodes based on two competitive ionophores for determining effective stability constants of ion–carrier complexes in solvent polymeric membranes, *Anal Chem*, Vol. 70, (1998): p. 295.

200. Bakker, E., M. Willer, M. Lerchi, K. Seiler, and E. Pretsch, Determination of complex-formation constants of neutral cation-selective ionophores in solvent polymeric membranes, *Anal Chem*, Vol. 66, (1994): p. 516.

201. Mi, Y., and E. Bakker, Ion binding properties of the lipophilic H^+-chromoionophore ETH 5294 in solvent polymeric sensing membranes as determined with segmented sandwich membranes, *Electrochem Solid-State Lett*, Vol. 3, (2000): p. 159.

202. Ceresa, A., and E. Pretsch, Determination of formal complex formation constants of various Pb^{2+} ionophores in the sensor membrane phase, *Anal Chim Acta*, Vol. 395, (1999): p. 41.

203. Szigeti, Z., A. Malon, T. Vigassy, V. Csokai, A. Grün, K. Wygladacz, N. Ye, et al., Novel potentiometric and optical silver ion-selective sensors with subnanomolar detection limits, *Anal Chim Acta*, Vol. 572, (2006): p. 1.

204. Stefanova, O. K., On the effect of coupling of ionic and neutral ligand fluxes in membrane potentials, *Elektrokhimiya*, Vol. 15, (1979): p. 1707.

205. Mokrov, S. B., and O. K. Stefanova, Manifestations of the coupling of the fluxes of ions and neutral chelating agent in the membrane potenital on the background of an ion exchanger in the membrane, *Elektrokhimiya*, Vol. 21, (1985): p. 540.

206. Shultz, M. M., O. K. Stefanova, S. B. Mokrov, and K. N. Mikhelson, Potentiometric estimation of the stability constants of ion–ionophore complexes in ion-selective membranes by the sandwich membrane method: Theory, advantages, and limitations, *Anal Chem*, Vol. 74, (2002): p. 510.

207. Mi, Y., and E. Bakker, Determination of complex formation constants of lipophilic neutral ionophores in solvent polymeric membranes with segmented sandwich membranes, *Anal Chem*, Vol. 71, (1999): p. 5279.

208. Qin, Y., and E. Bakker, Evaluation of the separate equilibrium processes that dictate the upper detection limit of ionophore-based potentiometric sensors, *Anal Chem*, Vol. 74, (2002): p. 3134.

209. Peshkova, M. A., A. I. Korobeynikov, and K. N. Mikhelson, Estimation of ion-site association constants in ion-selective electrode membranes by modified segmented sandwich membrane method, *Electrochim Acta*, Vol. 53, (2008): p. 5819.

210. Egorov, V. V., P. L. Lyaskovski, I. V. Il'inchik, V. V. Soroka, and V. A. Nazarov, Estimation of ion-pairing constants in plasticized poly(vinyl chloride) membranes using segmented sandwich membranes technique, *Electroanalysis*, Vol. 21, (2009): p. 2061.

211. Qin, Y., and E. Bakker, Quantitative binding constants of H^+-selective chromoionophores and anion ionophores in solvent polymeric sensing membranes, *Talanta*, Vol. 58, (2002): p. 909.

212. Reymond, F., G. Lagger, P.-A. Carrupt, and H. H. Girault, Facilitated ion transfer reactions across oil-water interfaces: Part II. Use of the convoluted current for the calculation of the association constants and for an amperometric determination of the stoichiometry of ML_j^{z+} complexes, *J Electroanal Chem*, Vol. 451, (1998): p. 59.

213. Lee, H. J., C. Beriet, and H. H. Girault, Amperometric detection of alkali metal ions on micro-fabricated composite polymer membranes, *J Electroanal Chem*, Vol. 453, (1998): p. 211.

2 Electrochemistry at Platinum Single Crystal Electrodes

Carol Korzeniewski, Victor Climent, and Juan M. Feliu

CONTENTS

2.1 INTRODUCTION

Single crystal electrodes have attracted practical interest for nearly 50 years (Refs. [1–8], and references therein). They are employed in experiments that benefit from surfaces that in principle are contaminant free, possess long range atomic order, and can be characterized down to the atomic length scale [1,2,4,9]. The latter properties provide knowledge of the surface structure and surface atom density. The surface atom density is an important quantity, because it enables adsorbate coverages and catalytic reaction rates to be expressed relative to the number of atoms present on the electrode surface [9]. The overwhelming majority of studies involving single crystals as working electrodes are based on the use of Pt-group (Pt, Ir, Rh, Pd, Ru) and coinage (Au, Ag, Cu, Ni) metals [1,2,4,6,8–10]. Carbon, mainly in the form of highly ordered pyrolytic graphite, also has been productively employed [11–13].

An early motivation for the use of single crystal electrodes was the opportunity to probe reactions in electrocatalysis that appeared sensitive to the atomic-level structure of the electrode surface [1,6,7]. Focus was on elucidating mechanisms for H_2 oxidation and O_2 reduction over Pt materials. Measurements were challenged by the tendency for Pt surfaces to easily poison and disorder [1,7]. To overcome these limitations, techniques and materials common to ultra high vacuum (UHV) surface science were adapted [1,2,4,9]. Single crystals that had been developed to support investigations of catalytic processes at gas–solid interfaces were applied as electrodes. Strategies were devised for the transfer of a single crystal between the UHV system, where the surface could be cleaned, ordered, and characterized in terms of structure and composition, and an electrochemical cell held in a chamber under inert atmosphere at ambient pressure [1,14–16]. The opportunity to perform fundamental surface science measurements on electrodes following manipulations in solution under potential control stimulated investigations along several lines aimed at gaining insights into metal/electrolyte solution interfacial structure and its relationship to reactivity and related phenomena in electrochemistry [2,4,9,10,17–21].

Shortly after applications of well-defined, UHV prepared, electrodes were first attempted, bench-top methods emerged for cleaning and ordering the surfaces of single crystal Pt, Rh, Au, and Ag [3,7,22–25]. In one approach, it was shown that heating the electrodes in a H_2, butane, or similar flame to just below the melting point removed surface impurities and annealed the surface atoms. The technique came to be known as "flame annealing" and was proposed and initially applied by Clavilier [7,22,23]. Pt electrodes were investigated first, and the development of procedures for other metals followed. The first voltammograms indicative of clean, ordered Pt single crystal surface planes, in fact, were obtained through the use of flame annealing [22,23], and the first successful transfer of a clean Pt surface from UHV was achieved a few years later [1,6,16]. The simplicity of flame annealing techniques stimulated greater use of single crystal materials in electrochemistry experiments [6,7,19,26]. As interest in single crystal electrochemistry grew, applications of bulk crystal and quasicrystalline metal films became more commonplace, and understanding of processes such as adsorption, electrodeposition, electrocatalysis, heterogeneous electron transfer, surface reconstruction, and thin film self-assembly more rapidly advanced (cf., Refs. [8,19,20,26–32]).

The focus of this chapter is on the preparation and application of Pt single crystal electrodes. In addition to their historical importance, Pt single crystals continue to provide new insights into interfacial processes and are of central importance in electrochemistry. The scope of the chapter, therefore, is limited to Pt materials to enable in-depth coverage. Methods for growing and orienting Pt single crystals and techniques for exposing, cleaning, and annealing Pt electrode surface planes are described. Applications of Pt single crystals in the study of fundamental double layer phenomena and electrocatalytic reaction steps and in the development of nanoscale catalyst materials are highlighted.

2.2 ELECTRODE PREPARATION

2.2.1 SINGLE CRYSTAL GROWTH AND ORIENTATION

The first electrochemical experiments that employed structurally characterized Pt single crystals utilized the bulk crystal materials (i.e., disks ~1 cm diameter by ~2 mm thick) common to UHV surface work. Although, in initial electrochemistry experiments, UHV techniques were used for electrode surface cleaning and ordering, the single crystal disks were easily adapted to flame annealing methods. Single crystal disks have continued to be used for electrochemistry experiments. They are typically obtained from specialty laboratories, and methods for their preparation have been described [3].

In the early 1980s, Clavilier demonstrated a simpler, less expensive approach to the preparation of Pt single crystal electrodes [7,22,23,33–35]. Crystals were grown from Pt wire and tended to be smaller but more robust than the commercially available single crystal disks. To begin, one end of a piece of Pt wire was carefully positioned in a fuel–O_2 flame until a spherical bead of molten Pt formed at the wire tip [36]. When carefully cooled, a single crystal displaying facets with (100) and (111) orientations formed from the molten Pt [37] (Figure 2.1). The single crystal nature of the beads was confirmed from diffraction patterns generated using the Laue x-ray back-reflection method [3,38] and by using visible light reflection [7] techniques. Following growth, a crystal bead was mounted in a goniometer, oriented and mechanically polished to expose an electrode surface plane with defined crystallographic orientation [7].

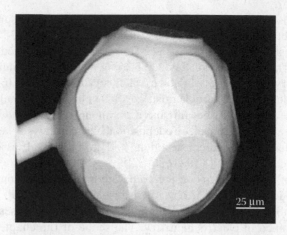

FIGURE 2.1   Scanning electron microscope image of a platinum single crystal prepared by melting the end of a platinum wire (25 μm diameter) in a CH_4/O_2 flame. To enhance the facets, the crystal was subsequently annealed in a CH_4 flame for 1 h. The (111) facets are oval shaped and the (100) facets are circular. (Adapted from Komanicky and Fawcett, *J Electroanal Chem*, Vol. 556, (2003): p. 109. With permission.)

Pt single crystal electrodes prepared by Clavilier's method have come to be known as bead-type electrodes, distinguishing them from the disk electrodes that developed from work in UHV. The size of a bead-type crystal is generally determined by the diameter of the wire from which it is grown. Frequently, 0.5 mm diameter wire is used to produce ~2.5 mm diameter quasispherical beads. However, larger beads (~6 mm diameter) are sometimes grown for use in spectroelectrochemical studies [28,39], and microelectrodes (~25 μm diameter) constructed from Pt single crystals are also of interest [37,40,41]. Since they are not available commercially, some details concerning the fabrication of bead-type single crystals follow.

Typically, single crystal beads are grown from high purity (>99.995%) Pt wire, although zone refining and further purification take place to some extent during bead formation. An experimental arrangement for single crystal Pt growth is diagramed in Figure 2.2a. The Pt wire is held fixed to avoid vibration during crystal cooling, while the torch is mounted on a stage that enables control of the flame position relative to the crystal. Initially, one forms a Pt bead by placing an end of the wire into a region of the flame where the temperature is sufficient to melt Pt (1772°C). Care should be taken, because if the bead becomes too large, it can easily detach from the wire. In addition, light emitted from Pt near its melting point is extremely bright. Therefore, the molten bead must be viewed through neutral density filters or welder's goggles. Once a bead has formed, it can undergo further flame treatment to increase the size and number of facets that appear on the surface after cooling. This stage requires fine manipulation of the flame position to allow only the bottom portion of the bead to become molten and to continuously adjust the height of the melt zone (i.e., the boundary between molten and solid phases) (Figure 2.2[inset]). By controlling the location of the bead in the

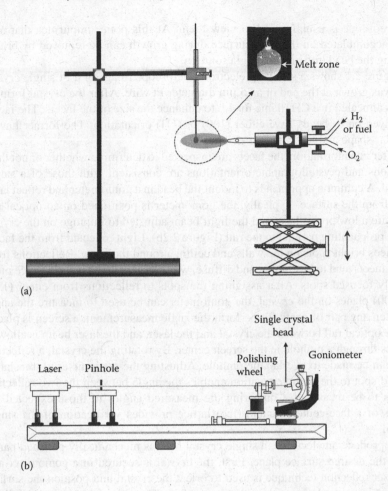

Melt zone

H₂ or fuel

O₂

(a)

Single crystal bead

Polishing wheel

Goniometer

Laser   Pinhole

(b)

FIGURE 2.2 (a) Experimental arrangement used to form a single crystal Pt bead from high purity Pt wire. The fuel/O₂ torch mounts to a stage that enables control of the flame position relative to the crystal. (inset) Image of a Pt bead within a H₂/O₂ flame recorded on a charge coupled device (CCD) detector through neutral density filters. The melt zone separates the molten (bright) portion of the crystal from the solid (dark) portion. (b) Apparatus for orienting a single crystal by laser back reflection and polishing to isolate the desired surface plane on the crystal.

flame, the melt zone is moved slowly, over a few minutes, from the bottom to near the top of the bead and then returned to the bottom. In performing these steps, optical filtering is essential in order to view the bead in the flame and discern the melt zone. The procedure concentrates impurities in the top portion of the bead and is repeated several times. During the cooling half of the cycle, flat facets can be observed to develop on the surface of the solid phase. After the last cycle, the bead can be removed from the flame and allowed to cool. For ~2.5 mm diameter spherical beads, the facets can be large enough to visualize unaided. However, a stereoscope is usually used to view them. At this point, impurities that may have accumulated on the bead surface during growth can be removed by briefly placing the bead into an aqua regia solution.

Figure 2.1 shows a scanning electron microscope image of a Pt single crystal that was grown at the end of a 25 μm diameter Pt wire. After the crystal formed, it was annealed in a CH_4 flame for 1 h to enhance the size of the facets. The facets displayed on Pt beads have either (100) or (111) orientation. The former have a circular shape and the latter are elliptical.

After bead formation, the facets are inspected to determine whether or not their positions and crystallographic orientations are consistent with those of a single crystal. A common approach is to mount the bead in a goniometer and reflect laser light from the surface. Typically, the goniometer is positioned on an optical rail opposite a low power laser and the light beam adjusted to impinge on the crystal while propagating parallel to the rail (Figure 2.2b). Light reflected from the facets produces bright spots on the walls and ceiling around the room. Reflections from (100) facets tend to be circular and diffuse, while larger (111) facets produce more sharply focused spots. After assigning the spots to reflections from either (111) or (100) planes on the crystal, the goniometer can be used to measure the angle between any pair of spots. To perform the angle measurement, a screen is placed on the optical rail between the crystal and the laser, and the laser beam is allowed to pass through a pinhole in the screen center. By rotating the crystal, a reflection spot can be made to overlap the pinhole. Adjusting the goniometer to translate a second spot to the pinhole position enables the angle between the two reflecting planes to be measured. Comparing the measured angles to those expected for planes of a face-centered-cubic (fcc) lattice provides verification of the single crystal.

To produce an electrode, a single crystal bead is mechanically polished parallel to the desired surface plane. First, the crystal is secured in a goniometer and the laser reflection technique is used to orient the crystal and position the surface plane of interest relative to the polishing surface. Then, a layer of epoxy is built up around the mounted crystal in stages. Small amounts of epoxy are applied to the sides of the crystal. The epoxy is allowed to harden and the alignment checked before the next layer is added. The procedure is repeated 2–3 times until the crystal is fully covered. At this point, the epoxy is usually allowed to harden overnight before polishing to expose the electrode surface plane. To polish the crystal, common approaches include securing the goniometer in a lapping fixture, which can be placed on a polishing wheel, and mounting a polishing wheel on the optical rail and bringing the wheel to the crystal. Coarse sandpaper (i.e., 320 grit) is used initially to remove close to a hemisphere of the crystal. Successively finer grades, down to at least 0.25 μm alumina or diamond paste, are used subsequently until a mirror finish is attained. Between grit changes, the crystal should be carefully rinsed to avoid carrying coarser particles into subsequent steps. In addition, it is useful to collect polishing debris to allow future recovery of Pt. After the final polishing, the epoxy resin is removed with a suitable solvent (chloroform) and the crystal is annealed in a fuel–air flame (see Section 2.2.2) for approximately 20 min.

2.2.2 Surface Annealing and Cleaning

Immediately prior to each electrochemical measurement, the surface of a Pt single crystal is treated to remove contaminants and order the top-most atoms. In studies aided by UHV techniques, the electrode undergoes cycles of Ar^+ sputtering, followed by thermal annealing using established surface science protocols [9]. The surface is characterized in terms of structure and composition before transfer to the chamber where electrochemistry is performed. For bench-top experiments, electrodes are thermally annealed for a few minutes in a fuel–air flame. Butane, CH_4, and H_2 are common fuels. The use of air as the oxidant gas maintains the flame temperature below the melting point of bulk Pt, but is sufficient to enable relaxation and rapid diffusion of atoms at the surface.

Flame annealed Pt electrodes must be cooled under conditions that protect the surface from contact with atmospheric contaminants and O_2. After removal from the flame, the crystal is moved to a vessel purged by an ultrapure $Ar + H_2$ (30 vol. % H_2) gas mixture [7,34,42–44]. The H_2 in the $Ar + H_2$ gas mixture reduces the likelihood that the atoms in the Pt surface will become oxidized and disordered as the crystal cools [7,44]. An example apparatus is shown in Figure 2.3. The bottom of the glass vessel is filled with ultrapure water. The purge gas is often bubbled through the water. The crystal is held above the water surface as it cools. After the redness disappears (a period of ~15–30 s for a 2.5 mm diameter bead-type crystal), the crystal is submerged in the ultrapure water. Successful removal of impurities from the Pt surface during annealing is signaled by the adherence of a large water droplet to the crystal as it is pulled from the water. The electrode can then be transferred from the cooling vessel to the electrochemical cell under the protection of this water droplet.

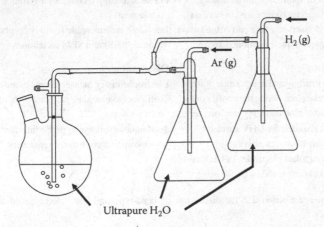

FIGURE 2.3 Apparatus for cooling single crystal electrodes in an atmosphere of $Ar + H_2$ (30 vol. % H_2) after flame annealing. The single crystal electrode is inserted through the side neck of the round bottom flask shown immediately after removal from the flame. The regulators for gas flow are not shown. It is important not to exceed 30 vol. % H_2 in Ar to prevent ignition of the H_2 at the flask outlet.

The bench-top flame annealing procedure described above is employed with both bead-type and disk single crystal Pt electrodes [24,45–47]. Due to the geometry of disk single crystal electrodes, though, they are sensitive to distortion from stress induced by cycles of heating and cooling [48]. In early studies, strategies were developed to heat crystals electrically [24] or in a furnace [49] and thereby provide a means to control temperature changes. Crystals were exposed to iodine vapor during annealing and cooling to protect the electrode surface from contamination and oxidation [24]. However, the complexity of the procedure relative to flame annealing has led to greater adoption of the flame annealing approach (cf., Refs. [45–47]. Recently, inductive heating methods have been employed with disk-type single crystals to reduce surface defects by providing more uniform heating conditions [50,51]. Inductive heating is a promising approach, because it can be performed in a controlled atmosphere and therefore can be applied to

---

TABLE 2.1
Approaches to Pt Single Crystal Electrode Characterization[a]

UHV[b]	In Situ
Advantages:	**Advantages:**
High sensitivity electron (i.e., LEED, AES, HREELS) and ion (i.e., ISS, SIMS) scattering techniques can be applied prior to and following electrochemical measurements to characterize the electrode structure and composition.[c]	Coulometry and voltammetry techniques are sensitive to surface structure, surface phase transitions, and the adsorption and desorption of species at submonolayer levels. The interface remains intact during analysis.
Through cycles of sputtering and annealing, an electrode surface can be prepared to have an initially well-defined structure and composition as verified by techniques mentioned above.	Flame annealing techniques are simple to implement. Interfaces are amenable to study by photon (IR, Raman, SXS) and SPM techniques.[d]
Disadvantages:	**Disadvantages:**
Changes in the structure and composition of the interface can take place during the transfer of the electrode from solution into UHV and vice versa.	Electrochemistry measurements alone are not sufficient to elucidate interface structure and composition.
The equipment required for UHV surface science experiments is expensive.	Techniques available to probe interface composition have limited sensitivity.
The need to manipulate electrodes in UHV adds complexity to experiments.	

[a] Additionally, see Section 2.2 for discussion of bead-type and disk electrode advantages and disadvantages.

[b] UHV = ultra high vacuum.

[c] LEED = low energy electron diffraction; AES = Auger electron spectroscopy; HREELS = high resolution electron energy loss spectroscopy; ISS = ion scattering spectroscopy; SIMS = secondary ion mass spectrometry.

[d] IR = infrared; SXS = surface x-ray scattering; SPM = scanning probe microscopy.

metals that absorb gases, such as palladium [52] and materials that easily oxidize (i.e., Cu, Ag). Another approach that leads to well-ordered surfaces uses a CO atmosphere to cool down the electrode after flame annealing [53].

Some key advantages and disadvantages of the UHV and *in situ* approaches to single crystal electrode characterization are summarized in Table 2.1. Furthermore, the small, bead-type single crystals offer advantages in that they can be grown and oriented in the laboratory and are easily annealed and cooled using bench-top methods; by contrast, the larger disk-shaped crystals require more careful handling during annealing and cooling. Nevertheless, more easily than the bead-type crystals, the disk-shaped crystals can be secured into a conventional working electrode holder to prevent the sides of the crystal from contacting solution and thereby eliminate the requirement to work in a meniscus configuration. This latter strategy has been used to advantage for rotating disk electrode measurements of dissolved gases under conditions that conform to simple theories of rotating disk electrode voltammetry [9,20].

2.3 EXPERIMENTAL TECHNIQUES FOR ELECTRODE CHARACTERIZATION

2.3.1 GENERAL EXPERIMENTAL CONSIDERATIONS

At the time the first single crystal electrochemistry experiments were performed, a controversy developed concerning differences in responses observed for flame annealed and UHV prepared single crystal Pt [7,48,54,55]. It was eventually shown, through a combination of electrochemical [7,43,56], UHV surface science [48,54,55], and later scanning tunneling microscope (STM) [57–59] and surface x-ray scattering (SXS) [60] experiments, that flame annealing produces clean, ordered Pt single crystal surfaces with less susceptibility to contamination during handling than electrodes prepared in UHV. Validation of the flame annealing approach opened opportunities in several areas to gain new insights through simple electrochemical measurements on single crystal Pt. The following sections discuss techniques for performing electrochemical measurements with single crystal electrodes and surface sensitive methods for electrode characterization.

2.3.1.1 Electrochemical Methods

In principle, the full range of voltammetry techniques can be applied to investigate processes at single crystal Pt electrodes. Cyclic voltammetry has probably been used most often. To perform voltammetry experiments, the prepared surface plane of a single crystal electrode is typically brought in contact with the electrolyte solution and positioned slightly above the liquid level such that a meniscus hangs from the face of the crystal (Figure 2.4) [7,61–63]. This so-called "meniscus configuration" prevents electrochemical reactions from taking place along the sides of the electrode. For applications that are not easily adapted to the meniscus configuration (i.e., spectroelectrochemistry,

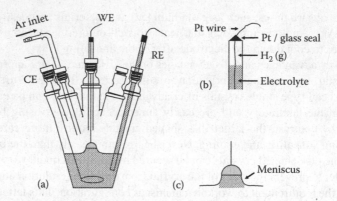

FIGURE 2.4 (a) Typical electrochemical cell employed for single crystal electrochemistry experiments. WE, CE, and RE refer to the working, counter, and reference electrodes. (b) Enlargement of the top portion of the RHE showing the Pt/glass seal and the H_2 gas segment. (c) Enlargement of the meniscus region.

hydrodynamic voltammetry), the sides of the crystal can be protected by wrapping in polytetrafluoroethylene (PTFE) tape (cf., [64,65]), or by pressure sealing in PTFE rod [45]. However, the meniscus configuration is simpler and reduces chances for surface contamination during electrode handling. Even approaches for performing rotating disk voltammetry in the meniscus configuration have been demonstrated [66–69]. It is common, although not required, to apply a few millivolts positive potential from the onset of hydrogen evolution to the single crystal electrode as it is brought in contact with the solution in the electrochemical cell. The applied potential can protect the Pt single crystal surface plane from oxidation during immersion into aqueous electrolyte solutions. Surface oxidation can cause disordering of the surface atoms through Pt–O place exchange reactions [58,70] and cause the formation of steps and nanoscale Pt deposits [58]. However, in cases where electroactive species are reduced at low potentials (e.g., under potential deposition [upd] processes), the immersion potential should be varied accordingly. In early experiments, electrodes were immersed at the open circuit potential.

To meet the experimental goals, electrolytes and gases employed for single crystal electrochemistry experiments should be of highest purity. Counter electrodes are typically constructed from Pt gauze or wire. A variety of reference electrodes can be employed, although, as discussed below, there are advantages to using a reversible hydrogen electrode (RHE) as the reference. In addition to the reference and counter electrodes, electrochemical cells need to accommodate the working electrode in the meniscus configuration and be equipped with ports for entry of gases to allow removal of dissolved O_2 from the electrolyte and blanketing of the electrolyte by an inert gas (i.e., Ar or N_2), as needed. An example electrochemical cell is shown in Figure 2.4. Several other cell types are possible, though, including those with separate compartments for the working, counter, and reference electrodes [24,48,49]. Cells are usually constructed from glass to prevent permeation of atmospheric gasses and enable decontamination by strong oxidant solutions. In some cases, a cell constructed from PTFE should be used if glass is not stable in the working solution. This is the case for hydrofluoric acid (HF) or strong alkaline solutions.

The assembled cell in the example in Figure 2.4 is shown with an RHE reference. The RHE is frequently employed in single crystal electrochemistry experiments. The internal electrolyte has the same composition as the electrolyte in the cell working electrode compartment, thereby eliminating the liquid junction potential and reducing the likelihood for contamination of the electrolyte by solution in the reference electrode. The potential of an RHE is in accord with the Nernst equation. In addition to temperature, the potential depends upon the H_2 partial pressure and internal electrolyte H^+ activity. The potential of an RHE can be checked by measuring with respect to a commonly available laboratory reference, such as a silver–silver chloride electrode, or standardized against a well-defined hydrogen reference electrode.

An RHE can be easily created by bubbling H_2 into a compartment where a platinum wire is immersed. Increasing the area of the platinum by depositing Pt black helps increase the stability and reproducibility of the potential. To prevent H_2 from reaching the working electrode, a Luggin capillary [71,72] and stopcock are used to bridge the reference and working electrode compartments. The electric contact between the compartments is achieved through the electrolyte film wetting the walls of the stopcock. This design, however, can result in relatively high impedance for the reference electrode and cause stability problems with some potentiostats.

An alternative RHE design traps a hydrogen bubble inside a soft glass tube that contains the platinum wire (~2.5 cm). Using a fuel–air flame, one end of the tube is softened and then sealed around the Pt wire as it is inserted about 1.5 cm into the tube. Just prior to electrochemical experiments, the tube is filled with electrolyte by turning the open end upward and dispensing the solution into the tube through a long pasture pipette. The tube should be filled completely and any entrapped air bubbles removed. Next, the tube is inverted over a small beaker (~50 mL) partially filled with electrolyte and the open end immersed in the solution. Care should be taken so as not to trap air in the tube. Then, H_2 gas is generated in the vicinity of the Pt wire by electrolysis. The Pt wire is connected to the negative terminal of a battery (9 V is convenient) or a power supply. A piece of Pt wire can be placed in the beaker to serve as the positive counter electrode. A plug of H_2 gas should be generated until about half the Pt wire inside the tube is exposed to the gas. The remainder of the Pt wire should remain in contact with electrolyte (Figure 2.4b). Finally, the RHE can be removed from the beaker and placed in the electrolyte inside the electrochemical cell, again being especially careful not to allow a bubble of air to enter the electrode. An air bubble entering the electrode will alter the partial pressure of the entrapped H_2 gas and thereby change the reference potential.

The cell and all components in contact with the working electrode compartment must be scrupulously cleaned. Glass components are soaked overnight in a bath of either strong acid (i.e., 1:3 by vol. $HNO_3 + H_2SO_4$) or a mixture of $KMnO_4$ (ca. 2 g/L) and dilute H_2SO_4. When more extensive cleaning is needed, for example in cases when new cells are used or difficult to remove contaminants are present, the components can also be boiled in concentrated H_2SO_4 for several hours. Just prior to experiments, glass parts are rinsed in cool ultrapure water. Those soaked in permanganate solution are additionally rinsed in a 1% $H_2O_2 + 0.1$ M H_2SO_4 solution to remove traces of manganese oxides. Components then undergo a rinsing cycle, which involves first boiling in ultrapure water for 30 min followed by flowing a stream of cool ultrapure water over the surfaces. The rinsing cycle is repeated at least three times, at the judgment of the experimentalist. If experiments are to be performed in $HClO_4$ electrolyte, several rinsing steps are required to remove strongly adsorbing anions, such as sulfate species. The procedure is also applied to remove contaminants from the inner surfaces of flasks used to prepare electrolyte solutions and glassware employed in the construction of the RHE.

2.3.1.2 Nonelectrochemical Surface Sensitive Methods

Several nonelectrochemical methods are employed to characterize single crystal electrodes. In early studies, UHV surface science techniques were used extensively (cf., Ref. [9] and references therein). These methods continue to be applied to assess electrode surface structure and composition. Recent examples can be found in studies that have guided the development of new electrocatalyst materials [20,73]. Because surface reconstruction and changes in adlayer structure can take place during the transfer of an electrode between UHV and the electrochemical cell, over the years, there has been considerable effort to develop techniques that enable electrodes to be characterized in solution under potential control. SXS [27,74] and STM [74,75] measurements have been demonstrated as powerful *in situ* probes of atomic-scale single crystal electrode surface properties. Optical reflection at ultraviolet and visible wavelengths [76,77] and second harmonic generation (SHG) [40,41,78–81] are sensitive to atomic-scale surface order and the presence of adsorbates on single crystal electrodes under potential control. Other *in situ* techniques, including radiotracer [82–84], infrared spectroscopy [85–88], and vibrational sum frequency generation (VSFG) spectroscopy [89,90], mainly interrogate the properties of the adsorbed species. Also, the chemical nature of volatile products produced on single crystal electrodes during an electrochemical reaction can be probed by differential electrochemical mass spectrometry (DEMS) [91–93]. While not directly applicable to single crystal electrode characterization, techniques of nuclear magnetic resonance (NMR) spectroscopy [94,95] and Raman [96,97] spectroscopy probe the interfacial properties of metal nanoparticles and provide results that guide understanding of chemical bonding at single crystal electrodes. Very recent developments have enabled Raman spectroscopy to be extended to single crystal electrodes [98].

Although detailed discussions of the techniques are beyond the scope of this chapter, additional information is available in the cited references.

2.3.2 Low Index Surface Planes

2.3.2.1 Pt(111)

Early applications of Pt single crystal electrodes focused on the surface electrochemistry of hydrogen reductive adsorption and oxidative desorption processes [7,22,23,54,55,99–103]:

$$H_3O^+ + Pt + e^- \leftrightarrow Pt - H + H_2O \qquad (2.1)$$

The low index surface planes were investigated in aqueous acid electrolyte with the aim of identifying the basic surface atom arrangements responsible for each wave appearing in the so-called hydrogen potential region, a range of potentials over which the reactions in Equation 2.1 take place, of cyclic voltammograms for polycrystalline Pt [7,100,102]. In addition to providing insights into the reactions in Equation 2.1, the experiments also revealed striking effects of electrolyte composition on the current in linear sweep voltammograms [23,34,48,55,104]. As discussed initially by Scherson and Kolb [104], responses were shown to be associated with anion adsorption on surface regions having long-range atomic order and to be sensitive to both the nature and the concentration of the supporting electrolyte [48,104–107].

Illustrative cyclic voltammograms for Pt(111) electrodes are shown in Figures 2.5 through 2.6. Figure 2.5 displays benchmark responses for Pt(111) in 0.5 M

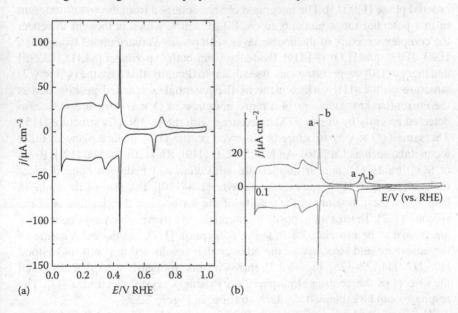

FIGURE 2.5 (a) Cyclic voltammogram of Pt(111) in 0.5 M H₂SO₄ recorded at a sweep rate of 50 mV/s. (From Climent et al., *Russ. J. Elec.*, Vol. 42, (2006): p. 1145. With permission.) (b) Same as (a), but with sweep rate 20 mV/s, from early work that examined the effects of cooling atmosphere. (From Clavilier et al., *J Electroanal Chem*, Vol. 295, (1990): p. 333. With permission.)

H_2SO_4. In Figure 2.5a, between 0.05 V and 0.25 V current in the positive and negative potential scan directions is nearly constant, and $i_{anodic} \approx - i_{cathodic}$. The current through the region results from the oxidation (positive scan) and reduction (negative scan) of hydrogen, in accord with Equation 2.1, mainly on the smooth (111) oriented surface plane [9,23,99,107,108]. Near 0.3 V, the changes in the current reflect a shift in the equilibrium in Equation 2.1 to the left, toward clean Pt. Integrating the voltammogram between 0.05 and 0.32 V and subtracting the double layer capacitive charging contribution (see Section 2.4.1.3 for additional details) give a charge associated with hydrogen desorption (or adsorption) of ca.160 μC cm^{-2} [107]. Referenced to the Pt atom density on the (111) surface plane (Table A.2), the hydrogen coverage inferred from the charge corresponds to 2/3 of a monolayer. The remaining 1/3 monolayer is probably not attained, because hydrogen evolution starts before the adlayer can be completed [109].

Scanning positive from 0.05 V to 0.6 V in Figure 2.5a, the surface transitions from hydrogen to SO_4^{2-} (or HSO_4^-) are covered. The oxidative desorption of Pt–H opens Pt sites, which become occupied by anions as excess positive surface charge develops (cf., Refs. [84,107,110–112], and references therein). The pair of spikes symmetric about the potential axis near 0.45 V signals the reversible transition of the SO_4^{2-} (or HSO_4^-) adlayer between disordered and ordered states [110,113]. In scanning positive from 0.3 V to 0.6 V, the anion coverage increases until the maximum ($\theta \approx 0.2$) [84,111] is attained, which drives the formation of an ordered phase [110,113]. The integrated charge obtained from the voltammogram in this potential range amounts to ca. 80 μC cm^{-2}, which is thought to reflect the complex structure of the double layer with possible contributions from SO_4^{2-}, HSO_4^-, OH^-, and H_3O^+ [84,114]. Evidence from both experiment [84,111,113,115] and theory [110] suggests the ions assemble into domains that have the ($\sqrt{3} \times \sqrt{7}$) structure on the (111) surface plane as the potential is scanned positive across the current spikes. Domains of a more compressed (3 × 1) structure were also detected recently by in situ STM coexisting with the ($\sqrt{3} \times \sqrt{7}$) structure [115]. The same ($\sqrt{3} \times \sqrt{7}$) structure is observed on the (111) surface plane of other fcc metals, such as Cu [116], Au [117,118], Ir [119], Rh [120], and Pd [121]. Rows of SO_4^{2-} (or HSO_4^-) appear to alternate with chains of hydrogen bonded water molecules and/or H_3O^+ ions in the adlayer [118–120]. The size of the spike at 0.45 V is a clear measure of the quality of the surface and the cleanliness of the solution [122]. Toward more positive potentials, the irreversible peaks near 0.7 V are known to be associated with anion adsorption [114], but the exact nature of the adsorbate and structure of the adlayer that results are not well understood [108,112–114,123–125]. Figure 2.5b shows results from early work that examined the effects of the cooling atmosphere on Pt single crystal electrodes [43]. The responses can be explained similarly to those in Figure 2.5a.

The symmetric response between 0.3 V and 0.6 V in Figure 2.5, and in general for Pt(111) in aqueous solution across the range of potentials where anion adsorption takes place, has often been referred to as the "butterfly" region [7,48,84,110,113,126,127], initially referred to as unusual states [23,54,55,128,129]. In sulfuric acid electrolyte, the nature of the adsorbed ions (SO_4^{2-} or HSO_4^-), their surface coordination geometry (bi-dentate, tri-dentate, etc.), and the adlayer structures that form been debated over the years [9,84,112,130]. The background leading to the present understanding in this area is discussed in detail in Section 2.4.1.4.

Figure 2.6 compares cyclic voltammograms of Pt(111) in different acid electrolytes. The variation in the chemical nature of the anion, $HClO_4$, H_2SO_4, or H_3PO_4 dramatically affects the responses. The voltammogram recorded in 0.1 M H_2SO_4 (Figure 2.6b) is similar to the voltammogram of the electrode in 0.5 M

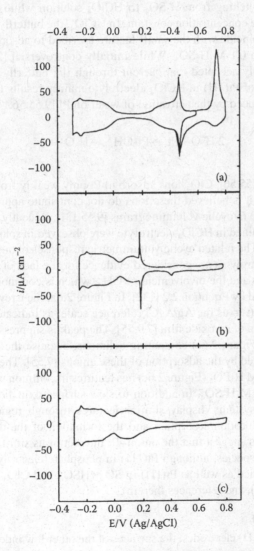

FIGURE 2.6 Cyclic voltammogram of Pt(111) in (a) 0.1 M $HClO_4$, (b) 0.1 M H_2SO_4, and (c) 0.1 M H_3PO_4. The voltammograms were recorded at a sweep rate of 50 mV/s. (From Herrero et al., *J Phys Chem*, Vol. 98, (1994): p. 5074. With permission.)

H_2SO_4 (Figure 2.5), as expected, but with subtle differences at potentials positive of the hydrogen region. In the range dominated by the formation of the disordered SO_4^{2-}/HSO_4^- adlayer (0.32 V–0.42 V in Figure 2.5), it is evident that the current changes are smoother in the lower concentration electrolyte. Also, the butterfly region, including the current spikes that signal the disorder-order phase transition of the adsorbed anions, shift positive with decreasing sulfate concentration [56,84,99,111,114]. The behavior reflects the sensitivity of anion adsorption to electrolyte concentration. Comparing the voltammograms in Figure 2.6a and b shows the effects of switching from H_2SO_4 to $HClO_4$ solution while maintaining the formal electrolyte concentration constant. In $HClO_4$, the butterfly peaks appear at potentials that coincide with the small feature ascribed to adsorbed OH^- formation on Pt(111) in 0.1 M H_2SO_4. While initially controversial [23,48,55,56,131], it is now generally accepted that current through the butterfly region in cyclic voltammograms of Pt(111) in $HClO_4$ electrolyte mainly results from the adsorption of OH^- produced by the activation of water on Pt [9,55,56,131]:

$$2\ H_2O + Pt \rightarrow Pt\text{-}OH_{ads} + H_3O^+ + e^- \qquad (2.2)$$

Similar to F^- [55,56], ClO_4^- ions adsorb to Pt only weakly from aqueous solutions. Therefore, it is believed these ions do not contribute appreciably to anion adsorption charge in cyclic voltammograms [9,55,131]. Recently, responses identical to those obtained in $HClO_4$ electrolyte were observed in solutions containing CF_3SO_3H [132]. The related cyclic voltammetric responses obtained in F^-, ClO_4^-, and $CF_3SO_3^-$ electrolytes are considered evidence for the lack of specific adsorption by these ions and the involvement of OH^-, which is common to all the solutions, as suggested by Equation 2.2 [132]. In Figure 2.6a, the irreversible peak just positive of 0.7 V (versus the Ag/AgCl reference scale, as indicated) is associated with the formation of an oxide film [7,9,55]. The peak is suppressed in the sulfate (and phosphate, Figure 2.6c) containing solutions, because the oxide formation reaction is inhibited by the adsorption of these anions [7,55]. The voltammogram for Pt(111) in 0.1 M H_3PO_4 (Figure 2.6c) has features in common with the response for Pt(111) in 0.1 M H_2SO_4. In addition to slow surface oxidation at the highest potentials, both systems display similar features through regions encompassing hydrogen adsorption/desorption and the formation of the disordered anion phase. The results suggest that the anion–Pt interaction is similar for phosphate and SO_4^{2-}/HSO_4^- species, although Pt(111) in phosphate electrolyte [63,133–135] has not been studied as well as Pt(111) in SO_4^{2-}/HSO_4^- and ClO_4^- solutions (Refs. [9,84,110,126,130], and references therein).

2.3.2.2 Pt(110)

In contrast to Pt(111) electrodes, the surfaces of the other low index surface planes undergo large structural changes associated with reconstruction that takes place during annealing and subsequent cooling steps [17,18]. Uncertainty in the surface structures that result for Pt(110) and Pt(100) electrodes has led to there being fewer studies performed with these surfaces relative to Pt(111). However, methods have emerged for stabilizing the unreconstructed (1 × 1) surfaces of Pt(110) and Pt(100) through control of the cooling conditions following annealing [53,136,137].

Markovic and coworkers confirmed by *in situ* SXS measurements that both Pt(110)-(1 × 1) and Pt(110)-(1 × 2) surfaces can be prepared by flame annealing procedures and maintained as stable under electrochemical potential control [136]. Pt(110)-(1 × 1) is the stable phase above 810°C and is therefore present during typical crystal annealing conditions, which usually reach to 1300°C or higher. When a crystal above the phase transition temperature is cooled in an inert gas containing about 30% H_2, the Pt(110)-(1 × 1) structure is preserved. On the other hand, slowly reducing the crystal temperature to below 810°C during annealing (a transformation in crystal color described as from "bright-yellow" to "orange" to "red" [136]) enables the transition to the Pt(110)-(1 × 2) surface structure, which

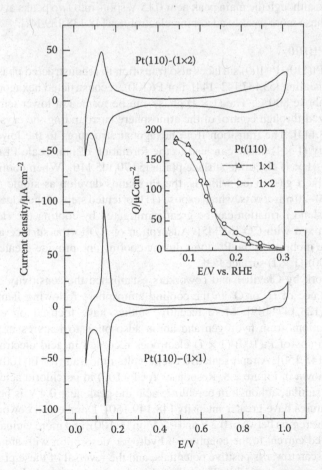

FIGURE 2.7　Cyclic voltammogram of Pt(110)-(1 × 1) and Pt(110)-(1 × 2) recorded in 0.5 M H_2SO_4 at a sweep rate of 20 mV/s. The inset reports the integrated charge as a function of potential. (From Markovic et al., *Surf. Sci*, Vol. 384, (1997): p. L805. With permission.)

is retained upon cooling. Cyclic voltammograms of Pt(110)-(1 × 1) and Pt(110)-(1 × 2) electrodes in 0.5 M H_2SO_4 are shown in Figure 2.7. For both surfaces, the pair of peaks near 0.15 V arises from the coupling of hydrogen desorption and anion adsorption on the scan toward positive potentials, and the reverse of these processes on the scan toward negative potentials [136,138–140]. The peaks are sharper and attain higher maximum current on the (1 × 1) surface. After subtraction of the capacitive component, the charge under the peak amounts to ca. 215 μC cm^{-2}. The contribution from adsorbed hydrogen, determined by performing charge displacement measurements (see Section 2.4.1.1) at 0.08 V, is ca. 150 μC cm^{-2}, which corresponds to a monolayer of hydrogen on the (1 × 1) surface [106]. Oxide formation (Equation 2.2) takes place above about 0.4 V and appears to commence at more positive potentials on Pt(110)-(1 × 1) [136]. Responses similar to those in Figure 2.7 are also observed using aqueous perchloric acid or HF electrolytes, although the main peak near 0.15 V splits into two peaks at 0.14 V and 0.24 V in these more weakly adsorbing electrolytes [48,138,139,141].

2.3.2.3   Pt(100)

Similar to Pt(110), Pt(100) surfaces also transition to reconstructed phases during thermal annealing [53,137,142–144]. For Pt(100), reconstructed hexagonal phases are stable above 800°C. The (1 × 1) phase can be formed at lower temperatures and stabilized through control of the atmosphere surrounding the crystal during cooling [53,144]. The transition from hexagonal structures to the lower surface atom density (1 × 1) phase can lead to the formation of nanoscale Pt islands on the Pt(100)-(1 × 1) electrode surface plane [53,70,145,146]. When cooling is performed in inert gas mixed with H_2, the Pt islands develop as single atom high squares, ~10–20 nm across, which expose (111) oriented steps at the edges [53,145]. However, island formation can be greatly mitigated by cooling the electrode in inert gas mixed with CO [53,145]. Adsorption of CO to the surface appears to increase the mobility of the Pt atoms during cooling and provide greater stability to the Pt(100)-(1 × 1) surface [53].

Early work by Clavilier and coworkers established the sensitivity of Pt(100) to the presence of H_2 or O_2 in the cooling atmosphere following flame annealing [33,34,128,147–149]. More recently, studies have focused on elucidating the contributions from hydrogen and anion adsorption to features in the cyclic voltammograms of Pt(100)-(1 × 1) electrodes recorded in acid electrolyte solutions [137,145,150]. A representative cyclic voltammogram for Pt(100) in 0.5 M H_2SO_4 is shown in Figure 2.8. Responses for Pt(100) in perchloric acid solutions are closely similar, although in perchloric acid the peak near 0.4 V is broader and the side features have greater intensity [48,149,150]. Domke and coworkers have examined the region between 0.2 V and 0.7 V on Pt(100), and more vicinal surfaces, and attributed current to the coupling of hydrogen desorption with anion adsorption on the scan towards positive potentials, and the reversal of these processes on the negative going scan [150]. The group also suggested current below 0.2 V may originate from hydrogen adsorption on (111) oriented step defects and assigned a peak at 0.25 V to contributions from hydrogen adsorption on the (100) terrace

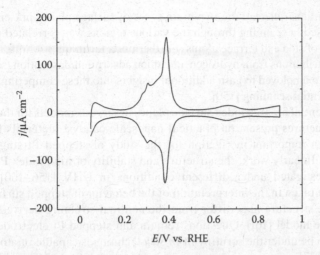

FIGURE 2.8 Cyclic voltammogram of Pt(100) in 0.1 M H_2SO_4 recorded at a sweep rate of 50 mV/s. (From Lopez-Cudero et al., *J Electroanal Chem*, Vol. 586, (2006): p. 204. With permission.)

sites at the step edge. However, Lopez-Cudero and coworkers recently pointed out that the defect density on state-of-the-art prepared Pt(100)-(1 × 1) electrodes is expected to give negligible hydrogen adsorption charge and instead suggested that responses below 0.2 V may arise mainly from adsorption of hydrogen to atop sites on (100) oriented terraces following filling of the fourfold sites [109,145].

2.3.3 HIGH INDEX SURFACE PLANES

Following early reports of responses for carefully selected stepped Pt electrodes [35,151,152], Motoo and Furuya studied in 0.5 M H_2SO_4 the voltammetry of 37 different Pt single crystal surface planes from zones that form the boarder of the unit stereographic triangle in Figure A.5 [153]. Soon after, Clavilier and coworkers scrutinized the electrochemistry of a series of Pt(s)-[n(111) × (111)] (or equivalently, Pt(s)-[(n−1)(111) × (110)] [28]) electrodes in 0.5 M H_2SO_4 [129]. Between about 0.05 V and 0.8 V (vs RHE), waves in linear scan voltammograms were characteristic of the (111) and (110) planes comprising the electrode surfaces. The charge determined by integration of the peak near 0.15 V associated with adsorption and desorption of hydrogen on (110) oriented step sites scaled linearly in proportion to the fraction of surface atoms in step sites. Similarly, the charge determined by integration of features characteristic of the (111) planes correlated directly with the density of terrace atoms. Analogous results were obtained in related experiments that employed a series of Pt(s)-[n(111) × (100)] electrodes [42].

More recently, surfaces with structure Pt(s)-[n(100) × (111)] were investigated in 0.1 M $HClO_4$ [150]. In keeping with earlier studies, features in cyclic voltammograms reflected the two main types of surface planes present, and the charge passed in scanning through the various regions was correlated directly to the density of step and terrace atoms. Furthermore, techniques for discriminating charge contributions from hydrogen and anion adsorption/desorption (see Section 2.4.1.1) were employed to gain additional insights into these competing processes during potential scanning [150].

The potential for high index surface planes to serve as models for the edge and terrace structures present on practical nanoscale catalyst materials [154–156] has been an important motivation for the study of stepped Pt single crystal electrodes. In early work, the structure and stability of high index Pt surfaces where investigated under different conditions in UHV [156–160]. Caution needs to be taken in the interpretation of the behavior of stepped surfaces since the surface structure may differ from the nominal structure derived from the hard sphere model [161]. Questions remain and stepped Pt electrode surfaces continue to be under the scrutiny of *in situ* techniques capable of atomic-scale resolution [27].

2.4 APPLICATIONS

2.4.1 DOUBLE LAYER PROCESSES

For any metal, the double layer region spans a range of potentials over which faradaic reactions do not take place and only the electrostatic charge on the metal is affected by changes in applied potential. The electrode is considered "ideally polarized" [3,71] at double layer potentials. Excess surface charge on the metal induces ordering of dipolar solvent molecules and accumulation of charge compensating ions in solution adjacent to the electrode. As potential is made to vary, the so-called double layer charging current flows in response to this atomic level restructuring. The double layer charging current provides insights into the structure and composition of the interface.

Basic theories of electrical double layer properties were developed through foundational experiments that employed liquid Hg electrodes. Studies were enabled by the availability of techniques for precise measurement of surface (interfacial) tension at Hg electrodes and the wide double layer potential range of Hg in aqueous solution (cf., Ref. [71]). Compared to Hg, the properties of the double layer formed by solid electrodes are more difficult to probe [3,71]. For many solid metals, the surface charge density at a given potential is sensitive to the crystallographic orientation of surface atoms and the presence of surface defects [3]. The emergence of single crystal electrochemistry techniques led to rapid developments in understanding double layer phenomena at solid electrodes. Au and Ag electrodes, which can readily be prepared and handled on the bench-top, have a relatively wide double layer potential region, and therefore have been at the focus of work on solid electrodes [3]. There have been fewer investigations of Pt group metals, due to their more complex surface chemistry [162].

The fundamental thermodynamic properties of the electrical double layer are described by the electrocapillary equation. At constant temperature and pressure, the electrocapillary equation can be expressed as:

$$-d\gamma = \sigma_M dE + \sum_i \Gamma_i d\mu_i \qquad (2.3)$$

where γ is the interfacial tension, E is the electrode potential (measured with respect to a constant reference), Γ_i and μ_i are the relative Gibbs excess and the chemical potential, respectively, of species i, present at the metal–solution interface, and σ_M is the excess charge density at the metal surface. When ion adsorption involves charge transfer (the so-called specific adsorption), the equilibrium between solution phase anions (A^{n_a-}) or cations (C^{n_c+}) and the corresponding adsorbed anion (A) or cation (C) species can be described by the following equations:

$$A^{n_a-} \leftrightarrow A + n_a e^- \qquad (2.4a)$$

$$C^{n_c+} + n_c e^- \leftrightarrow C \qquad (2.4b)$$

where n_a and n_c represent the magnitude of the respective anion and cation charges. The chemical potentials of the adsorbed and solution free ions are related by:

$$d\mu_{A^{n_a-}} = d\mu_A - n_a F dE \qquad (2.5a)$$

$$d\mu_{C^{n_c+}} - n_c F dE = d\mu_C \qquad (2.5b)$$

where F is the Faraday constant. Expanding Equation 2.3 to account for specific adsorption of ions gives:

$$-d\gamma = \sigma_M dE + \Gamma_{A^{n_a-}} d\mu_{A^{n_a-}} + \Gamma_A d\mu_A + \Gamma_{C^{n_c+}} d\mu_{C^{n_c+}} + \Gamma_C d\mu_C$$
$$+ \sum_{i \neq A, A^{n_a-}, C, C^{n_c+}} \Gamma_i d\mu_i \qquad (2.6)$$

and substituting expressions for $d\mu_A$ and $d\mu_C$ from Equations 2.5a and 2.5b leads to

$$-d\gamma = (\sigma_M + n_a F \Gamma_A - n_c F \Gamma_C) dE + (\Gamma_{A^{n_a-}} + \Gamma_A) d\mu_{A^{n_a-}}$$
$$+ (\Gamma_{C^{n_c+}} + \Gamma_C) d\mu_{C^{n_c+}} + \sum_{i \neq A, A^{n_a-}, C, C^{n_c+}} \Gamma_i d\mu_i \qquad (2.7)$$

The term that multiplies dE in Equation 2.7 can be defined as the total charge density (Q) and in general will be expressed as:

$$Q = \sigma_M - \sum_k z_k F \Gamma_k \qquad (2.8)$$

where the summation extends to all terms that involve charge transfer [114,163,164] and z_k is the charge transferred by species k with its sign.

Commonly encountered ions that undergo specific adsorption at Pt include OH^-, the halides Cl^-, Br^-, and I^-, and sulfate species (HSO_4^- and SO_4^{2-}). The reduction of H_3O^+ in accord with Equation 2.1 is sometimes also treated as a specific adsorption process. By contrast, the distribution at the Pt–solution interface of nonspecifically adsorbed ions (i.e., F^-, ClO_4^-) is affected primarily by long-range electrostatic interactions. When the electrode surface is free of specifically adsorbed ions, $Q = \sigma_M$ in Equation 2.8 and represents the classical capacitive charging of the electrical double layer [71].

From the standpoint of thermodynamics, Equation 2.7 and Equation 2.3 are equivalent with the difference being that Equation 2.7 breaks out separate terms for the various interfacial charge contributions. A more detailed theory for the charge transferred upon ionic adsorption would require knowledge of the position of the charges at the interface and extend beyond the macroscopic scope of thermodynamics. However, the presence or absence of charge transfer has important consequences for the structure of the double layer and the magnitude of the electric field at the interphase, which is governed by the true charge on the metal, σ_M. Also important is that the experiment does not provide a means to determine the separate terms in Equation 2.7 and only total charge (Q) can be measured.

In aqueous electrolyte solutions, the double layer region for Pt roughly spans potentials between hydrogen adsorption (Equation 2.1) and oxide formation (Equation 2.2). Depending upon the nature of the electrolyte, though, H_3O^+ adsorption (Equation 2.1), OH^- adsorption (Equation 2.2), or both can take place in the same potential region as anion specific adsorption, making the potential limits of the double layer region difficult to define [162,165]. There has been a great deal of interest in understanding processes that occur at double layer potentials on single crystal Pt electrodes. An early motivation was the need to determine origins of the unexpected features in linear scan voltammograms that have come to be associated with adsorption of common electrolyte anions (Figures 2.5 through 2.8). However, double layer phenomena have broad importance with impact on processes in electrochemistry that include electrodeposition, electrocatalysis, and heterogeneous electron transfer, to name a few. The following sections discuss strategies for investigating double layer properties of Pt electrodes and some important applications.

2.4.1.1 Charge Displacement Technique

For Pt, the charge displacement technique is used to determine the excess surface charge on an electrode held at a fixed potential in the double layer or hydrogen adsorption region. When ions and polar solvent initially present at the interface are displaced by neutral nonpolar species, the integral of the transient current that flows is equal to the interfacial charge removed. CO has typically been used as a neutral probe for displacement experiments with Pt single crystal electrodes [107,150,165–168]. It has been shown that formation of the Pt–CO bond contributes negligible charge to the measurements [166,167], and the current arises from the charge transfer processes that accompany the loss of adsorbed hydrogen or anions. Additional advantages of CO as a displacing agent include the fact that, first, as a gas it is simple to introduce into the cell without atmospheric oxygen interference and, second, dissolved CO is easily removed from the solution by Ar purging. The latter allows consecutive experiments to be performed without the need to change the solution. Another displacing agent used, although with more limited applicability, has been iodide [169].

Results of charge displacement measurements are shown in Figure 2.9. The current–time traces in parts a–c were recorded while an initially clean, ordered Pt(111) electrode held at a fixed potential in 0.5 M H_2SO_4 was exposed to CO [107]. The CO was admitted to the atmosphere in the cell through a purge tube placed

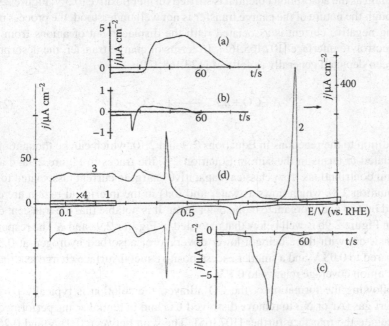

FIGURE 2.9 Current density–time transients recorded during the adsorption of CO onto a well-ordered Pt(111) electrode poised at (a) 0.08 V, (b) 0.30 V, and (c) 0.50 V in 0.5 M H_2SO. (d) Voltammograms for the same electrode in the CO-free electrolyte: (1) a cyclic scan recorded just after CO adsorption showing the effect of CO blocking (scan rate = 50 mV/s); (2) a linear scan that records the CO stripping peak (scan rate = 20 mV/s); (3) a cyclic scan showing the recovery of the initial voltammetric profile for clean Pt(111) (scan rate = 50 mV/s). The potentials are reported with respect to the RHE reference. (From Feliu et al., *J Electroanal Chem*, Vol. 372, (1994): p. 265. With permission.)

near the meniscus formed by the Pt(111) electrode. In this arrangement, CO diffuses across the gas–solution interface into the meniscus region. Changes in current that accompany CO exposure reflect charge transfer processes associated with the electrode/solution interface transforming from its initial state to a state in which the Pt surface is covered by CO. For the measurements at 0.08 V in Figure 2.9a, between about 20 to 80 s the current flow is in response to the development of the CO monolayer. The positive sign indicates that an oxidation process is occurring. The following reaction has been proposed as contributing to the measured current [107,150,165–167]:

$$Pt\text{-}H + CO + H_2O \longrightarrow Pt\text{-}CO + H_3O^+ + e^- \tag{2.9}$$

Figure 2.9b shows results for a CO displacement measurement performed at a more positive potential (0.3 V). A cathodic current spike precedes the anodic current excursion. The response indicates that a fast process associated with cathodic current flow occurs ahead of Pt–H oxidation. The cathodic peak becomes more dominant as the adsorption potential is shifted further positive (0.5 V, Figure 2.9c). Although the nature of the charge transfer is not well understood, the process producing negative current is associated with the displacement of anions from the Pt–electrolyte interface [107,165,168]. In terms of charge transfer, the desorption has been depicted generally as follows [123,168,170]

$$Pt\text{-}A + CO + n_a e^- \longrightarrow Pt\text{-}CO + A^{n_a-} \tag{2.10}$$

In addition to the reactions in Equations 2.9 and 2.10, which can be thought of as associated to terms in the sum in Equation 2.8, the traces in Figure 2.9a–c also contain contributions from classical capacitive charging current (associated to σ_M in Equation 2.8), which flows as water and ions in the interfacial region are displaced by CO and reorganize in the outer layer. It is notable that the transient current in Figure 2.9b is well below that recorded in Figures 2.9a and c. The response is consistent with there being a lower coverage of adsorbed hydrogen at 0.3 V compared to 0.08 V and a smaller excess positive metal surface charge leading to lower anion coverage relative to 0.5 V.

Following the formation of the CO adlayer, the solution is typically purged by inert gas (Ar or N_2) to remove dissolved CO and potential scans performed to interrogate the interface further [107,165]. The scan between 0.05 V and 0.25 V in Figure 2.9d (labeled "1") contains information about the extent to which CO has blocked the surface. In the scan to 1.0 V (Figure 2.9d, "2"), CO is stripped from the surface by oxidation, and the charge under the stripping peak, located at ~0.85 V in Figure 2.9d, is proportional to the CO coverage [168,171]. The final scan, labeled "3" in Figure 2.9d, is the voltammogram of the electrode in blank electrolyte and enables the electrode to be checked for any surface contamination and disorder that may have developed during charge displacement experiments.

2.4.1.2 Potential of Zero Charge

The potential of zero charge (pzc) is the potential at which the excess surface electronic charge on the metal side of the electrode solution interface is zero. In aqueous solution, metals such as Hg, Ag, and Au are ideally polarizable over a wide potential range, and values of their pzc are often accessible from interfacial tension, in the case of Hg, or from the identification of the differential capacity minimum characteristic in dilute solutions [3,71]. The situation is more complicated for Pt group metals. Although hydrogen adsorption (Equation 2.1) and the initial stages of surface oxide formation (Equation 2.2) are limited to a monolayer and do not progress with sustained faradaic current, charge transfer takes place and limits the potential range over which the electrode is ideally polarizable [162,165,172]. As a consequence, Frumkin suggested two types of pzc be identified for Pt group metals [162]. One is the potential of zero free charge (pzfc), defined as the potential at which the free excess surface charge on the metal (σ_M in Equation 2.8) is zero [162,165]. The second is the potential of zero total charge (pztc), which accounts for contributions from reversible adsorption processes where electron transfer plays a role [162,165]. With reference to Equation 2.8, the pztc is the potential at which $Q = 0$ and provides insights into both the free excess surface charge on the metal (σ_M term) and the charge associated with chemisorption steps (terms in the summation). For Pt electrodes, the pztc is more accessible experimentally than the pzfc [173].

Values for pztc have been reported for Pt low index surface planes, and stepped surfaces with structure Pt(s)-[$n(111) \times (111)$] and Pt(s)-[$n(100) \times (111)$], in 0.1 M HClO$_4$ [150,165]. These pztc determinations employed linear scan voltammetry and charge displacement measurements. The total interfacial charge at potential E, Q_E, was calculated for a range of potentials encompassing the pztc. Then, the value for E at which $Q_E = 0$ (the pztc) was identified from a plot of Q_E versus E. Q_E was evaluated from

$$Q_E = \frac{1}{\nu} \int_{E^*}^{E} j\,dE - q_{d,E^*} \qquad (2.11)$$

where j is the linear scan voltammogram current density, ν is the voltammetric scan rate, E^* is the potential at which the charge displacement measurement was carried out, and q_{d,E^*} is the charge displaced by CO at E^* [150,165,168].

For the series of Pt(s)-[$n(111) \times (111)$] and Pt(s)-[$n(100) \times (111)$] electrodes investigated, in progressing through a zone, the pztc values showed the general trend of shifting toward less positive potentials as the step density increased [150,165]. The behavior has been observed also for Ag and Au electrodes in aqueous electrolyte [3,174]. The work function values for the metals, which represent the energy required to move an electron from the bulk metal into vacuum and are measured with the crystals in UHV, change in an analogous manner [175,176]. The trends in work function for metals in UHV have been explained in terms of the surface dipole formed by the spillover of electronic charge at the metal–vacuum interface [3,176]. On a flat metallic surface, the surface dipole created by the spillover of electrons is negative toward the vacuum and creates an energy barrier for the removal of an electron. Near steps and low coordination atoms, spreading, or smoothing, of the electronic charge lead to the accumulation of positive charge at the low coordination atom and the formation of a local dipole that opposes the dipole moment of the flat surface. This so-called Smoluchowski smoothing [3,176,177] lowers the surface work function of stepped surfaces relative to the (111) plane.

The pztc values for Pt surfaces are not always consistent with the expectations based on work function trends, though. A notable exception is Pt(100), which has a more positive pztc than Pt(111) in $HClO_4$ electrolyte solutions [150,165,173]. Differences can be traced to the effects of ions and dipolar solvent molecules in altering the charge on the metal surface relative to UHV [26,178]. On Pt electrodes, the possibility for interactions with water, electrolyte ions, hydrogen, and oxides complicates the relationship between the total surface charge (Q in Equation 2.8) and the metal work function, which is more closely related to the free charge residing on the metal (σ_M in Equation 2.8) [173] (see Section 2.4.1.5).

In general, there has been a great deal of interest in identifying the effects of electrolyte on the surface charge of single crystal metals relative to UHV [26]. An informative strategy has been to construct model electrochemical double layers by stepwise dosing of solvent and ionic components onto single crystal surfaces in UHV followed by interrogation of the interfaces using work function and other techniques of UHV surface science [26,179,180]. Another approach providing valuable insights into the double layer properties of Pt electrodes is an *in situ* technique that examines the transient potential changes that accompany a sudden heating of the interface by a short (~5 ns) laser pulse [173,181–185]. The entropy changes associated with the formation of the double layer can be calculated from the potential transients. Adsorption processes that affect charging of the interface are probed with submillisecond time resolution, and with careful control of electrolyte composition, the reorganization of water molecules can be distinguished from the adsorption of ions [173,185]. The response of water molecules sheds light on the metal electrode pzfc, while responses of the ions give information about the total interfacial charge and pztc [173]. The coulostatic laser induced temperature jump experiments are discussed in greater detail in Section 2.4.1.5.

It is notable that in early studies, trends in the vacuum work function for surfaces of single crystal fcc metals were invoked to explain very simply the potentials at which features for hydrogen desorption and anion adsorption were observed on Pt low index surface planes [48,104]. The trend in work function shifting to higher (more positive) values in the order Pt(110) < Pt(100) < Pt(111) was correlated to the relative potentials of the voltammetric waves for hydrogen desorption, or the onset of anion adsorption, which was indicated to follow the same trend [48,104]. These relationships are considered further in Section 2.4.1.5.

2.4.1.3   Coulometric Determination of Surface Coverages

A limitation in determining the coverage of adsorbate species on single crystal Pt electrodes based on coulometric measurements is the need to correct for contributions associated with double layer reorganization and the gain, or loss, of specifically adsorbed ions. In general, the charge measured during a coulometric determination ($q_{stripping}$) will be given by

$$q_{stripping} = q_{far} + (q_f - q_i) \qquad (2.12)$$

where q_{far} stands for the faradaic charge, and q_i and q_f are the total double layer charges at the beginning and end of the electrochemical transformation, respectively. The term in parentheses corresponds to the contribution due to the change in double layer structure and represents a correction that should be applied to $q_{stripping}$ to obtain q_{far}. It happens often during the electrochemical stripping of an adsorbed layer that the charge on the adsorbate covered surface, q_i, can be considered negligible ($q_i \ll q_f$). Furthermore, q_f can be evaluated from measurements on the electrode in the electrolyte solution, when the solution and the electrode are free of the redox active adsorbate, by using Equation 2.11 ($q_f \equiv Q_E$ in Equation 2.11). Once q_{far} is obtained, the coverage (θ) of the redox active adsorbate is easily calculated as:

$$\theta = \frac{q_{far} \times N_A}{nFN_{(hkl)}} \qquad (2.13)$$

where N_A is Avogadro's number, n is the number of electrons transferred per adsorbate species, and $N_{(hkl)}$ is the surface atom density. The approach and results for a few common adsorbate systems are discussed below.

Voltammetric stripping measurements have often been used to estimate the coverage of adsorbed CO (CO_{ads}) on the surface of Pt (and Pt group metal) electrodes. The faradaic charge for CO_{ads} oxidation in aqueous acid media arises from the processes depicted in the following reaction:

$$CO_{ads} + 3\,H_2O \longrightarrow CO_2 + 2\,H_3O^+ + 2e^- \qquad (2.14)$$

When carried out in electrolyte solution that has been purged of dissolved CO, the oxidative removal of adsorbed CO is accompanied by the adsorption of anions on vacant Pt sites in accord with Equation 2.4a. Using the linear scan voltammograms in Figure 2.9d as an example, the CO coverage can be determined from the sweeps labeled "2" and "3." Integration of the CO stripping peak, located near 0.85 V in the scan labeled "2," gives a charge ($q_{stripping}$) of 437 $\mu C/cm^2$ [168,171]. This charge contains a faradaic component from the oxidation of CO (Equation 2.14) and an additional double layer contribution (($q_f - q_i$) $\approx q_f$), which is dominated by adsorption of anions (Equation 2.4a) as the redox active adsorbate is stripped. The double layer charge is evaluated from the linear scan labeled "3" in Figure 2.9d by applying Equation 2.11 [165,168,171]. Setting E^* in Equation 2.11 to the pztc (0.32 V for Pt(111) in 0.5 M H_2SO_4) and E to the upper limit used for integration of the CO stripping peak (1.0 V in this example) gives $Q_E \equiv q_f = 129$ μC/cm^2 [168]. The difference between the stripping charge and the double layer charge, 308 μC cm^{-2}, is the faradaic charge associated with CO oxidation (q_{CO}) depicted by Equation 2.14. The coverage of CO molecules (θ_{CO}) can then be expressed as the number of CO molecules oxidized per Pt surface atom ($N_{(111)} = 1.5 \times 10^{15}$ atoms/cm^2) from Equation 2.13. In this example, for CO on Pt(111), $\theta_{CO} = 0.64$.

It should be noted that in evaluating the double layer charge, E^* in Equation 2.11 can be set to values other than the pztc, as long as the appropriate charge displacement measurements needed to evaluate q_{d,E^*} are performed [171]. Moreover, it can be shown that the correction can be exact in this case, without the necessity of neglecting q_i in Equation 2.12, by choosing a cycle where the initial and final states are equal [186]. Saturation coverages for CO on a variety of Pt, Rh, and Ir single crystal electrodes in different electrolyte solutions have been assessed using the general approach [28,165,168,171,187,188]. Uncertainties in the range of ± 0.02 to ± 0.1 have been reported for the coverage of CO on Pt single crystal electrodes [28,187].

NO is another neutral adsorbate that has been successfully quantified on Pt (and Rh) electrodes by coulometry with account for background charge corrections [168,189]. Adsorbed NO is stripped through a reductive process, and the final product in acid solutions has been identified spectroscopically as NH_4^+ [190]. The double layer corrections evaluated through the use of Equation 2.11 account for the charge passed as NO is replaced by adsorbed hydrogen at metal sites that become vacant as NO undergoes reaction. In a similar manner, reductive stripping has been used to determine the coverage of submonolayers formed by the anions Br^- and I^-, which reversibly adsorb and desorb at Pt surfaces in accordance with Equation 2.4a [168,191,192]. For adlayers of CO, NO, and Br^- formed on electrodes, their coverages measured in coulometry experiments with account of background corrections have been in good agreement with the expectations based on structures of the adlayers known to form in UHV [168,189,191]. Urea (($H_2N)_2CO$), although neutral, displays anion-like characteristics in its adsorption and desorption behavior at Pt [168,193–195].

Hydrogen coverage on Pt also is frequently studied by coulometry, in this case it is based on the integration of cyclic voltammograms over a range encompassing features associated with reactions in Equation 2.1. For Pt(111) in $HClO_4$ or H_2SO_4 electrolyte solutions, charge displacement measurements indicate that hydrogen adsorption (or desorption) and anion desorption (or adsorption) occur in fairly well-separated potential regions, with the transition between the predominately hydrogen and anion covered states occurring very near the pztc [107,165,168]. Additionally, the pztc for these systems is close to a point of minimum current between cyclic voltammetric waves [107,165,168]. In 0.5 M H_2SO_4, the charge measured in scanning between this minimum current point (0.32 V in Figures 2.5 and 2.9) and the potential just prior to the onset of H_2 evolution (0.06 V) is ca.

170 $\mu C/cm^2$. A correction for capacitive charging is usually estimated by assuming that a small amount of background current, of the magnitude near 0.6 V in Figures 2.5 and 2.9, flows across the range between 0.32 V and 0.06 V. This double layer charge typically amounts to about 13 μC cm^{-2}. Applying the correction to the total charge measured in scanning negative (or positive) through the hydrogen region gives a value close to 160 $\mu C/cm^2$ as the charge that flows in response to processes depicted in Equation 2.1. Calculating the hydrogen coverage (θ_H) from Equation 2.13 gives $\theta_H = 0.67$, or about 2/3 monolayer. In contrast to Pt(111), the determination of hydrogen coverage based on stripping charge is not as clear-cut for other surface planes of Pt single crystals, because the potential regions where hydrogen and anion adsorption takes place overlap more strongly (cf., Ref. [150]).

2.4.1.4 Thermodynamic Analysis of Ion Adsorption

An important consequence of the electrocapillary equation (Equation 2.3) is that it allows the interfacial excess of a species to be determined at constant potential by differentiation of γ with respect to μ_i. For solid electrodes, the interfacial tension cannot typically be measured. However, for constant solution composition, cyclic voltammetry and chronocoulometry measurements provide a means to access Q as E is varied, thereby enabling determination of the total differential of the interfacial tension, $d\gamma$.

In studies of Pt single crystal electrodes, the electrocapillary equation has been employed most often to determine the Gibbs excess of and charge transferred by specifically adsorbed anions as a function of E [111,114,164,196–200]. Sulfate has received a great deal of attention [111,114,164,196] and will be used as an example in the discussion that follows to describe the analysis method. When experiments are performed in sulfuric acid electrolyte under conditions that maintain the pH and ionic strength constant (see below), Equation 2.3 can be reduced to:

$$-d\gamma = QdE + \Gamma_-RTd\ln\left(c_{HSO_4^-} + c_{SO_4^{2-}}\right) \tag{2.15}$$

where Γ_- is the sum of the HSO_4^- and the SO_4^{2-} surface excesses, R is the universal gas constant, T is the temperature, and $c_{HSO_4^-}$ and $c_{SO_4^{2-}}$ represent the electrolyte solution concentrations of bisulfate and sulfate, respectively. To obtain Equation 2.15 from Equation 2.3, the relationship between the chemical potentials imposed by the acid–base equilibrium has to be used. The Gibbs excess of sulfate species can be assessed at fixed E from Equation 2.15 by differentiation of γ with respect to $RT\ln\left(c_{HSO_4^-} + c_{SO_4^{2-}}\right)$ [111,114,164].

Experimentally, the Gibbs excess of sulfate species has been determined by first performing cyclic voltammetry [111,114], or chronoamperometry [111,164] measurements with the electrode of interest in solutions of approximately constant pH and ionic strength, but with varying concentration of sulfate species. These conditions have been achieved by adding small amounts of sulfate to an

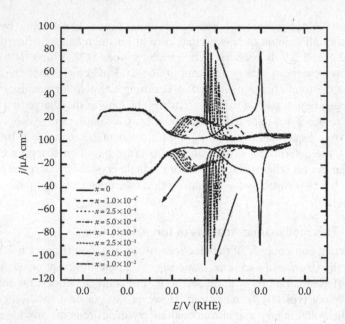

FIGURE 2.10 Cyclic voltammograms recorded at a Pt(111) electrode in solutions of 0.1M $HClO_4$ + x M H_2SO_4, where x equals 0, 1.0×10^{-4}, 2.5×10^{-4}, 5.0×10^{-4}, 1.0×10^{-3}, 2.5×10^{-3}, 5.0×10^{-3}, and 1.0×10^{-2}, as indicated. (From Herrero et al., *J Electroanal Chem*, Vol. 534, (2002): p. 79. With permission.)

$HClO_4$ electrolyte solution (cf., Refs. [111,114]). To illustrate, Figure 2.10 shows cyclic voltammograms of a Pt(111) electrode in 0.1 M $HClO_4$ + x M H_2SO_4, with x ranging from 0 to 10^{-2} M. The large excess of $HClO_4$ relative to sulfate species keeps the variation in solution ionic strength and H_3O^+ activity negligible with increasing sulfate content, at least up to 5×10^{-3} M H_2SO_4 [201]. The voltammograms progress from that expected for Pt(111) in $HClO_4$ electrolyte to responses that display features characteristic of sulfate adsorption. Next, the current response (either current–potential scans from cyclic voltammograms [111,114], or current–time transients from chronoamperometry measurements [111,164]) are integrated to enable a plot of charge density versus potential to be constructed for each sulfate concentration in the series. Figure 2.11 demonstrates a charge density versus potential plot derived from the cyclic voltammetry data in Figure 2.10. The absolute charge was calculated by using the pztc as one limit of integration, although use of the pztc is not a requirement for the thermodynamic analysis [164]. From each curve in Figure 2.11, which represents a constant concentration of sulfate species, a quantity proportional to $d\gamma$ in Equation 2.3 can be determined by integration. The interfacial pressure ($\Pi = \gamma_{\theta=0} - \gamma_\theta$) is calculated as follows:

$$\Pi = \int_{E^*}^{E_\theta} \left(Q_\theta - Q_{\theta=0}\right) dE \qquad (2.16)$$

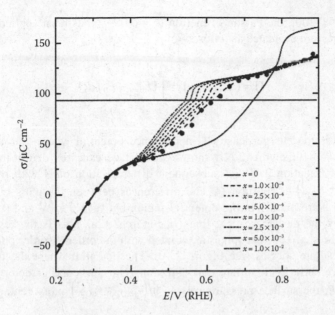

FIGURE 2.11 Lines show the total charge density versus E relationship derived from the cyclic voltammetry data in Figure 2.10 by integration according to Equation 2.11 with $E^* = 0.32$ V. The filled circles represent data obtained from the analogous potential step measurements for the 0.1 M $HClO_4$ + 1.0×10^{-4} M H_2SO_4 solution. All other experimental parameters are the same as given in the legend to Figure 2.10. (From Herrero et al., *J Electroanal Chem*, Vol. 534, (2002): p. 79. With permission.)

where the subscripts θ and $\theta = 0$ refer to quantities measured in the presence and absence of H_2SO_4, respectively. The lower limit of integration in Equation 2.16, E^*, can be any potential where anion adsorption is negligible [164]. From the interfacial pressure data, the Gibbs excess of sulfate species is obtained from:

$$\Gamma_- = \frac{1}{RT} \left[\frac{\partial \Pi}{\partial \ln(c_{HSO_4^-} + c_{SO_4^{2-}})} \right]_E \qquad (2.17)$$

Figure 2.12a shows a plot of Γ_- versus E derived from the data in Figures 2.10 and 2.11. The coverage of sulfate species increases as E becomes more positive and reaches a maximum near 3.0×10^{14} ions/cm^2, which corresponds to a coverage of $\theta \approx 0.2$, in good agreement with independent STM [113] and radiotracer measurements [84].

Although the approach outlined above follows logically from Equation 2.3, it has been shown that smaller uncertainties in values for the surface excess are obtained when Q instead of E is the independent variable [111]. In recent studies

[111,114,198–200], the Parson's function ($\xi = QE + \gamma$) has been applied and the surface pressure evaluated as follows:

$$\Pi = \left(\xi_{\theta=0} - \xi_{\theta}\right) = \int_{Q^*}^{Q} \left(E_{\theta=0} - E_{\theta}\right) dQ \qquad (2.18)$$

where Q^* is the charge density at a reference potential where ions are completely desorbed. Figure 2.12b shows the surface excesses determined from the use of Equation 2.18 and subsequent differentiation of Π with respect to $\ln\left(c_{HSO_4^-} + c_{SO_4^{2-}}\right)$ at constant Q. The differences between the plots in Figures 2.12a and b are mainly in the potential region between 0.55 V and 0.65 V, as marked by the dashed vertical lines in Figure 2.12a. This is the region that includes the sharp current spikes associated with the order-disorder phase transition of sulfate species (see Figure 2.10). The potential where the transition takes place shifts with sulfate concentration. Thus, when Π is determined at constant E, the surface pressure versus $\ln\left(c_{HSO_4^-} + c_{SO_4^{2-}}\right)$ plots contain points

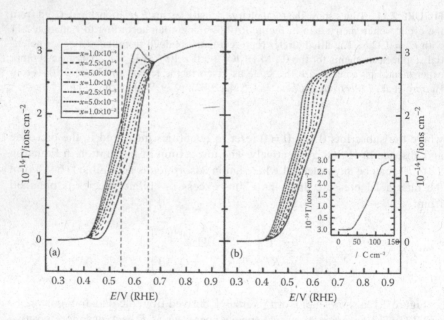

FIGURE 2.12 Plot of the Gibbs excess of sulfate species (Γ_-) against electrode potential for solutions of 0.1M $HClO_4 + x$ M H_2SO_4, where x equals 0, 1.0×10^{-4}, 2.5×10^{-4}, 5.0×10^{-4}, 1.0×10^{-3}, 2.5×10^{-3}, 5.0×10^{-3}, and 1.0×10^{-2}, as indicated. Data were calculated by differentiation of Π with respect to $RT \ln \left(c_{HSO_4^-} + c_{SO_4^{2-}}\right)$ determined at: (a) constant potential and (b) constant charge. Potentials were measured versus the RHE reference. Inset: plot of the surface excess of sulfate species (Γ_-) versus Q. (From Herrero et al., *J Electroanal Chem*, Vol. 534, (2002): p. 79. With permission.)

corresponding to both the disordered and ordered states of the adsorbed anions. By contrast, when Π is calculated at constant Q, points in the plots correspond to a single state, because the phase transition occurs at constant charge [111]. Data affected by a mix of anion adsorption states are not well fit by simple adsorption models and contribute error to the determination of surface excess. The peaked features near 0.6 V in Figure 2.12a are an artifact resulting from this error [111]. The traces in Figure 2.12b reflect more accurately the dependence of surface excess on E.

By cross differentiation of Equation 2.3, two numbers can be defined related to the quantity of charge exchanged between the electrode and adsorbed ion [111,114,200,201]. The number of electrons flowing per adsorbed species at constant potential is commonly known as the electrosorption valency (IUPAC recommends the name "formal partial charge number" [202]) and given the symbol l. The number of electrons flowing per adsorbed species at constant chemical potential is the inverse of the Esin–Markov coefficient [71] and is given by n'. For an adsorbing anion:

$$l = -\frac{1}{e}\left(\frac{\partial Q}{\partial \Gamma}\right)_E = \frac{1}{e}\left(\frac{\partial \mu}{\partial E}\right)_\Gamma = \frac{1}{F}\left(\frac{\partial \Delta G^\circ_{anion}}{\partial E}\right)_\Gamma \tag{2.19}$$

where e is the magnitude of the elementary charge and ΔG°_{anion} is the standard Gibbs free energy for adsorption of the anion species, and

$$n' = -\frac{1}{e}\left(\frac{\partial Q}{\partial \Gamma}\right)_\mu = \frac{1}{e}\left(\frac{\partial \mu}{\partial E}\right)_Q \tag{2.20}$$

Experimentally, l is determined from the slope of plots for Q versus Γ measured at constant E and n' is calculated from the slope of plots for E versus μ at constant Q [111,114,197,200].

Charge numbers for ion adsorption to Pt have been reported recently for H_3O^+, Cl^-, Br^-, OH^-, and sulfate species [111,114,197–200]. For the halides, in the potential range where hydrogen adsorption does not compete, l and n' are consistent with the transfer of the full charge on the solution free ions to form zero valent adatoms. At hydrogen adsorption potentials, l and n' are less than -1, reflecting the charge contribution from the displacement of adsorbed hydrogen [198,200]. For sulfuric acid electrolyte, l and n' fall between -1 and -2 over a narrow potential range (0.40V–0.55 V vs standard hydrogen electrode (SHE)), but vary a great deal outside this range due to competition with H_3O^+ and OH^- adsorption and other double layer effects [114].

Equation 2.19 shows that the thermodynamic analysis provides access to the standard Gibbs free energy of adsorption for ions. These adsorption energies are usually determined from a fit of the Gibbs excess, or film pressure data to the equation of an adsorption isotherm [111,114,196,198–200]. The empirical "square root" isotherm [203] has often been used [111,114,196,198,200], but other isotherms also have been applied [199]. Similar to the charge numbers, values of the standard Gibbs free energy for adsorbing anions also are affected by the competition for surface sites with other species in the electrolyte solution [111,114,196,198–200].

In considering the results obtained relating to ion adsorption based on a thermodynamic analysis of voltammograms, it should be kept in mind that the thermodynamic approach does not yield direct information about interfacial structure. Additional measurements involving spectroscopic or scanning probe techniques are required to confirm structure. The case of SO_4^{2-} (or HSO_4^-) on (111) surface planes provides an excellent example of the types of complementary measurements that advance understanding of electrode–solution interfacial properties (see Section 2.3.2.1).

2.4.1.5 Coulostatic Temperature Jump Measurements

As discussed in Section 2.4.1.2, coulostatic laser induced temperature jump measurements probe the entropy changes associated with double layer formation [173,181–184,204]. The response to the thermal perturbation is dominated by the reorganization of dipolar solvent species at the electrode–solution interface [173,181,184,185]. The reorganization of dipoles in the interphase is mainly determined by electrostatic interactions, and therefore the technique is responsive to the pzfc at Pt electrodes [173,185].

In probing double layer structure at single crystal electrodes, the experimental approach typically employed involves recording the relaxation of a small potential excursion (~5 mV) induced by a temperature perturbation applied at constant interfacial charge [173,181–185,204]. A pulse of approximately 5 ns duration from a frequency doubled Nd-YAG laser (532 nm) is used to deposit about 8 mJ/ cm^2 of energy at the working electrode–solution interface. For Pt in aqueous electrolyte solutions, the maximum temperature change caused by the laser pulse has been estimated to be <2 K at 0.5 µs [173,183,184]. At the start of an experiment, the working electrode is held at a constant potential. About 200 µs before the laser firing, the electrode is disconnected from the potentiostat and the potential changes in response to the laser pulse are measured. Potentiostatic control is reestablished before the next laser pulse. The experiment is typically repeated at a rate of 10 Hz, which allows sufficient time for the electrode to return to its original temperature and applied potential [173,183–185], and the data are averaged over 128 or 256 cycles. In one measurement configuration, a second Pt working electrode is used (cf., Ref. [184]). Both working electrodes are held at the same initial potential and released from the potentiostat at the same time. After the laser firing (the laser fires on only the first working electrode), a differential amplifier connected between the two working electrodes records the potential excursion. The differential measurement reduces interferences from noise sources that affect both electrodes, but do not arise from the laser illumination [173,183–185].

The temperature relaxation following a laser pulse, for $t \gg t_o$, is in accordance with:

$$\Delta T (t) = \frac{1}{2} \Delta T_o \sqrt{\frac{t_o}{t}} \tag{2.21}$$

where ΔT is the temperature change, t is time, and t_o is the time at which the maximum temperature (ΔT_o) is reached (i.e., the laser pulse duration, ~5 ns). When the potential change is small enough to allow a linear variation of potential with temperature, then:

$$\Delta E = \left(\frac{\partial E}{\partial T}\right)_Q \Delta T = \left(\frac{\partial E}{\partial T}\right)_Q \frac{\Delta T_o}{2} \sqrt{\frac{t_o}{t}} \tag{2.22}$$

Furthermore, from the electrocapillary equation it is possible to show [205]:

$$\left(\frac{\partial E}{\partial T}\right)_Q = -\left(\frac{\partial \Delta S}{\partial Q}\right)_T \tag{2.23}$$

where ΔS is the interfacial entropy of formation of the interface. When the interfacial response to the temperature perturbation is purely capacitive, Equations 2.21 through 2.23 can be used to determine ΔS as a function of the interfacial charge [181,204]. For each temperature induced ΔE-t transient, plotting ΔE versus $t^{-0.5}$ for the decaying portion of the transient gives a linear response, the slope of which is the temperature coefficient of the potential, $(\partial E/\partial T)_Q$. Values for ΔS are obtained by measuring ΔE-t transients over a range of potentials, with each potential corresponding to a constant interfacial charge, and extracting the $(\partial E/\partial T)_Q$ quantities. The ΔS values are determined by integrating a plot of $(\partial E/\partial T)_Q$ versus Q [204]. Steps in the procedure have been demonstrated for Au(111) in 0.1 M $HClO_4$ [204]. For Pt electrodes, the possibility for hydrogen and OH^- specific adsorption can complicate the analysis and require that relationships in addition to Equations 2.21 through 2.23 are considered [173,181].

The potential at which ΔS reaches a maximum is significant and known as the potential of maximum entropy (pme) [173,184,185,204]. At the pme, the net dipolar contribution is zero, suggesting an average dipolar orientation parallel to the surface and a maximum degree of disorder in the interfacial solvent network [173,181,184,185]. The pme is closely related to the pzc [184], or more specifically for Pt electrodes, the pzfc [173,181,185]. From Equation 2.23, it can be seen that the pme is also the point where $(\partial E/\partial T)_Q$ is zero, and it follows that the ΔE-t transient response is zero at the pme [184].

A series of ΔE-t transients is shown Figure 2.13 for Pt(111) in 0.1 M $KClO_4$ + 10^{-3} M $HClO_4$ electrolyte. At the lowest potentials studied, the direction of ΔE is toward

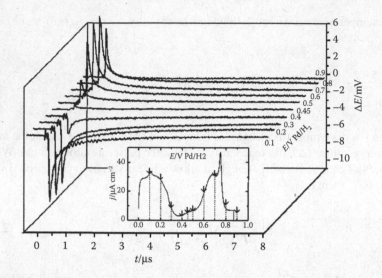

FIGURE 2.13 Laser induced potential transients for Pt(111) in 0.1 M $KClO_4$ + 10^{-3} M $HClO_4$ at the indicated potentials. The inset shows the potentials corresponding to the laser induced transients relative to the cyclic voltammogram of the electrode. (From Climent et al., *J Phys Chem B,* Vol. 106, (2002): p. 5988. With permission.)

negative values. Then, as the electrode potential is made more positive, the transient ΔE becomes smaller and attains a value close to zero before changing sign. The responses have been interpreted in terms of the effect of heating on the structure of the water layer at the metal–solution interface [173,181–185,204]. At low potentials, the electric field at the interface orients water molecules with the positive (hydrogen) end of the dipole toward the surface. Heating the interface disrupts the solvent dipole layer, causing the potential at constant charge to become more negative until the equilibrium interfacial structure is reestablished. At the most positive potentials depicted in Figure 2.13, the surface has net excess positive charge and the electric field orients the water dipoles with the negative (oxygen) end toward the surface. In this case, the thermal disruption of the solvent dipole layer leads to positive excursions in ΔE. For the transient recorded at 0.45 V in Figure 2.13, ΔE (and hence $(\partial E/\partial T)_Q$) approaches zero, indicating 0.45 V is near to the pme for the system.

Furthermore, the electrolyte used for the experiments in Figure 2.13 was selected to minimize the effects of anion specific adsorption, through the use of ClO_4^-, and slow the kinetics of hydrogen adsorption and desorption, by adjustment of the pH to 3 (instead of ~1 typical for 0.1 M $HClO_4$). The conditions enable the ΔE-t transient responses to be interpreted in terms of thermal perturbations to the solvent dipole layer [114,173,181,185]. Thus, it was also concluded that pme ≈ pzfc for Pt(111) in 0.1 M $KClO_4$ + 10^{-3} M $HClO_4$ [181].

The ability to locate approximate pme values for Pt by inspection of ΔE-t transients recorded over a range of potentials with careful control of electrolyte composition has led to the discovery of multiple pme values for stepped Pt(111) surfaces [185]. In experiments with Pt(s)-[$(n − 1)(111) \times (110)$] and Pt(s)-[$(n)$ $(111) \times (100)$] electrodes, two potentials of zero transient were observed and attributed to local effects of steps and terraces on the orientation of water dipoles. Some of the results are summarized in Figure 2.14. At low potentials (Figure 2.14a), the pme values correlate to the cyclic voltammetric peaks for hydrogen adsorption/desorption at step sites on the surfaces. The responses were discussed in terms of changes in hydrogen coverages at step sites triggering a reorientation of water dipoles across the electrode surface [185]. The second pme (Figure 2.14b), which shifts to more positive potentials with decreasing terrace length, was attributed to the disruption of the water layer on the terraces [185].

For the low index surface planes, Figure 2.15 displays pme and pztc values determined from ΔE-t transients and CO displacement experiments, respectively, measured as a function of the electrolyte composition [173]. The graphs suggest that the values are sensitive to the proton concentration in solution. For Pt(111) the shift in the pme and pztc values with the logarithm of the proton concentration is about −0.060 V/decade and reflects the effect of pH on the RHE reference [173]. Since the pztc lies

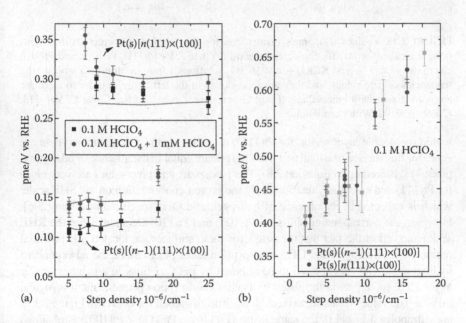

FIGURE 2.14 Plots showing pme values determined from coulostatic laser induced temperature jump measurements for a series of stepped Pt(111) electrodes. (a) pme values associated with step sites as a function of step density in 0.1 M $HClO_4$ (squares) and 0.1 M $KClO_4$ + 10^{-3} M $HClO_4$ (circles) for Pt(s)-[$(n−1)(111) \times (110)$] and Pt(s)-[(n) $(111) \times (100)$] electrodes, as indicated. Dotted lines show the potential of the voltammetric peak associated with hydrogen adsorption/desorption at steps. (b) pme values associated with terrace sites as a function of step density in 0.1 M $HClO_4$ for Pt(s)-[$(n−1)$ $(111) \times (110)$] (squares) and Pt(s)-[$(n)(111) \times (100)$] (circles). (From Garcia-Araez et al., *Electrochim Acta*, Vol. 54, (2009): p. 966. With permission.)

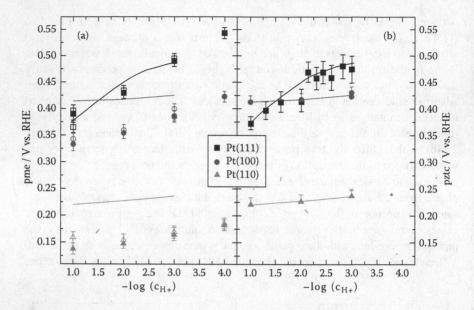

FIGURE 2.15 Values of (a) pme's, uncorrected (open symbols) and corrected (filled symbols) from the thermodiffusion potential; and (b) pztc's, for Pt(111), Pt(100), and Pt(110) electrodes in $(0.1 - x)$ M $KClO_4 + x$ M $HClO_4$ solutions. Lines are drawn to indicate the tendencies of pztc values, and they are reproduced in the left figure in order to facilitate the comparison with pme values. (From Garcia-Araez et al., *J Phys Chem C*, Vol. 113, (2009): p. 9290. With permission.)

within the double layer region for Pt(111), it can be assumed that pztc and pzfc are equal for this surface (the total and free charges are equal in the absence of adsorption processes, according to Equation 2.8). The pztc, pzfc, and pme values are very close for Pt(111) and essentially unaffected by the proton concentration on the SHE scale, which is expected in the absence of hydrogen and OH^- specific adsorption [206]. However, the corresponding shifts for Pt(100) and Pt(110) are smaller (on the RHE scale) and reflect the fact that the pztc's for these surfaces are located in a potential region where hydrogen and OH^- coadsorption takes place. Since the adsorption of hydrogen and OH^- is expected to be affected by pH variations in a manner similar to the RHE reference, in the absence of other coadsorption phenomena, it is logical that the pztc values for these surfaces show little change. If the curves in Figure 2.15b are extrapolated to pH 0, the same order (Pt(110) < Pt(111) < Pt(100)) is obtained as that discussed previously (Section 2.4.1.2) for the pztc of the low index planes [150,165,173]. However, the pme values follow the trend Pt(110) < Pt(100) < Pt(111), in accordance with work function values, reinforcing the idea that the reorientation of water dipoles at the interphase is mainly governed by the free charge density and the finding that pzfc values are closer to pme values than pztc values.

Also notable, in Figure 2.15a the pme values are shown before and after correction for the thermodiffusion potential. The thermodiffusion potential arises from the temperature difference between the heated solution at the surface of the electrode and the cooler solution near the reference electrode [173,181,204]. Figure 2.15a indicates the correction results in a small decrease in the pme in the more acidic solutions, but is essentially negligible for pH > 3.

It is important to point out that when conditions are such that hydrogen or OH⁻ adsorption is fast enough to compete with the reorientation of water dipoles at the interface, interpretations of ΔE-t transients are not as simple as discussed previously [173,181]. In addition, the chemical interactions that tend to orient water molecules at metal electrodes in the absence of an electric field at the interface also require consideration in the determination of pme values [114,173,181]. The preferential orientation of solvent dipoles through surface–adsorbate interactions will shift the pme slightly relative to the value expected in the absence of the chemical effects [114,173,181].

2.4.1.6 Metal Adatom Deposition Processes

The modification of a Pt electrode through the deposition of a partial or a full monolayer of a foreign metal is generally carried out through one of two processes. The most widely studied of these is the underpotential deposition (UPD) [27,207–208]. UPD is chemically reversible and takes place at potentials positive of the Nernst potential for bulk deposition of the foreign metal. Stronger chemical bonding interactions between the electrode substrate and adatoms of the foreign metal relative to the bonds formed within the bulk foreign metal provide the driving force for UPD. In addition, the UPD layer is stable only when in equilibrium with the precursor ions in solution. As a consequence of the equilibrium requirement, the adlayer properties can be understood and characterized in terms of thermodynamic relationships. In contrast to UPD, some metals form ad-species on Pt through a process that involves spontaneous and irreversible deposition [184,209]. The adlayers formed are stable in solutions that lack the precursor ion. Following modification, electrodes can be transferred to an electrolyte that is free of the precursor ion and the potential varied, within the range of chemical stability of the adlayer, without changing the adatom surface coverage. The ability to control adlayer coverage and electrode potential independently is an advantage of spontaneous, irreversible deposition relative to UPD processes for the modification of Pt surfaces [209].

Although awareness and understanding of UPD have been growing for many years [9,27,207–209], studies of spontaneous, irreversible deposition are relatively more recent [9,148,184,209–212]. The majority of investigations into spontaneous, irreversible deposition have focused on elements of the p-block of the periodic table. Work has been motivated mainly by the possibility for improving the electrocatalytic properties of Pt through simple surface modification procedures [209,213], although the approach is also proving useful in the study of electrical double layer structures [183,184] and for the identification of crystal planes in heterogeneous Pt materials [213–215]. Since metal UPD has received coverage in several reviews [9,27,207,208], the following discussion will focus on bringing to light important aspects of spontaneous, irreversible deposition.

A representative example is the irreversible adsorption of bismuth on Pt(111) and stepped Pt(111) surfaces [183,209,210,216]. Bismuth adsorption to Pt has been investigated for its promotional effects on electrocatalytic reactions (cf., Refs. [9,209,217–224]), and as a probe of electrical double layer properties [183,184] and atomic level surface structure [213,216]. These applications benefit from the quantitative information contained in the charges for the adatom redox processes and the hydrogen and anion adsorption steps. A characteristic of bismuth is that the charge associated with oxidation and reduction of the adatoms on extended (111) oriented domains selectively reports on the coverage of (111) surface sites [213,216].

Figure 2.16 shows cyclic voltammograms for Pt(s)-[n(111) × (100)] electrodes following the irreversible adsorption of a complete monolayer of bismuth. In the potential window between 0.06 V and 0.56 V, there is a small and almost constant current, which has been attributed to double layer charging. Particularly notable is the absence of features from hydrogen adsorption/desorption, since the presence of a full monolayer of bismuth adatoms completely blocks the adsorption of hydrogen. At potentials more positive than 0.56 V, the reversible peaks are associated with charge transfer processes that arise from the presence of bismuth adatoms on (111) oriented terraces. The stoichiometry of the charge transfer is consistent with the formation of adsorbed oxides as follows [216]:

$$Pt_3Bi + 4H_2O \leftrightarrow Pt_3Bi(OH)_2 + 2H_3O^+ + 2e^- \tag{2.24}$$

It is difficult to discern details of the OH adsorption site, whether on Bi or on neighboring Pt atoms [224–226], from voltammetry measurements alone. As discussed by Markovic and coworkers [224–226], the local environment in the vicinity of the bismuth adatoms may be more complex than suggested by Equation 2.24. The analogous peaks for bismuth adatoms on sites with (100) and (110) orientation appear above 0.8 V [9,209,210,225] and, thus, are well separated from the peaks for bismuth adatoms on (111) oriented sites.

The stoichiometry in Equation 2.24 in relation to the number of Pt sites occupied per bismuth atom and the number of electrons transferred in the bismuth redox process was demonstrated in early studies of bismuth adsorption to Pt single crystal electrodes [210]. These quantities may be obtained by measuring the experimental charge through the hydrogen adsorption region and the bismuth redox charge as a function of bismuth coverage [209]. A linear relationship exists between the charge under the bismuth adatom redox peaks and the charge associated with uncovered platinum surface atoms [209]. The available Pt can be determined by integration of the voltammetric features that arise from hydrogen and anion adsorption/desorption processes, as described in Section 2.4.1 [209].

In Figure 2.16, the morphology and integrated charge of the peaks for bismuth adatoms change with the length of the terraces. The plot in Figure 2.16c shows that the integrated charge becomes smaller as the (111) terraces shorten. The fitted line indicates the bismuth redox charge scales linearly with $\cos \beta \times (n - 1/3)^{-1}$ in accordance with the relationship suggested by Equation A.3. Instead of the

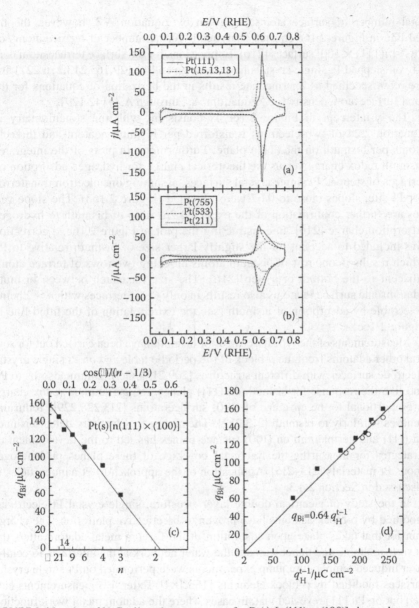

FIGURE 2.16 (a and b) Cyclic voltammograms for Pt(s)-[n(111) × (100)] electrodes covered by irreversibly adsorbed bismuth in 0.5 M H₂SO₄. The scan rates were 50 mV/s. (c) Charge density values for the bismuth redox process plotted versus $\cos\alpha \times (n - 1/3)^{-1}$, where α is the angle between the surface and the terrace planes (β in Appendix D and in the text), for Pt(s)-[n(111) × (100)] electrodes. (d) Charge density values for the bismuth redox peak versus the charge associated with the (111) sites on the terrace for Pt(s)-[n(111) × (100)] (filled squares) and Pt(s)-[n(111) × (111)] (open circles) electrodes. (Adapted from Rodriguez et al., *Anal Chem*, Vol. 77, (2005): p. 5317. With permission.)

total number of surface atoms, as given by Equation A.3, however, the fit-ted line in Figure 2.16c equates analytically to the number of *terrace* atoms on Pt(s)-[n(111) × 100] surfaces [216]. Expressions for the surface terrace atom den-sity on stepped Pt single crystals have been given [42,129,165,213,216,227] and are often specified as intermediate results in the derivation of equations for the total surface atom density (i.e., Equations A.2 through A.4) [42,129].

The y-intercept in Figure 2.16c is consistent with the stoichiometry in Equation 2.24 of two electrons transferred per bismuth adatom and three Pt atoms per bismuth on the (111) plane. Furthermore, in plots of the integrated bismuth redox charge versus the theoretical charge for hydrogen adsorption on terraces of stepped Pt single crystal surfaces, assuming one electron transferred per Pt site, slopes close to 0.67 were obtained (Figure 2.16d). The slope val-ues are further confirmation of the expected 2/3 ratio of bismuth to hydrogen adsorption charge [216]. In constructing the plot in Figure 2.16c, a correction was included to account for the slightly larger size of bismuth relative to Pt, which results in one row of bismuth atoms blocking two rows of terrace atoms adjacent to the bottom of a step [216]. The size mismatch between bismuth adatoms and surface Pt atoms also results in only (111) terraces with $n > 2$ being detectable by adsorption of bismuth (see the extrapolation of the fitted line in Figure 2.16c).

Measurements of the type described for bismuth have been carried out for sev-eral other adatoms from the p-block of the periodic table and on Pt single crystal electrode surfaces with different structures [209,213]. Germanium adsorbs to Pt, and like the response for bismuth on (111) planes, the germanium redox charge is proportional to the coverage of (100) surface atoms [213,228,229]. Tellurium behaves similarly to bismuth [213,227]. The ability to selectivity detect bismuth on (111) and germanium on (100) surface planes has led to the development of strategies for measuring the fractional coverage of these planes on heteroge-neous Pt materials [213–215]. Application of the approach to Pt nanoparticles is discussed in Section 2.4.3.

In the study of electrical double layer structure, single crystal Pt electrodes modified by p-block adatoms have proven to be effective platforms. The charge transfer that takes place upon adsorption of the foreign metal adatom alters the Pt surface dipole and thereby affects the water network at the interface. Recently, laser induced temperature jump experiments were performed on Pt single crystal surfaces modified by p-block elements [183,184]. Extensive measurements car-ried out on Pt(111) revealed that, in cases where the adatom metal work function was lower than the work function for Pt (i.e., bismuth and lead adatoms), the pme shifted negative with increasing adatom coverage [184]. By contrast, the pme shifted positive in cases where the work function for the adatom metal was higher than Pt (i.e., selenium and sulfur adatoms) [184]. The findings are sum-marized in Figure 2.17. The dominating influence was suggested to be the surface dipole changes that take place at the Pt electrode upon adatom adsorption (see the illustrative insets to Figure 2.17). The direction of the dipole is a function of the electronegativity differences between the adatom and the Pt surface, and the

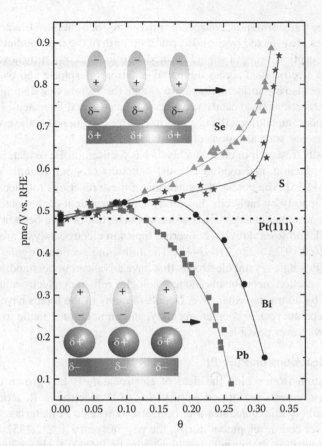

FIGURE 2.17 Plot showing pme values for adatom modified Pt(111) as a function of the adatom coverage in a 0.1 M KClO$_4$ + 1 mM HClO$_4$ solution: Lead is represented by squares, bismuth is represented by circles, selenium is represented by triangles, and sulfur is represented by stars. The inset illustrations show the interpretation of the effect of the adatoms at high coverage on the potential transients measured in the laser induced temperature jump experiments. The dotted, zero-slope line corresponds to the averaged reference pme value of unmodified Pt(111). (From Garcia-Araez et al., *J Am Chem Soc*, Vol. 130, (2008): p. 3824. With permission.)

resulting dipole influences the ordering of water molecules and the pme at the interface [184].

2.4.2 ELECTROCATALYSIS

2.4.2.1 Definition and Overview of Electrocatalysis

Electrocatalytic reactions are catalyzed by atoms at the surface of an electrode and are accompanied by faradaic charge transfer across the electrode–electrolyte interface [9,230–232]. The geometric arrangement and electronic properties of surface atoms lower the activation energy, but the electrode surface is unchanged following the transformation, consistent with the role of a catalyst. Electrocatalytic reaction rates are affected by the adsorption strength of the intermediates. In general, the adsorption strength should be moderate for maximum turnover [9,233]. The applied electric field across the metal–solution interphase and the presence of electrolyte also are influential and are among the factors that distinguish electrocatalytic reactions from catalytic processes at gas–metal interfaces. In the case of Pt electrodes, an additional and sometimes strong influence is the coverage by spectator species, such as anions or adsorbed hydrogen.

Historically, research in electrocatalysis has focused on H_2 oxidation, H_2 evolution, O_2 reduction, and oxidative transformations of small organic molecules [6,9,232,234,235]. The potential application of these reactions for electric power generation, mainly by fuel cells, has motivated a great deal of the interest. Other small molecules that undergo dissociation at Pt electrodes have been investigated [6,236–240], and several reviews covering topics in electrocatalysis are available (cf., Refs. [6,9,19,231–236,240–244]). The following sections highlight studies enabled by Pt single crystal electrodes that have advanced understanding of electrocatalytic reaction mechanisms and technologies reliant on electrocatalysis. The coverage is by no means exhaustive. Details of many of the topics have been discussed in separate review chapters recently, and an attempt is made to cite these references wherever possible.

2.4.2.2   Reactions on Pure Pt

The foundation from which the field of electrocatalysis has grown comprises largely measurements performed with the use of Pt electrodes. Reactions of H_2, O_2, and small organic molecules over bulk and nanoscale Pt materials have been central to fuel cell development during the past 60 years [9,232,235]. The well-defined composition and atomic-level structure of Pt(hkl) surfaces have guided the study of reaction mechanisms and the development of models for active sites on practical catalysts [6,9,243].

The H_2 evolution and H_2 oxidation reactions were among the first to be studied at single crystal electrodes that met modern standards for surface cleanliness and structural order. The net reactions in aqueous acid solutions can be represented by:

$$H_3O^+ + e^- \leftrightarrow \tfrac{1}{2}H_2 + H_2O \tag{2.25}$$

However, the mechanistic steps for the transformations are considerably more complicated and involve adsorbed hydrogen species [9,234,245–249] as depicted by Equation 2.1.

The exchange current densities at Pt for the reactions in Equation 2.25 are high, which has made it challenging to perform kinetic measurements and discern effects of surface atomic structure on the reactions [234,246]. The earliest studies employed conventional acid electrolytes and showed similar rates among the low index Pt surface planes [220,250,251], but subsequent experiments that varied temperature and solution pH revealed structure sensitivity [245–249]. Moving from acid into alkaline electrolyte slowed the rates of H_2 evolution and

H_2 oxidation such that differences became evident among the different surfaces across kinetically limited regions of voltammograms [245,247,249]. Reducing the temperature below ambient, down to 274 K, uncovered similar surface structure effects in both acid and base solutions [9,246,247]. The response to temperature was considered to be consistent with the expected changes in the rate constants, while the slowing of the reactions with increasing pH was attributed to the blocking of hydrogen adsorption sites by OH^- [9,247]. Mechanisms that have been proposed to explain the structure sensitivity of the H_2 oxidation and H_2 evolution reactions on Pt single crystal electrodes consider possible differences in the coverage and adsorption strength for hydrogen atoms at different symmetry sites on the surfaces and have been discussed in detail [9,234].

The reduction of O_2 on Pt electrocatalyst materials has received at least as much, if not more, attention than the reactions of H_2 [9,234,241,244,252]. The complete reduction of O_2 to H_2O is a multielectron process. High overpotentials are required to obtain appreciable turnover, even on Pt, which is among the most active catalyst for O_2 reduction. For low temperature fuel cells, slow kinetics of O_2 reduction limit the voltage and are a major obstacle to advancing performance [20,244,252–254].

Similar to reactions of H_2, early studies of O_2 reduction at Pt single crystal electrodes reported a lack of sensitivity to the atomic level structure of the surface [255]. With the development of techniques for adapting Pt single crystals for use as rotating disk electrodes, mass transport limitations were overcome and the sensitivity of O_2 reduction kinetics to the surface atom geometry on Pt single crystals was demonstrated [45,256,257]. Analysis of early hydrodynamic voltammetry data showed that O_2 reduction mainly progresses through the four electron pathway to water on all three low-index Pt surfaces in the range of potentials between hydrogen adsorption and the onset of O_2 reduction [45,256]. Since anion adsorption takes place at potentials where O_2 reduction occurs at a practical rate, the presence of adsorbed anions and the structure of the anion layer as influenced by the underlying Pt surface atoms have a pronounced effect on the O_2 reduction kinetics [9,45,68,69,244,258–260]. The anions attenuate the O_2 reduction rate by blocking the sites for O_2 adsorption [9,45,68,69,259]. In acid solutions, HSO_4^- and Cl^- are among the anions that commonly obstruct O_2 adsorption [45,69], whereas OH^- is the dominant inhibitor of O_2 reduction in basic pH electrolyte [9,259]. Recent studies that examined O_2 reduction on a series of stepped Pt single crystal electrodes showed that the rate of O_2 reduction generally increases with the step density; however, it was found that the role of steps in lowering the anion coverage is more significant than the effects of surface atom geometry on the Pt–oxygen adsorption energy [68,69]. Advances in O_2 reduction electrocatalysis are discussed further in the following sections that address adlayer modified Pt and Pt alloys (Sections 2.4.2.3 and 2.4.2.4).

Among the small molecules that undergo electrocatalytic reactions at Pt, CO is the best understood in terms of its atomic scale interactions with the electrode. In addition to electrochemical measurements, CO molecules adsorbed to single crystal Pt electrodes have been detected in *in situ* STM experiments [31,261] and by *in situ* infrared, SXS, and VSFG spectroelectrochemical techniques

[9,26,27,87,89,90,236,262,263]. The multiplicity of probes suitable for the study of CO on electrodes has enabled correlations to be made between the electrochemical kinetics of CO oxidation and factors such as the CO adlayer structure, the Pt–CO chemical bonding, and lateral interactions among adsorbates [27,31,264–266].

Early experiments that probed CO electrochemistry at Pt single crystals have been discussed in several reviews [9,26,236,267]. CO is important for its role as an intermediate and surface poison in many electrocatalytic reactions. The electrochemical oxidation of CO at Pt low index surface planes was investigated soon after the demonstration of flame annealing techniques [24,268–270]. An initial direction explored the potential to draw connections between responses recorded in electrochemistry experiments during CO adsorption and oxidation at Pt single crystals and measurements performed on Pt–CO in UHV [24,26,180,269,271].

Combining data from electrochemical and UHV measurements on CO proved particularly fruitful as applied to *in situ* infrared spectroelectrochemistry experiments [26,87,180,262–264,272–274]. For example, Weaver and coworkers [26] showed that the arrangement of molecules in a CO monolayer on a single crystal electrode can be suggested based on infrared spectra by comparing with the results for the related system in UHV, where infrared spectral measurements are often accompanied by more direct probes of structure, such as electron diffraction and ion scattering techniques [4,271,275]. The group demonstrated that hydrophobic interactions with water can promote the formation of densely packed CO islands when the CO coverage is below saturation [264–266], a finding that has impacted understanding of adsorbed CO oxidation kinetics [122,264,265]. The group also showed, with confirming information from *in situ* STM studies (next paragraph), that the electrode potential and resulting electric field across the metal–solution interface affect the CO adlayer structure [26,31,180,276–278] and the chemical bonding interaction between CO molecules and the electrode [26,180,264,279]. This foundational work focused on low index surface planes. The strategies developed subsequently were used to investigate CO adsorption and models for CO oxidation at stepped Pt single crystal electrodes [280–283].

A breakthrough was made when the crystallographic structure of the adlayer formed by CO on Pt(111) electrodes was determined *in situ* using an STM [31,276]. Both *in situ* STM and *in situ* SXS techniques revealed that the CO adlayer transforms from the (2×2)-3CO to the $(\sqrt{19} \times \sqrt{19})$-R23.4° structure as potential is scanned from the hydrogen adsorption through the double layer region [9,27,31,276]. The factors that control the onset of the phase transition continue to be identified [9,27,47,51,80,284,285]. Since the area has been reviewed recently, details will not be discussed further [9,27].

Understanding of CO oxidation mechanisms has advanced through investigations of the reaction kinetics on single crystal Pt electrodes [28,122, 243,283,286,287]. Lebedeva and coworkers showed that the stripping peak in linear scan voltammograms resulting from the oxidation of a CO monolayer initially at saturation coverage has a strong dependence on the crystallographic orientation of the electrode surface (Figure 2.18) [122]. CO adsorbs to Pt irreversibly. Therefore, it is possible to form an adlayer by exposing the electrode to CO dissolved in the electrolyte and retain the adlayer after displacing CO from solution by purging with inert gas, provided the electrode is poised negative of the CO oxidation potential and within the limits of solvent stability [9,26,28,122]. The voltammograms in Figure 2.18 were recorded after dosing the electrode to full CO coverage. The potential of the CO oxidation peak shifts negative with increasing density of steps on the surfaces. The net reaction for CO oxidation is shown in Equation 2.14. Although there have been refinements to the basic mechanism [28,243], the elementary step for CO oxidation, which is often rate determining, is given by

$$CO_{ads} + OH_{ads} + H_2O \rightarrow CO_2 + H_3O^+ + e^- \qquad (2.26)$$

Equation 2.2 depicts the elementary step leading to the formation of OH_{ads} from water. The structure sensitivity evident in the voltammograms in Figure 2.18 has been discussed as arising from the preference of OH_{ads} to form at step sites, and therefore increase in coverage in proportion to the density of steps on the surface, and the lower CO_{ads} packing densities attained as the surface step site density increases [122].

In the CO_{ads} oxidation reaction, OH_{ads} serves as both a spectator species and a reactive intermediate [9,287]. When present, strongly adsorbing supporting electrolyte anions compete with OH_{ads} for reaction sites and affect the reaction rate [288]. However, unlike other adsorbed ions, in its role as a reactant the loss of OH_{ads} frees an additional surface site by consuming CO_{ads}, which enables further OH_{ads} formation (Equation 2.2).

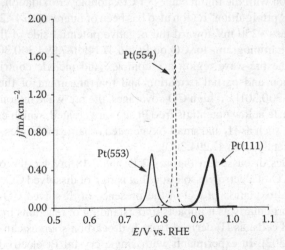

FIGURE 2.18 Linear scan voltammograms for the oxidation of CO adlayers at saturation coverage on Pt(111) and Pt(s)-[n(111) × (111)] electrodes where $n = 5$ and 10, for Pt(553) and Pt(554), respectively. (From Lebedeva et al, *J Electroanal Chem*, Vol. 487, (2000): p. 37. With permission.)

Another approach that has revealed mechanistic details of CO oxidation on Pt single crystal electrodes is based on the use of chronoamperometry [28,188,243,286,289–291]. The method involves forming a CO adlayer on the electrode, as described above in the discussion of Figure 2.18, and initiating the reaction by stepping to a potential where CO oxidation takes place. Processes that include the reactions in Equations 2.2 and 2.26 limit the current that flows in response to the potential step. Example current–time transients recorded during the oxidation of saturated CO adlayers on Pt(s)-[n(111) × (111)] electrodes are displayed in Figure 2.19. The central bell-shaped region reflects conditions where Equation 2.26 is limiting and can be fit to a mean-field kinetic model based on the Langmuir–Hinshelwood (LH) mechanism [28,170,188,243,286]. The responses at early times do not fit the LH model and are likely limited, at least in part, by the formation of adsorbed or hydroxide species (Equation 2.2). The development of the theory that governs the fitted lines in Figure 2.19 has been reviewed recently [243]. The adherence of the central portion of the transients to mean-field LH kinetics implies that under the conditions of the experiment the CO adsorbed on the electrode surface undergoes rapid diffusion as the adlayer is depleted [28,243,286].

The extent to which the transient responses for CO monolayer oxidation can be fit to the LH model is highly sensitive to the cleanliness and structural order of the electrode. The presence of contaminants [187,286] and disordering to form nanoscale Pt deposits [70,187] can cause the peak potential to shift and the transient shape to deviate from LH kinetics [187,286]. In chronoamperometry studies of CO monolayer oxidation on Pt nanoparticle catalyst, nonideal responses of this type often have been reported and provide insights into the effects of catalyst electronic properties and crystallite structure on the reaction [292–297].

In connection with the initial stages of CO monolayer oxidation, the so-called "pre-wave," or "pre-ignition" region also has been of interest [46,47,294,298–300]. The region spans ~250 mV toward the negative potential side of the main peak in stripping voltammograms for CO on Pt [46,47,145,187,294,300,301]. Processes occurring in the pre-wave region are complex and include contributions from anion adsorption and partial oxidation and rearrangement of the CO adlayer [46,47,145,294,300,301]. At high CO coverages, the pre-wave region has practical importance, since at low potentials free Pt sites are created, which enable the oxidation of fuels, such as H_2 and small oxygenated organic molecules, on otherwise CO poisoned catalyst [46,47,294,300].

The examples discussed in Figures 2.17 and 2.18 involve the oxidation of a monolayer of CO in electrolyte solution that is free of dissolved CO. However, the electrochemical oxidation of CO in the presence of dissolved CO has practical importance, particularly as it impacts understanding of reactions in low temperature H_2/O_2 fuel cells, as H_2 derived from hydrocarbon sources can contain trace CO [9,46,300,302]. In experiments with single crystal Pt electrodes, the onset potential for CO oxidation in the presence of bulk CO shifts positive relative to the oxidation of a CO monolayer in CO free electrolyte [9,46,122,243,245,300]. The continuous supply of CO to the electrode surface from the bulk solution blocks sites for OH_{ads}, which suppresses CO_2 formation at low potentials.

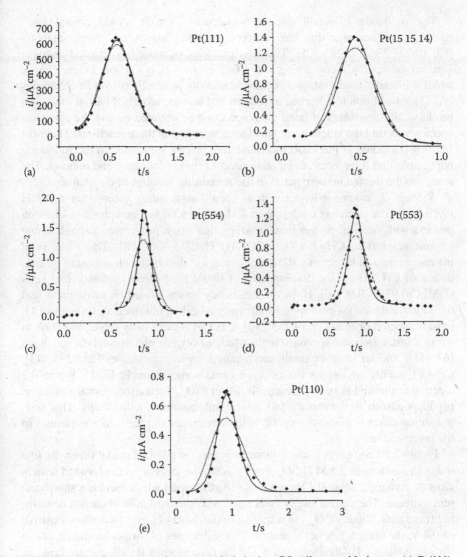

FIGURE 2.19 Current transients recorded during CO adlayer oxidation on (a) Pt(111) at 0.905 V, (b) Pt(15 15 14) at 0.88 V, (c) Pt(554) at 0.83 V, (d) Pt(553) at 0.805 V, and (e) Pt(110) at 0.755 V. Experimental data (diamonds), fit by the mean-field LH model for CO oxidation (solid line), fit by the progressive nucleation and growth model (dashed line). Only a few experimental data points are depicted for the sake of clarity. (From Lebedeva et al., *J Phys Chem B*, Vol. 106, (2002): p. 12938. With permission.)

The development of rate equations for CO oxidation in the presence of bulk CO requires account of (i) CO transport from solution to the electrode surface and (ii) Pt–CO bond formation, in addition to Equations 2.2 and 2.26 [243,303]. Just as in CO free solution, CO oxidation in the presence of dissolved CO is sensitive to the metal crystallographic orientation [9,122,300,304].

The oxidation of small organic molecules over Pt single crystal electrodes is another area that has received attention in several review chapters [6,9,19,235,236,242,243,305]. The vast majority of work has focused on oxygenated compounds containing one (C_1) or two (C_2) carbon atoms. These small molecules tend to show structure sensitivity in reactions over Pt electrodes [9,243]. In addition to adsorbed hydrogen and anions, adsorbed partial oxidation products and fragments of bond cleavage must be counted among the spectator species affecting the kinetics and interfacial structure in these reactions. The most frequently studied of the small molecules are briefly mentioned in the following paragraphs and references are given to recent primary literature and reviews. The reactions are treated in more detail in the remaining sections of the chapter.

Among C_1 compounds, there has been longstanding interest in CH_3OH [9,19,243]. The complete oxidation of CH_3OH to CO_2 is a complex, six-electron process with coupled proton transfer steps that often progresses through stable intermediates (i.e., CH_2O, CO, HCOOH) [19,243,306–308]. There has been intense interest in harnessing the energy released during the electrocatalytic oxidation of CH_3OH for the development of liquid feed direct methanol fuel cells (DMFCs) [9,19,309,310]. However, efficiency losses caused by incomplete fuel oxidation and catalyst poisoning by strongly adsorbed intermediates, such as CO, have hampered progress [9,19,235,243]. CH_3OH was an early target of efforts to adopt a surface science approach to the study of organic electrocatalytic reactions [63,311]. Similar to other small, oxygenated organic molecules [235,236,243], CH_3OH readily undergoes dissociative chemisorption on Pt [311]. When the electrode potential is below the threshold for OH_{ads} formation, partial oxidation products adsorb to the surface and inhibit subsequent reaction steps. This self-poisoning effect is evident in cyclic voltammograms and leads to a hysteresis in the reverse scan.

Figure 2.20 compares cyclic voltammograms for CH_3OH oxidation on the low index Pt surfaces in 0.5 M H_2SO_4. For Pt(100), the current on the forward scan is close to zero until about 0.4 V (versus Ag/AgCl), upon which there is a sharp current increase. The current rise is associated with the oxidation of carbon containing fragments, notably CO_{ads}, in response to the onset of OH_{ads} formation. Positive of 0.4 V, the current reaches a peak and then declines as further oxidation of the surface inhibits CH_3OH dissociation. On the reverse scan, *anodic* current flows as the steady-state coverage of inhibiting oxides becomes lower and opens sites for CH_3OH dissociation. The anodic current remains high on the reverse scan through intermediate potentials, reflecting the presence of reactive OH_{ads} species and rapid oxidation of CO_{ads}, and other poisoning carbon species. In Figure 2.20, Pt(110) shows the largest hysteresis, or disparity in anodic current between the forward and reverse scans. The low current on the forward scan has been attributed to extensive poisoning due to the high density of step sites available to catalyze C–H bond cleavage and formation of CO_{ads}. Pt(111) appears to poison least of the low index Pt surfaces, but the lower overall current density for CH_3OH oxidation on Pt(111) reflects slower reaction kinetics compared to the other surfaces, likely because the coverage of defects and other low coordination atomic sites is small.

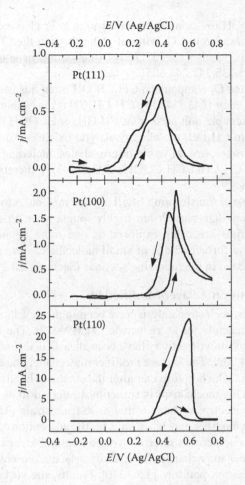

FIGURE 2.20 Cyclic voltammograms for Pt(111), Pt(100), and Pt(110) electrodes, as indicated, in 0.2 M CH₃OH + 0.5 M H₂SO₄. The scan rate was 50 mV/s. (Adapted from Koper et al., *Fuel Cell Catalysis: A Surface Science Approach*, Wiley: Hoboken, New Jersey (2009). p. 159, and Herrero et al., *J Phys Chem*, Vol. 98, (1994): p. 5074. With permission.)

In recent years, there also has been growing interest in HCOOH as a fuel [19,243,312]. In contrast to CH_3OH, the conversion of HCOOH to CO_2 over Pt progresses through fewer intermediates and can be more complete [9,243,305]. Combined with other advantages, such as lower toxicity and reduced ability to transport across fuel cell ionomer membranes, there is renewed interest in HCOOH for its potential use in direct formic acid fuel cells (DFAFCs) [312,313]. HCOOH also has become a model for mechanistic studies of small molecule oxidation reactions, because it has the simplest oxidation mechanism of C_n compounds. Cyclic voltammograms of electrolyte solutions containing HCOOH show hysteresis effects similar to those discussed for CH_3OH. Although CO_{ads} is formed readily from HCOOH dissociative chemisorption at Pt electrodes, HCOOH has the potential to transform to CO_2 without going through the CO_{ads} intermediate. This "dual path" mechanism for HCOOH oxidation has been under scrutiny for many years [19,235,236,242,243,305].

Of the oxygenated C_2 compounds, CH_3CH_2OH often has been investigated as an alternative to C_1 fuels [243,314,315]. CH_3CH_2OH is renewable and has a slightly higher energy content per unit mass than CH_3OH, or HCOOH [243,309,310,312]. However, similar to CH_3OH, the electrocatalytic oxidation of CH_3CH_2OH over Pt can progress through several stable intermediates, including C_2 partial oxidation products (i.e., CH_3CHO, CH_3COOH) [243,314–318], thereby limiting energy conversion efficiency.

Catalysts capable of transforming small organic molecules to CO_2 with greater speed and efficiency than pure Pt are highly sought after. Pt alloy and adlayer modified Pt materials are being explored as one route to meeting this need. These materials and further details of small molecule electrocatalytic oxidation reactions are discussed in the following sections (Sections 2.4.2.3 and 2.4.2.4).

2.4.2.3   Reactions on Adlayer-Modified Pt

Some key advances in electrocatalysis have been in the development of modified Pt surfaces, most recently for O_2 reduction [20,319–323]. The effects of surface modifiers on electrocatalytic activity have been classified according to three main mechanisms [9,184,209]. The surface modifier may change the electronic properties of the substrate, which in turn can alter the chemical bonding interactions of reacting species. This mechanism is sometimes referred to as the ligand effect [324] and includes electronic changes that arise from strain [325] in the surface layers. The surface modifier may block sites for the adsorption of reactive species. This mechanism is sometimes referred to as an ensemble, or third-body effect, because the modifier may change the size of surface atom ensembles available for a particular reaction pathway [326–330]. Finally, the surface modifier may provide sites that are more energetically favorable than the substrate for the adsorption of a second species necessary for the reaction to take place [331–334]. CO oxidation (Equation 2.26) over PtRu electrode surfaces, wherein the OH_{ads} forms at lower potentials on Ru than Pt, is a frequently cited example of the latter case, sometimes called the bifunctional mechanism [331,332,334].

Adatom modified single crystal surfaces have been studied for applications in electrocatalysis since the early developments of flame annealing techniques [9,209,217,335–337]. Initial studies focused on the promotion of HCOOH oxidation by bismuth [217,219,335,338,339] and other elements from the p-block of the periodic table (Section 2.4.1.6) [336,337]. On Pt(111), enhancements in the rates of HCOOH and CO oxidation were reported in the presence of even small amounts of bismuth ($\theta_{Bi} \geq 0.05$) [209]. For CO, the increases were traced to a bifunctional mechanism associated with the low potential for oxide formation at bismuth sites (Equation 2.24) on (111) planes of Pt [39,223]. Similar rate increases were not as evident for CO oxidation at bismuth modified Pt(100) electrodes, and Pt(s)-[n(111) × (110)] electrodes with bismuth selectively adsorbed to the steps, since the redox processes related to bismuth on (100) and (110) planes take place at potentials more positive than CO oxidation [39,148,209,223]. A bifunctional mechanism also was put forward to explain faster rates for CO oxidation on Pt single crystal electrodes modified by other p-block elements (i.e., As, Sb, Sn) that form oxides at potentials coincident with CO oxidation [9,39,209]. In relation to the bifunctional model, the possibility for enhanced adsorption of OH⁻, or other anions, caused by changes in local work function at surface sites near adatoms has long been proposed as a potential mechanism affecting small molecule electrocatalytic oxidation rates [9,124,224].

For HCOOH, bismuth and other p-block adatoms generally result in faster oxidation rates relative to Pt [9,209]. However, the mechanism is more complicated than for CO, because HCOOH has the potential to progress to CO_2 without the intermediate formation of CO [9,209,224,340]. Electronic, surface blocking and bifunctional effects all have been suggested to explain enhancements in the electrocatalytic oxidation of HCOOH on adatom modified Pt single crystal surfaces [9,209,224,340–343]. On stepped surfaces, the possibility for blocking of active sites for CO poison formation by bismuth adsorption to steps also has been discussed [305,341–343]. Despite its apparent simplicity, a great deal remains to be learned about the mechanism of HCOOH electrocatalytic oxidation [9,209,340]. In the case of CH_3OH, bismuth has a blocking effect on the reaction, disrupting the ensembles of contiguous Pt required for the initial stages of the reaction (see Section 2.4.2.4) [219,339,344]. Of the p-block elements, Sn has been of greatest interest for its catalytic effect on the oxidation of CH_3OH and larger organic molecules [345–347].

The strategy of assembling transition metal atom components onto the surface of a single crystal electrode also has been used to prepare new electrocatalyst materials [319,322,323,348–354]. The single crystal surface can affect both the spatial structure and the electronic properties of atoms in the overlayer [322,323,325,353,354], or serve as a support for nanoscale deposits of the second catalytic material in a bifunctional catalyst [348–352]. An early approach involved the modification of Pt low index surface planes by spontaneous deposition of Ru [348–352,355,356], or Os [355,357]. The transition metal adatoms formed nanoscale islands [348,355] on the Pt surfaces. The modified surfaces were investigated mainly for their promotional effects on CO and CH_3OH oxidation and found to facilitate the reactions through a bifunctional mechanism involving the activation of H_2O to form reactive OH_{ads} species on the adatom sites [349,351,352,355].

For the reduction of O_2, there has been interest in Pt(111) modified by Pd adatoms [353,358]. In 0.1 M KOH, the O_2 reduction rate was enhanced over the modified surfaces relative to Pt(111) and increased with Pd coverage (θ_{Pd}) up to a full monolayer. The O_2 reduction rate decreased as the Pd coverage exceeded one monolayer, and fell below the rate measured over Pt(111) for $\theta_{Pd} = 1.5$ [353]. In acid electrolyte, Pd adatoms did not enhance the kinetic current for O_2 oxidation relative to Pt(111) [358]. Collectively, the results were explained in terms of (i) the electronic properties of Pd adatoms on Pt(111), providing a more optimum adsorption energy for O_2 and its reduction intermediates compared to Pt(111) and Pd multilayers; and (ii) stronger blockage of O_2 adsorption sites by anions in acid than base electrolyte [353]. Using a related approach, a Pd(111) electrode was modified by the adsorption of Pt, or a mixed monolayer of Pt and another late transition metal (i.e., Au, Pd, Rh, Ru, Ir, Re, Os) [322]. The trend in the O_2 reduction rates for different adlayer compositions is shown in Figure 2.21. Increasing activity for O_2 reduction was attributed to the effect of oxides on the late transition metal atoms lowering the coverage of inhibiting OH_{ads} on Pt sites through lateral repulsive forces and thereby freeing Pt sites for adsorption of O_2 and its reaction intermediates.

2.4.2.4 Reactions on Pt Alloy and Pt Intermetallic Electrodes

As adlayer modified Pt(hkl) electrodes were being investigated for applications in electrocatalysis, related experiments that employed well-defined bulk alloy and intermetallic materials were underway [20,73,74,302,320,330,334,359–363]. UHV techniques and *in situ* SXS measurements were essential to characterizing the structure and composition of the electrode surfaces. In UHV, low energy

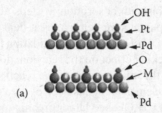

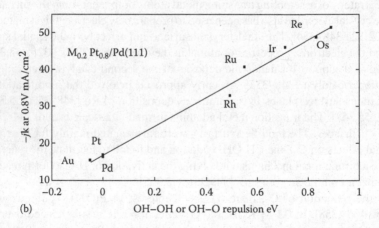

FIGURE 2.21 (a) Model for a Pd(111) electrode modified by Pt adatoms (top) and a mixture of Pt and a late transition metal, M (bottom) illustrating the decrease of the OH coverage on Pt, caused by a high OH or O coverage on M. (b) Kinetic current measured at 0.80 V (vs RHE) for the monolayers indicated as a function of the calculated interaction energy between two OHs, or OH and O in a (2 × 2) unit cell. Positive energies indicate more repulsive interaction compared to a full monolayer of Pt on Pd(111). Measurements were performed in 0.1 M $HClO_4$ and at a Pt:M ratio in the monolayer of 80:20. (From Zhang et al., *J Am Chem Soc*, Vol. 127, (2005): p. 12480. With permission.)

ion scattering (LEIS) [20,73,330,362] was particularly valuable for assessing the atomic ratios of different types of metal atoms present at the surface. Similar to Pt and adlayer modified Pt electrodes, spectator species, in particular forms of OH_{ads}, have a strong effect on the kinetics of electrocatalytic reactions at surfaces of Pt alloy and Pt intermetallic materials.

PtRu was investigated initially. Electrodes spanning a range of Pt:Ru atomic ratios were prepared as random, bulk alloys by arc-melting [330]. The electrodes were applied mainly for the oxidation of CO [334], CH_3OH [330,359,364], and H_2/CO mixtures [302,360]. In acid electrolyte, a composition of 50 atomic % Ru was found to be optimal for promoting CO oxidation, whereas 10–30 atomic % Ru was most active for CH_3OH oxidation [330,334]. It was determined that Ru sites produce reactive surface oxides from H_2O at a lower potential than pure Pt, thereby enhancing at low overpotentials the rate of oxidation for CO and fragments from CH_3OH dissociative chemisorption. For CH_3OH, it was found that at ambient temperature an ensemble of Pt sites (~3 Pt atoms per ensemble) are needed to carry out the C–H bond cleavage steps [330,359]. The high activity of Pt-rich alloys for CH_3OH oxidation could be explained by a statistical model that considered the need for Pt ensembles and access to Ru sites that supply OH_{ads} [330].

In bulk crystals of the pure metals, Pt adopts an fcc and Ru a hexagonal closed packed (hcp) arrangement of atoms in the lattice. PtRu alloys have the fcc structure when the Pt composition is greater than 40% and adopts the hcp structure for Pt compositions below about 10% [330]. The arrangement of Pt and Ru atoms in the alloy lattice is random. By contrast, many fcc metals, when alloyed with Pt, can produce an fcc crystal with well-defined metal atom composition within the unit cell. These so-called intermetallic compounds [74,363] have been investigated for applications in electrochemistry. Intermetallic materials employed as electrodes include alloys formed by Pt and one of the following: Ni, V, Fe, Co, Ti, Pb, Sn, In, or Bi [20,73,74,320,321,363,365].

In experiments that utilize Pt alloy electrodes, it is important to determine any differences that exist between the surface and bulk compositions. UHV surface science and *in situ* SXS techniques have been vital in this regard and in shedding light on the role of the surface in promoting reactions [20,73,74,320,321,330,362,365–367]. For bimetallic alloys formed by Pt and Ru, or the 3d transition metals Ni, V, Fe, Co, and Ti, the bulk and surface compositions are the same after cycles of annealing followed by ion sputtering in UHV. However, the surfaces become enriched in Pt after the annealing (~700°C–800°C) step. For the Pt_3M (where M is one of the 3d transition metals mentioned above), annealing produced a Pt top layer and a second layer enriched in the 3d metal [20,320,321,366,367]. This "Pt skin" structure [20,320,321,362,365] has electronic properties different from the surface of pure Pt and has been shown to play a role in the promotion of O_2 reduction electrocatalysis [20,74,320,321,362,365].

Results of LEIS measurements that employed 1 keV Ne^+ ions to interrogate the surface of a Pt_3Fe alloy electrode are displayed in Figure 2.22. The LEIS spectrum of the lightly sputtered surface contains peaks indicating the presence of Pt and

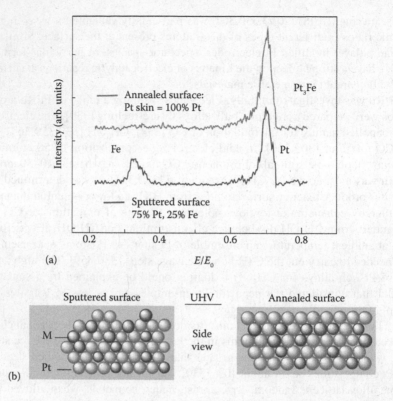

(a)

(b)

FIGURE 2.22 (a) Low energy ion scattering (LEIS) spectra, collected using 1 keV Ne$^+$ ions, of a Pt$_3$Fe alloy electrode surface after mild sputtering (bottom) and mild sputtering followed by annealing at 700°C (top). (b) Cartoon of the Pt$_3$Fe alloy viewed from the side depicting both Pt (light) and Fe (M, dark) on the sputtered surface (left) and the Pt enriched surface that develops after annealing (right). (Adapted from Stamenković et al., *Nature Mater*, Vol. 6, (2007): p. 241. With permission.)

Fe atoms, whereas the spectrum of the annealed surface contains a strong peak for Pt, but no evidence of surface Fe. The cartoon in Figure 2.22b depicts the effects of annealing on the surface roughness and composition, including the formation of the smooth Pt skin. The Pt skin develops as a phase separation takes place near the alloy surface, which leads to the formation of a second layer enriched in the alloying metal [20,366,367]. LEIS was used in a similar manner to demonstrate Pt enrichment at surfaces of PtRu and the other Pt$_3$M alloys. In the studies of O$_2$ reduction, Pt skins formed on Pt$_3$M electrodes remained stable during potential scans. In acid solutions, however, the less noble metal was soluble, and Pt rich surfaces were observed to form through the loss of M by dissolution [320,365]. These atomically rough, Pt-rich surfaces were referred to as having a "Pt skeleton" structure [320,365]. Pt skeleton structures on electrodes displayed enhancements in O$_2$ reduction kinetics similar to Pt skins, although with slightly lower activity [74,320,365].

Recently, the variation in metal composition within the first few surface layers of a bimetallic electrode containing a Pt skin was assessed by *in situ* SXS measurements [20] through the analysis of crystal truncation rods [27]. Following cycles of sputtering and annealing, a single crystal $Pt_3Ni(111)$ electrode held at 0.05 V in 0.1 M $HClO_4$ was confirmed to have a Pt top layer, a Ni enriched (52% relative to 25% Ni in the bulk) second layer, and a Pt enriched (87%) third layer [20]. Furthermore, the composition and structure of the surface and near surface layers were shown to remain fixed following potential scans across the range of 0.05 V to about 1.0 V (vs RHE). Similar results were obtained for the other low index surface planes of Pt_3Ni crystals [20].

The rotating ring-disk electrode voltammetry results in Figure 2.23 demonstrate the enhancement in current through the kinetically limited potential region for O_2 reduction on a Pt skin bimetallic electrode. In Figure 2.23 (bottom), positive shifts in the half wave potential in the 0.8 V–0.9 V region are evident in progressing from polycrystalline Pt and Pt(111) to $Pt_3Ni(111)$ electrodes. A figure of merit frequently used to describe electrocatalytic O_2 reduction activity is the *kinetic current* assessed from Koutecky–Levich plots derived from rotating disk electrode measurements [45,68,69,320–322,362,368]. Approaches for extracting the kinetic current and related information for O_2 reduction from hydrodynamic voltammetry measurements have been described in several reports [45,68,69,353,362,369]. Figure 2.23 (top) also includes the current response at the ring electrode arising from H_2O_2 generated at the catalyst samples. On pure Pt electrodes, in acid electrolyte H_2O_2 production from O_2 becomes significant at potentials in the hydrogen adsorption region [9,45,68,69]. However, on Pt skin

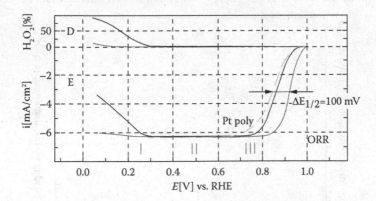

FIGURE 2.23 Voltammograms from rotating ring-disk electrode experiments performed in O_2 saturated 0.1 M $HClO_4$ at 60°C. Top: Response from H_2O_2 detected at the ring electrode. Bottom: Linear scan voltammograms recorded during O_2 reduction on 700°C annealed $Pt_3Ni(111)$ (light, solid), Pt(111) (dark, solid) and polycrystalline Pt (light, dashed) disk electrodes. (Rotation rate = 1600 rpm; Potential scan rate = 50 mV/s). The arrows indicate the values of the half-wave potential, which on $Pt_3Ni(111)$ is shifted positively relative to polycrystalline Pt by about 100 mV. (From Stamenkovic et al., *Science*, Vol. 315, (2007): p. 493. With permission.)

surfaces, the H_2O_2 producing pathway is less significant and is almost negligible for the $Pt_3Ni(111)$ electrode [20,320,362].

The kinetic current has been shown to correlate with electronic properties of the metal electrode [20,74,244,320–322,365]. Figure 2.24 demonstrates the relationship between the kinetic current for O_2 reduction measured under the conditions specified and the energy of the metal d-band center determined by synchrotron-based ultraviolet photoemission spectroscopy (UPS) for the electrodes in UHV [320]. The d-band center is the average energy of the d states on the metal surface atoms to which the adsorbate of interest binds and can be correlated to the surface–adsorbate bond energy [321]. For the Pt skin surfaces considered in Figure 2.24, the energy of the d-band center is reported relative to the Fermi level and is closest to the Fermi level for pure Pt. Among the alloys, the d-state energy is shifted down relative to Pt in amounts that increase in progressing from Ni to Ti across the 3d row of the periodic table. Furthermore, it has been demonstrated that the Pt–O surface bond energy follows a related trend in that it is strongest for pure Pt and weakens in progressing from Ni to Ti [321]. The "volcano" shaped plot in Figure 2.24 reflects the Sabatier principle [9,320,321], which predicts that the fastest electrocatalytic reaction rates take place over metals that bind intermediates moderately. In relation to O_2 electrocatalysis, when the catalyst binds oxygen too strongly, O_2 reduction is slowed by the presence of inhibiting oxides and anions on the electrode, and when the surface binds oxygen too weakly, O_2 reduction is limited by the coupled electron and proton transfer steps associated with O_2 dissociative chemisorption [320,321]. Of the surfaces considered in Figure 2.24, the near surface layers of the annealed Pt_3Co alloy produce a moderate d-band center energy, which implies an intermediate Pt–O bond

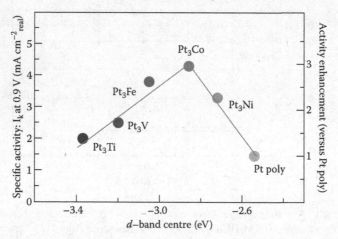

FIGURE 2.24 Relationships between the experimentally measured kinetic current for O_2 reduction on Pt_3M (Pt skin) surfaces in 0.1M $HClO_4$ at 60°C versus the energy of the metal d-band center position determined for the electrodes in UHV by synchrotron-based ultraviolet photoemission spectroscopy (UPS). (From Stamenkovic et al., *Nature Mater*, Vol. 6, (2007): p. 241. With permission.)

strength at the Pt skin surface. Hence, the annealed polycrystalline Pt_3Co alloy displayed maximum activity for O_2 reduction among the samples investigated.

The trends in Figure 2.24 have been corroborated by periodic density functional theory (DFT) calculations [20,321]. In general, experiments performed on single crystal Pt and well-characterized Pt alloy electrodes have laid a foundation for the investigation of O_2 electrocatalytic oxidation mechanisms by electronic structure methods [20,74,244,320–322]. The calculations provide insights into relationships between the turnover rate and kinetic current for O_2 reduction and properties of metal–adsorbate chemical bonding interactions [20,74,244,320–322].

2.4.3 NANOPARTICLE SURFACE CHARACTERIZATION

Understanding that has developed through the study of electrochemistry at Pt single crystals has been extended to measurements involving Pt nanoparticle materials [213–215,370–376]. Pt nanoparticles have practical impact in applied areas of electrochemistry, notably as fuel cell catalyst [74,253,292,377]. Particles ranging in size from a few to tens of nanometers can be spherical with a polyoriented surface, or can be prepared as small crystallites that have a preferential shape [213,375]. The surfaces of Pt nanoparticles comprise small ensembles of atoms that form (111) and (100) planes connected by low coordination atoms [213]. On preferentially oriented Pt nanoparticles, ideally, the number and spatial arrangement of these planes are well-defined. In practice, samples contain a distribution of particle sizes, and slight irregularities in shape develop due to differences in the growth rate among the crystal planes [378,379]. A collection of preferentially oriented Pt nanoparticles reflects dominant growth of either the (100) planes, resulting in mainly cubic shapes, or (111) planes, resulting in mainly tetrahedral or octahedral shapes [213,375,378,379]. Intermediate growth conditions are also possible, and produce particles that display both (111) and (100) planes [375].

Procedures for characterizing the surface crystallographic composition of Pt nanoparticles have been developed based on knowledge of voltammetric responses for bulk Pt single crystals in aqueous electrolytes and the discovery of adatom deposition processes selective for different low index Pt surface planes [213–215,375]. Features through the hydrogen and anion adsorption regions of voltammograms for Pt nanoparticles in contact with aqueous electrolytes guide the identification of dominant crystallographic planes. From this base voltammetry and measurements of the bismuth and germanium adsorption charge on, respectively, (111) and (100) surface planes (Section 2.4.1.6), the fractional coverage of low index planes on ensembles of Pt particles can be determined [213–215,375]. An important advantage of this voltammetric approach is that it enables the surface planes on Pt nanoparticles to be characterized *in situ* under conditions relevant to the experiment.

Before discussing voltammetric responses for Pt nanoparticles, it will be useful to mention related studies of bulk polycrystalline Pt. Figure 2.25 shows a voltammogram of bulk polycrystalline Pt in comparison to other single crystal Pt surface planes. The two peaks in the voltammogram for polycrystalline Pt have been traced to contributions from hydrogen adsorption at step sites of (100) and

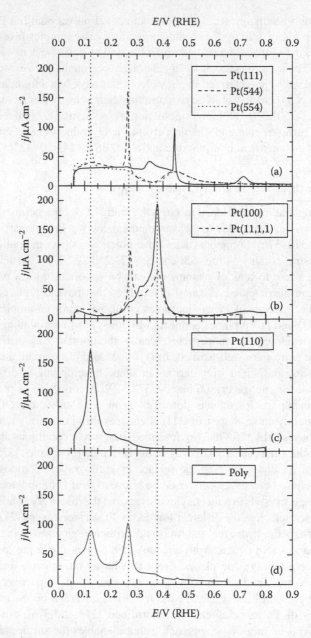

FIGURE 2.25 The anodic scans from cyclic voltammograms of different Pt electrodes in 0.5 M H_2SO_4. (a) Pt(111), Pt(544) ≡ Pt(s)-[9(111) × (100)] and Pt(554) ≡ Pt(s)-[9(111) × (110)]; (b) Pt(100) and Pt(11, 1, 1) ≡ Pt(s)-[6(100) × (111)]; (c) Pt(110) (d) a polyoriented Pt sphere. The sweep rate was 50 mV/s. (From Solla-Gullon et al., *Phys Chem Chem Phys*, Vol. 10, (2008): p. 1359. With permission.)

(110) symmetry with low bidimensional order [213,214]. The peaks at 0.125 V and 0.26 V are consistent with the oxidative desorption of hydrogen from steps on (111) terraces that have (110) and (100) orientation, respectively. The assignments are evident by comparison to the voltammograms in Figure 2.25a. The Pt(s)-[9(111) × (110)] and Pt(s)-[9(111) × (100)] electrodes display the corresponding step site peaks, and the responses for all three electrodes include the constant current component below 0.3 V associated with hydrogen on (111) terraces. Other related matches can be identified with peaks in the voltammograms for Pt(110) (Figure 2.25c) and Pt(100) (Figure 2.25b) [213].

Although the peak comparison illustrated in Figure 2.25 enables the presence of dominant surface crystallographic structures on a heterogeneous Pt sample to be identified, additional information is required to quantify the fractional coverage of the low index surface planes present. It is not sufficient to integrate the current in the region of hydrogen and anion adsorption/desorption features in base voltammograms, because the peaks, which result from surface chemical processes, have complicated shapes that make deconvolution difficult [213]. A solution has been to determine the coverage of atoms in (111) and (100) planes independently, in separate measurements, based on the bismuth and germanium adsorption charge, as described in Section 2.4.1.6.

Returning to Pt nanoparticles, voltammograms of samples in 0.5 M H_2SO_4 are displayed in Figure 2.26. Responses for polyoriented Pt and three types of preferentially oriented Pt nanoparticles are compared. As an initial step in the analysis, particles were immobilized at the surface of a polished Au disk electrode, which served as a conductive, catalytically inert support over the potential range investigated. An aliquot of particles suspended in solution was cast onto the electrode surface and dried under a gentle stream of Ar. Evaporation of the solution left behind an adherent layer of particles a few monolayers in thickness. Surface impurities were removed by cycling the electrode through the hydrogen and anion adsorption regions in 0.5 M H_2SO_4. This general approach has been adopted widely for investigations of Pt nanoparticle electrochemistry [213,380–383]. In these broader applications, Au or glassy carbon typically has been employed as the support electrode. In addition to physical adsorption of particles, ionomer materials sometimes have been used to ensure adherence of the particles to the support [381–383]. CO adsorption and stripping is also effective as a final cleaning step [213,371–374].

In the voltammograms of Figure 2.26, the hydrogen and anion adsorption peaks are sharply defined and reflect the distribution of Pt surface atoms in different coordination sites on the Pt nanoparticles. The sharpness and symmetry of the peaks indicate that a high level of cleanliness has been attained. Figure 2.26a corresponds to the cyclic voltammogram for the polycrystalline Pt nanoparticles. The features present are very similar to those of bulk polycrystalline Pt (Figure 2.25d). For the preferentially oriented particles, features dominant in the voltammogram of the cubic particles are characteristic of (100) planes (Figure 2.26b), while the sample containing a mixture of octahedral and tetrahedral shaped particles displays prominent features for (111) planes (Figure 2.26d). The particles that contain a combination of (100) and (111) planes show behavior intermediate to the cubic and tetrahedral plus octahedral nanoparticle samples (Figure 2.26c).

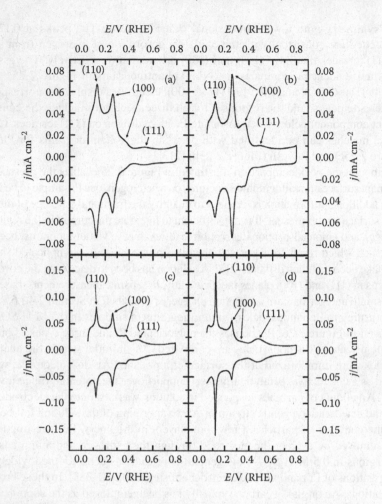

FIGURE 2.26 Voltammograms corresponding to (a) polyoriented Pt nanoparticles, and (b) (100), (c) (100)-(111), and (d) (111) Pt nanoparticles. The solution was 0.5 M H_2SO_4 and the sweep rate was 50 mV/s. (From Solla-Gullon et al., *Phys Chem Chem Phys*, Vol. 10, (2008): p. 3689. With permission.)

The average relative percentage of ordered (111) and (100) domains present on the surface of the nanoparticles in each sample is given in Table 2.2. The determinations were based on the oxidation charge for a monolayer of bismuth, or germanium adsorbed to the particles on the electrode. A calibration curve was used to convert the adatom oxidation charge to the coverage of sites on the (100), in the case of germanium, or (111) for bismuth (Figure 2.16d), surface planes. The results in Table 2.2 were found to agree with those expected from inspection of the hydrogen and anion adsorption regions in cyclic voltammograms and analysis of the particle samples by transmission electron microscopy [375].

TABLE 2.2

Fraction of the (111) and (100) Ordered Domains Determined for the Different Nanoparticle Samples [375]

Sample[a]	(111) Sites/%	(100) Sites/%
Pt_{poly}	9	18
$Pt_{(100)}$	18	40
$Pt_{(100)-(111)}$	30	32
$Pt_{(111)}$	42	3

[a] Pt_{poly} = polyoriented Pt nanoparticles; $Pt_{(100)}$, $Pt_{(100)-(111)}$ and $Pt_{(111)}$ are preferentially oriented nanoparticles.

Practical applications of Pt nanoparticle materials in electrochemistry are motivating continued study of particles that have a preferential orientation. The ability to perform experiments on particles that have predictable shapes and a predominance of either (100) or (111) crystal surface planes is helping to bridge the gap in understanding the differences in the catalytic activity of bulk single crystals and practical catalysts. The Pt nanoparticle samples prepared for the experiments displayed in Figure 2.26 were also employed in studies of HCOOH and CH_3OH oxidation [375]. The measurements were performed in parallel with bulk Pt(100), Pt(111), Pt(110), and polycrystalline Pt electrodes. Responses recorded in experiments with the shape-controlled particles showed evidence of the dominant low index surface planes present. However, relative to the bulk single crystals, responses for the particles were more complex and reflected the catalytic activity of low coordination atoms on the particles. Recently, HCOOH oxidation was investigated at shape-controlled Pt nanoparticles modified by bismuth [376]. Samples dominated by (111) planes were more active than polycrystalline particles in linear scan voltammetry measurements, consistent with expectations based on studies of bismuth modified bulk Pt single crystal electrodes [376]. The studies are an example of the steps being taken to use preferentially oriented Pt nanoparticles and findings from studies with bulk single crystal electrodes to guide the development of new high surface area catalysts.

2.5 SUMMARY

The ability to characterize the structure and composition of Pt single crystal surfaces down to the atomic scale has led to the development of more detailed microscopic models for electrical double layer phenomena and electrocatalytic reaction mechanisms, topics that have broad importance. For a single crystal Pt electrode in contact with an aqueous electrolyte solution, the energies and shapes of voltammetric peaks recorded while scanning across the hydrogen adsorption and double layer potential regions are profoundly affected by the surface crystallographic structure of the Pt electrode. The responses reflect the competition for different adsorption sites among solvent molecules, electrolyte anions, hydrogen, and OH⁻. In recent years, it has been observed that on Pt(111) electrodes the potential regions for hydrogen and OH⁻ adsorption are well separated, whereas these regions are not as distinct on surfaces with steps and other low coordination sites [173]. The Pt metal surface atomic structure affects the composition of the surface adlayer, which in turn influences the charge distribution within the electrical double layer region. These details of the interface are of fundamental interest, but have more general importance in impacting processes such as pathways and rates of electrocatalytic reactions. The microscopic details of the metal–solution interface revealed by experiments with Pt single crystal electrodes assist in the identification of factors that limit interfacial charge transfer [243].

Trends in electrocatalysis toward the study of modified Pt surfaces and nanoscale metal particles have built on the foundations of Pt single crystal electrochemistry. Pt single crystal electrodes are often employed as standards for benchmarking reaction rates and calibrating adsorbáte surface coverages in support of these applications. Work on Pt_3M alloys provides an example of the power of adopting an electrochemical surface science approach to understand structure–function relationships and define principles that guide the design of new catalysts [20,320]. Pt single crystal electrochemistry experiments also are inspiring the use of shape controlled metal nanoparticles as steps toward understanding the reaction mechanisms over practical catalysts and testing strategies for catalyst preparation [213,375].

The demand for new materials and tailored interfaces for applications such as efficient electrical energy production, corrosion prevention, and chemical sensing, will continue to motivate the use of well-defined surfaces to advance understanding of interfacial properties. There is still a great deal to learn about the structure, stability, and composition of Pt electrode surfaces, particularly under conditions relevant to practical applications. Pt single crystals have played an important role in electrochemistry and many more exciting avenues for investigation lie ahead.

REFERENCES

1. Ross, P. N., Surface crystallography at the metal-solution interface, in: *Structure of Electrified Interfaces*, J. Lipkowski, and P. N. Ross (eds.), VCH: New York (1993), p. 35.
2. Hubbard, A. T., Electrochemistry of well-defined surfaces, *Acc Chem Res*, Vol. 13, (1980): p. 177.
3. Hamelin, A., Double-layer properties at sp and sd metal single-crystal electrodes, in: *Modern Aspects of Electrochemistry*, R. E. White, J. O. M. Bockris, and B. E. Conway (eds.), Plenum Press: New York (1985), Vol. 16, p. 1.
4. Hubbard, A. T., Electrochemistry of well-defined surfaces, *Chem Rev*, Vol. 88, (1988): p. 633.
5. Clavilier, J., Electrochemical surface characterization of platinum electrodes using elementary electrosorption processes at basal and stepped surfaces, in: *Electrochemical Surface Science: Molecular Phenomena at Electrode Surfaces*, M. P. Soriaga (ed.), American Chemical Society: Washington, D.C. (1988), ACS Symposium Series 378, p. 202.

6. Adzic, R., Reaction kinetics and mechanisms on metal single-crystal electrode surfaces, in: *Modern Aspects of Electrochemistry*, R. E. White, J. O. M. Bockris, and B. E. Conway (eds.), Plenum Press: New York (1990), Vol. 21, p. 163.

7. Clavilier, J., Flame-annealing and cleaning technique, in: *Interfacial Electrochemistry: Theory, Experiment, and Applications*, A. Wieckowski (ed.), Marcel Dekker: New York (1999), p. 231.

8. Dakkouri, A. S., and D. M. Kolb, Reconstruction of gold surfaces, in: *Interfacial Electrochemistry: Theory, Experiment, and Applications*, A. Wieckowski (ed.), Marcel Dekker: New York (1999), p. 151.

9. Markovic, N. M., and P. N. Ross, Surface science studies of model fuel cell electrocatalysts, *Surf Sci Reports*, Vol. 45, (2002): p. 117.

10. Soriaga, M. P., Molecular adsorption at single-crystal electrode surfaces, in: *Structure of Electrified Interfaces*, J. Lipkowski, and P. N. Ross (eds.), VCH: New York (1993), p. 103.

11. McCreery, R. L., and R. T. Packard, Raman monitoring of dynamic electrochemical events, *Anal Chem*, Vol. 61, (1989): p. 775A.

12. Kneten, K. R., and R. L. McCreery, Effects of redox system structure on electron-transfer kinetics at ordered graphite and glassy carbon electrodes, *Anal Chem*, Vol. 64, (1992): p. 2518.

13. Tao, N. J., Potential controlled ordering in organic monolayers at electrode-electrolyte interfaces, in: *Imaging of Surfaces and Interfaces*, J. Lipkowski, and P. N. Ross, (eds.), Wiley-VCH: New York (1999), p. 211.

14. Hubbard, A. T., J. L. Stickney, M. P. Soriaga, V. K. F. Chia, S. D. Rosasco, B. C. Schardt, T. Solonum, D. Song, J. H. White, and A. Wieckowski, Electrochemical processes at well-defined surfaces, *J Electroanal Chem*, Vol. 168, (1984): p. 43.

15. Homa, A. S., E. Yeager, and B. Cahan, LEED-AES thin-layer electrochemical studies of hydrogen adsorption on platinum single crystals, *J Electroanal Chem*, Vol. 150, (1983): p. 181.

16. Aberdam, D., R. Durand, R. Faure, and F. El-Omar, Structural changes of a Pt(111) electrode induced by electrosorption of oxygen in acidic solutions: A coupled voltammetry, LEED and AES study, *Surf Sci*, Vol. 171, (1986) p. 303.

17. Kolb, D. M., Surface reconstruction at metal-elecrolyte interfaces, in: *Structure of Electrified Interfaces*, J. Lipkowski, and P. N. Ross (eds.), VCH: New York (1993), p. 65.

18. Kolb, D. M., Reconstruction phenomena at metal-electrolyte interfaces, *Prog Surf Sci*, Vol. 51, (1996): p. 109.

19. Jarvi, T. D., and E. M. Stuve, Fundamental aspects of vacuum and electrocatalytic reactions of methanol and formic acid on platinum surfaces, in: *Electrocatalysis*, J. Lipkowski, and P. N. Ross (eds.), Wiley-VCH: New York (1998), Vol. Ch. 3, p. 75.

20. Stamenkovic, V. R., B. Fowler, B. S. Mun, G. Wang, P. N. Ross, C. A. Lucas, and N. M. Markovic, Improved oxygen reduction activity on $Pt_3Ni(111)$ via increased surface site availability, *Science*, Vol. 315, (2007): p. 493.

21. Strmcnik, D., K. Kodama, D. van der Vliet, J. Greeley, V. R. Stamenkovic, and N. M. Markovic, The role of non-covalent interactions in electrocatalytic fuel-cell reactions on platinum, *Nature Chemistry*, Vol. 1, (2009): p. 466.

22. Clavilier, J., R. Faure, G. Guinet, and R. Durand, Preparation of monocrystalline platinum microelectrodes and electrochemical study of the plane surfaces cut in the direction of the {111} and {110} planes, *J Electroanal Chem*, Vol. 107, (1980): p. 205.

23. Clavilier, J., The role of anion on the electrochemical behavior of a {111} platinum surface: An unusual splitting of the voltammogram in the hydrogen region, *J Electroanal Chem*, Vol. 107, (1980): p. 211.

24. Zurawski, D., L. Rice, M. Hourani, and A. Wieckowski, The in-situ preparation of well-defined, single crystal electrodes, *J Electroanal Chem*, Vol. 230, (1987): p. 221.

25. Hourani, M., and A. Wieckowski, Electrochemistry of the ordered rhodium (111) electrode. Surface preparation and voltammetry in perchloric acid electrolyte, *J Electroanal Chem*, Vol. 227, (1987): p. 259.

26. Chang, S. C., and M. J. Weaver, In situ infrared spectroscopy at single-crystal metal electrodes: An emerging link between electrochemical and ultrahigh-vacuum surface science, *J Phys Chem*, Vol. 95, (1991): p. 5391.

27. Lucas, C. A., and N. M. Markovic, In-situ x-ray diffraction studies of the electrode-solution interface, in: *Diffraction and Spectroscopic Methods in Electrochemistry*, R. Alkire, D. M. Kolb, J. Lipkowski, and P. N. Ross (eds.), Wiley-VCH: Weinheim, 2006, Vol. 9, p. 1.

28. Lebedeva, N. P., M. T. M. Koper, J. M. Feliu, and R. A. van Santen, Role of crystalline defects in electrocatalysis: Mechanism and kinetics of CO adlayer oxidation on stepped platinum Electrodes, *J Phys Chem B*, Vol. 106, (2002): p. 12938.

29. Chidsey, C. E. D., C. R. Bertozzi, T. M. Putvinski, and A. M. Mujsce, Coadsorption of ferrocene-terminated and unsubstituted alkanethiols on gold: electroactive self-assembled monolayers, *J Am Chem Soc*, Vol. 112, (1990): p. 4301.

30. Chidsey, C. E. D., Free energy and temperature dependence of electron transfer at the metal-electrolyte interface, *Science*, Vol. 251, (1991): p. 919.

31. Villegas, I., and M. J. Weaver, Carbon monoxide adlayer structures on platinum (111) electrodes: A synergy between in-situ scanning tunneling microscopy and infrared spectroscopy, *J Chem Phys*, Vol. 101, (1994): p. 1648.

32. Nichols, R. J., Imaging metal electrocrystallization at high resolution, in: *Imaging of Surfaces and Interfaces*, J. Lipkowski, and P. N. Ross (eds.), Wiley-VCH: New York (1999), p. 99.

33. Clavilier, J., R. Durand, G. Guinet, and R. Faure, Electrochemical adsorption behavior of platinum(100) in sulfuric acid solution, *J Electroanal Chem*, Vol. 127, (1981): p. 281.

34. Clavilier, J., D. Armand, and B. L. Wu, Electrochemical study of the initial surface condition of platinum surfaces with (100) and (111) orientations, *J Electroanal Chem*, Vol. 135, (1982): p. 159.

35. Clavilier, J., D. Armand, S. G. Sun, and M. Petit, Electrochemical adsorption behavior of platinum stepped surfaces in sulphuric acid solutions, *J Electroanal Chem*, Vol. 205, (1986): p. 267.

36. Kaishev, R., and B. Muftaftschiew, Electrolytic nucleus formation on globular platinum single-crystal electrodes, *Z Phys Chem*, Vol. 204, (1955): p. 334.

37. Komanicky, V., and W. R. Fawcett, Fabrication of an annealable platinum (111) single crystal ultramicroelectrode, *J Electroanal Chem*, Vol. 556, (2003): p. 109.

38. Nuffield, E. W., *X-ray Diffraction Methods*, Wiley, New York, 1966.

39. Herrero, E., A. Rodes, J. M. Perez, J. M. Feliu, and A. Aldaz, CO adsorption and oxidation on Pt(111) electrodes modified by irreversibly adsorbed arsenic in sulphuric acid medium. Comparison with bismuth-modified electrodes, *J Electroanal Chem*, Vol. 393, (1995): p. 87.

40. Pozniak, B., Y. Mo, I. C. Stefan, K. Mantey, M. Hartmann, and D. A. Scherson, In situ time-resolved second-harmonic generation from Pt(111) microfacetted single-crystal platinum microspheres, *J Phys Chem B*, Vol. 105, (2001): p. 7874.

41. Pozniak, B., and D. A. Scherson, Dynamics of a surface phase transition as monitored by in situ second harmonic generation, *J Am Chem Soc*, Vol. 125, (2003): p. 7488.

42. Rodes, A., K. El Achi, M. A. Zamakhchari, and J. Clavilier, Hydrogen probing of step and terrace sites on Pt(s)-[n(111)x(100)], *J Electroanal Chem*, Vol. 284, (1990): p. 245.

43. Clavilier, J., K. E. Achi, M. Petit, A. Rodes, and M. A. Zamakhchari, Electrochemical monitoring of the thermal reordering of platinum single-crystal surfaces after metallographic polishing from the early stage to the equilibrium surfaces, *J Electroanal Chem*, Vol. 295, (1990). p. 333.

44. Lebedeva, N. P., M. T. M. Koper, J. M. Feliu, and R. A. van Santen, The effect of the cooling atmosphere in the preparation of flame-annealed Pt(111) electrodes on CO adlayer oxidation, *Electrochem Commun*, Vol. 2, (2000): p. 487.

45. Markovic, N. M., H. A. Gasteiger, and P. N. Ross, Jr., Oxygen reduction on platinum low-index single-crystal surfaces in sulfuric acid solution: Rotating ring-Pt(hkl) disk studies, *J Phys Chem*, Vol. 99, (1995): p. 3411.

46. Markovic, N. M., B. N. Grgur, C. A. Lucas, and P. N. Ross, Electrooxidation of CO and H_2/CO mixtures on Pt(111) in acid solutions, *J Phys Chem B*, Vol. 103, (1999): p. 487.

47. Strmcnik, D. S., P. Rebec, M. Gaberscek, D. Tripkovic, V. Stamenkovic, C. Lucas, and N. M. Markovic, Relationship between the surface coverage of spectator species and the rate of electrocatalytic reactions, *J Phys Chem C*, Vol. 111, (2007): p. 18672.

48. Markovic, N., M. Hanson, G. McDougall, and E. Yeager, The effects of anions on hydrogen electrosorption on platinum single-crystal electrodes, *J Electroanal Chem*, Vol. 214, (1986): p. 555.

49. Motoo, S., and N. Furuya, Electrochemistry of platinum single crystal surfaces. Part I. Structural change of the Pt(111) surface followed by an electrochemical method, *J Electroanal Chem*, Vol. 172, (1984): p. 339.

50. Lima, F. H. B., J. Zhang, M. H. Shao, K. Sasaki, M. B. Vukmirovic, E. A. Ticianelli, and R. R. Adzic, Catalytic activity – d-band center correlation for the O_2 reduction reaction on platinum in alkaline solutions, *J Phys Chem C*, Vol. 111, (2007): p. 404.

51. Lucas, C. A., P. Thompson, M. Cormack, A. Brownrigg, B. Fowler, D. Strmcnik, V. Stamenkovic, et al., Temperature-induced ordering of metal/adsorbate structures at electrochemical interfaces, *J Am Chem Soc*, Vol. 131, (2009): p. 7654.

52. Cuesta, A., L. A. Kibler, and D. M. Kolb, A method to prepare single crystal electrodes of reactive metals: Application to Pd(hkl), *J Electroanal Chem*, Vol. 466, (1999): p. 165.

53. Kibler, L. A., A. Cuesta, M. Kleinert, and D. M. Kolb, In-situ STM characterisation of the surface morphology of platinum single crystal electrodes as a function of their preparation, *J Electroanal Chem*, Vol. 484, (2000): p. 73.

54. Wagner, F. T., and P. N. Ross, Jr., LEED analysis of electrode surfaces. Structural effects of potentiodynamic cycling on Pt single crystals, *J Electroanal Chem*, Vol. 150, (1983): p. 141.

55. Wagner, F. T., and P. N. Ross, Jr., Long-range structural effects in the anomalous voltammetry ultra-high vacuum prepared Pt(111), *J Electroanal Chem*, Vol. 250, (1988): p. 301.

56. Al Jaaf-Golze, K., D. M. Kolb, and D. Scherson, On the voltammetry curves of Pt(111) in aqueous solutions, *J Electroanal Chem*, Vol. 200, (1986): p. 353.

57. Itaya, K., S. Sugawara, and K. Higaki, In situ scanning tunneling microscopy for platinum surfaces in aqueous solutions, *J Phys Chem*, Vol. 92, (1988): p. 6714.

58. Itaya, K., S. Sugawara, K. Sashikata, and N. Furuya, In situ scanning tunneling microscopy of platinum (111) surface with the observation of monatomic steps, *J Vac Sci Technol*, Vol. 8, (1990): p. 515.

59. Tanaka, S., S.-L. Yau, and K. Itaya, In-situ scanning tunneling microscopy of bromine adlayers on Pt(111), *J Electroanal Chem*, Vol. 396, (1995): p. 125.

60. Tidswell, I. M., N. M. Markovic, and P. N. Ross, Potential dependent surface structure of the Pt(111) electrolyte interface, *J Electroanal Chem*, Vol. 376, (1994): p. 119.

61. Dickertmann, D., F. D. Koppitz, and J. W. Schultze, Eine methode zum ausschluss von randeffekten bei elektrochemischen messungen an einkristallen: Test anhand der adsorptionssysteme Ag/Pb^{2+} und Au/Cu^{2+}, *Electrochim Acta*, Vol. 21, (1976): p. 967.

62. Herrero, E., J. Clavilier, J. M. Feliu, and A. Aldaz, Influence of the geometry of the hanging meniscus contact on the hydrogen oxidation reaction on a Pt(111) electrode in sulphuric acid, *J Electroanal Chem*, Vol. 410, (1996): p. 125.

63. Herrero, E., K. Franaszczuk, and A. Wieckowski, Electrochemistry of methanol at low index crystal planes of platinum: An integrated voltammetric and chronoamperometric study, *J Phys Chem*, Vol. 98, (1994): p. 5074.

64. Leung, L.-W. H., A. Wieckowski, and M. J. Weaver, In situ infrared spectroscopy of well-defined single-crystal electrodes: Adsorption and electrooxidation of CO on Pt(111), *J Phys Chem*, Vol. 92, (1988): p. 6985.

65. Chang, S.-C., and M. J. Weaver, Coverage and potential dependent binding geometries of carbon monoxide at ordered low index platinum and rhodium aqueous interfaces: Comparisons with adsorption in corresponding metal-vacuum environments, *Surf Sci*, Vol. 238, (1990): p. 142.

66. Villullas, H. M., and M. Lopez Teijelo, The hanging-meniscus rotating disk (HMRD) Part 1. Dependence of hydrodynamic behavior on experimental variables, *J Electroanal Chem*, Vol. 384, (1995): p. 25.

67. Villullas, H. M., and M. Lopez Teijelo, Meniscus shape and lateral wetting at the hanging meniscus rotating disc (HMRD) electrode, *J Appl Elect*, Vol. 26, (1996): p. 353.

68. Macia, M. D., J. M. Campina, E. Herrero, and J. M. Feliu, On the kinetics of oxygen reduction on platinum stepped surfaces in acidic media, *J Electroanal Chem*, Vol. 564, (2004): p. 141.

69. Kuzume, A., E. Herrero, and J. M. Feliu, Oxygen reduction on stepped platinum surfaces in acidic media, *J Electroanal Chem*, Vol. 599, (2007): p. 333.

70. Strmcnik, D. S., D. V. Tripkovic, D. van der Vliet, K.-C. Chang, V. Komanicky, H. You, G. Karapetrov, J. P. Greeley, V. R. Stamenkovic, and N. M. Markovic, Unique activity of platinum adislands in the CO electrooxidation reaction, *J Am Chem Soc*, Vol. 130, (2008): p. 15332.

71. Bard, A. J., and L. R. Faulkner, *Electrochemical Methods: Fundamentals and Applications*, 2nd Edition, Wiley: New York, 2001.

72. Greef, R., R. Peat, L. M. Peter, D. Pletcher, and J. Robinson, *Instrumental Methods in Electrochemistry*, Ellis Horwood Ltd.: Chichester, 1985.

73. Stamenkovic, V. R., M. Arenz, C. A. Lucas, M. E. Gallagher, P. N. Ross, and N. M. Markovic, Surface chemistry on bimetallic alloy surfaces: Adsorption of anions and oxidation of CO on Pt$_3$Sn(111), *J Am Chem Soc*, Vol. 125, (2003): p. 2736.

74. Stamenkovic, V., and N. M. Markovic, Electrochemistry at well-characterized bimetallic surfaces, in: *Fuel Cell Catalysis: A Surface Science Approach*, M. T. M. Koper (ed.), Wiley: Hoboken, New Jersey (2009), p. 245.

75. Moffat, T. P., Scanning tunneling microscopy studies of metal electrodes, in: *Electroanalytical Chemistry: A Series of Advances*, A. J. Bard, and I. Rubinstein (eds.), Dekker: New York (1999), Vol. 21, p. 211.

76. Kolb, D. M., UV-visible reflectance spectroscopy, in: *Spectroelectrochemistry: Theory and Practice*, R. J. Gale (ed.), Plenum: New York (1988), p. 87.

77. Sagara, T., UV-visible reflectance spectroscopy of thin organic films at electrode surfaces, in: *Diffraction and Spectroscopic Methods in Electrochemistry*, R. Alkire, D. M. Kolb, J. Lipkowski, and P. N. Ross (eds.), Wiley-VCH: Weinheim (2006), p. 47.

78. Richmond, G. L., Optical second harmonic generation as an in situ probe of electrochemical interfaces, in: *Electroanalytical Chemistry: A Series of Advances*, A. J. Bard (ed.), Dekker: New York (1991), Vol. 17, p. 87.

79. Corn, R. M., Optical second harmonic generation studies of adsorption, orientation and order at the electrochemical interface, *Anal Chem*, Vol. 63, (1991): p. 285A.

80. Akemann, W., K. A. Friedrich, and U. Stimming, Potential-dependence of CO adlayer structures on Pt(111) electrodes in acid solution: Evidence for a site selective charge transfer, *J Chem Phys*, Vol. 113, (2000): p. 6864.

81. Pozniak, B., and D. A. Scherson, In situ dual-beam coincidence second harmonic generation as a probe of spatially resolved dynamics at electrochemical interfaces, *J Am Chem Soc*, Vol. 126, (2004): p. 14696.

82. Zelenay, P., and A. Wieckowski, Radioactive labeling: Toward characterization of well-defined electrodes, in: *Electrochemical Interfaces: Modern Techniques for In-Situ Surface Characterization*, H. D. Abruna (ed.), VCH: New York (1991), p. 479.

83. Balashova, N. A., and V. E. Kazarinov, Use of the radioactive-tracer method for the investigation of the elecric double-layer structure, in: *Electroanalytical Chemistry: A Series of Advances*, A. J. Bard (ed.), Dekker: New York (1969), Vol. 3, p. 135.

84. Kolics, A., and A. Wieckowski, Adsorption of bisulfate and sulfate anions on a Pt(111) electrode, *J Phys Chem B*, Vol. 105, (2001): p. 2588.

85. Iwasita, T., and F. C. Nart, In-situ Fourier transform infrared spectroscopy: A tool to characterize the metal-electrolyte interface at a molecular level, in: *Advances in Electrochemical Science and Engineering*, H. Gerischer, and C. Tobias (eds.), VCH Publishers: New York (1995), Vol. 4, p. 123.

86. Iwasita, T., and F. C. Nart, In situ infrared spectroscopy at electrochemical interfaces, *Prog Surf Sci*, Vol. 55, (1997): p. 271.

87. Korzeniewski, C., Recent advances in in situ infrared spectroscopy and applications in single crystal electrochemistry and electrocatalysis, in: *Diffraction and Spectroscopic Methods in Electrochemistry*, R. Alkire, D. M. Kolb, J. Lipkowski, and P. N. Ross (eds.), Wiley-VCH: Weinheim (2006), p. 233.

88. Osawa, M., In-situ surface-enhanced infrared spectroscopy of the electrode/solution interface, in: *Diffraction and Spectroscopic Methods in Electrochemistry*, R. Alkire, D. M. Kolb, J. Lipkowski, and P. N. Ross (eds.), Wiley-VCH: Weinheim (2006), p. 269.

89. Lu, G. Q., A. Lagutchev, T. Takeshita, R. L. Behrens, D. D. Dlott, and A. Wieckowski, Sum frequency generation studies of surface intermediates involved in fuel cell electrocatalysis, in: *Fuel Cells: A Surface Science Approach*, M. T. M. Koper (ed.), Wiley: Hoboken, New Jersey (2009), p. 375.

90. Baldelli, S., and A. A. Gewirth, Sum frequency generation studies of the electrified solid/liquid interface, in: *Diffraction and Spectroscopic Methods in Electrochemistry*, R. Alkire, D. M. Kolb, J. Lipkowski, and P. N. Ross (eds.), Wiley-VCH: Weinheim (2006), p. 163.

91. Hartung, T., U. Schmiemann, I. Kamphausen, and H. Baltruschat, Electrodesorption from single-crystal electrodes: Analysis by differential electrochemical mass spectrometry, *Anal Chem*, Vol. 63, (1992): p. 44.

92. Wang, H., T. Loffler, and H. Baltruschat, Formation of intermediates during methanol oxidation: A quantitative DEMS study, *J Applied Electrochemistry*, Vol. 31, (2001): p. 759.

93. Schmiemann, U., and H. Baltruschat, The adsorption of ethene at Pt single crystal electrodes. Desorption products and observation of multiple adsorption states by DEMS, *J Electroanal Chem*, Vol. 340, (1992): p. 357.

94. Tong, Y. Y., E. Oldfield, and A. Wieckowski, Exploring electrochemical interfaces with solid-state NMR, *Anal Chem*, Vol. 70, (1998): p. 518A.

95. Tong, Y. Y., Coupling interfacial electrochemistry with nuclear magnetic resonance spectroscopy: An electronic perspective, in: *In-situ Spectroscopic Studies of Adsorption at the Electrode and Electrocatalysis*, S.-G. Sun, P. A. Christensen, and A. Wieckowski (eds.), Elsevier: Amsterdam (2007), p. 441.

96. Tian, Z.-Q., and B. Ren, Adsorption and reaction at electrochemical interfaces as probed by surface-enhanced Raman spectroscopy, *Annu Rev Phys Chem*, Vol. 55, (2004): p. 197.

97. Pettinger, B., Tip-enhanced raman spectroscoy – Recent developments and future prospects, in: *Diffraction and Spectroscopic Methods in Electrochemistry*, R. Alkire, D. M. Kolb, J. Lipkowski, and P. N. Ross (eds.), Wiley-VCH: Weinheim (2006), p. 377.

98. Li, J. F., Y. F. Huang, Y. Ding, Z. L. Yang, S. B. Li, X. S. Zhou, F. R. Fan, et al., Shell-isolated nanoparticle-enhanced Raman spectroscopy, *Nature*, Vol. 464, (2010): p. 392.

99. Jerkiewicz, J., Hydrogen sorption at/in electrodes, *Prog Surf Sci*, Vol. 57, (1998): p. 137.

100. Hubbard, A. T., R. M. Ishikawa, and J. Katekaru, Study of platinum electrodes by means of electrochemistry and low energy electron diffraction. Part II. Comparison of the electrochemical activity of Pt(100) and Pt(111) surfaces, *J Electroanal Chem*, Vol. 86, (1978): p. 271.

101. Yeager, E., W. E. O'Grady, M. Y. C. Woo, and P. Hagans, Hydrogen adsorption on single crystal platinum, *J Electrochem Soc*, Vol. 125, (1978): p. 348.

102. Ross, P. N., Jr., Structure sensitivity in the electrocatalytic properties of Pt. I. Hydrogen adsorption on low index single crystals and the role of steps, *J Electrochem Soc*, Vol. 126, (1979): p. 67.

103. Yamamoto, K., D. M. Kolb, R. Kotz, and G. Lehmpfuhl, Hydrogen adsorption and oxide formation on platinum single crystal electrodes, *J Electroanal Chem*, Vol. 96, (1979): p. 233.

104. Scherson, D. A., and D. M. Kolb, Voltammetric curves for Au(111) in acid media: A comparison with Pt(111) surfaces, *J Electroanal Chem*, Vol. 176, (1984): p. 353.

105. Adzic, R. R., F. Feddrix, B. Z. Nikolic, and E. Yeager, The elucidation of hydrogen and anion adsorption on Pt(111) through the co-adsorption of metal adatoms and carbon monoxide, *J Electroanal Chem*, Vol. 341, (1992): p. 287.

106. Clavilier, J., J. M. Orts, R. Gómez, J. M. Feliu, and A. Aldaz, On the nature of the charged species displaced by CO adsorption from platinum oriented electrodes in sulphuric acid solution, in: *Electrochemistry and Materials Science of Cathodic Hydrogen Absorption and Adsorption*, B. E. Conway, and G. Jerkiewicz (eds.), The Electrochemical Society Proceedings Series: Pennington, New Jersey (1995), PV 94–21, p. 167.

107. Feliu, J. M., J. M. Orts, R. Gomez, A. Aldaz, and J. Clavilier, New information on the unusual adsorption states of Pt(111) in sulfuric acid solutions from potentiostatic adsorbate replacement by CO, *J Electroanal Chem*, Vol. 372, (1994): p. 265.

108. Koper, M. T. M., J. J. Lukkien, N. P. Lebedeva, J. M. Feliu, and R. A. van Santen, Adsorbate interactions and phase transitions at the stepped platinum/electrolyte interface: Experiment compared with Monte Carlo simulations, *Surf Sci*, Vol. 478, (2001): p. L339.

109. Strmcnik, D., D. Tripkovic, D. van der Vliet, V. Stamenkovic, and N. M. Markovic, Adsorption of hydrogen on Pt(111) and Pt(100) surfaces and its role in the HOR, *Electrochem Comm*, Vol. 10, (2008): p. 1602.

110. Hermse, C. G. M., A. P. van Bavel, M. T. M. Koper, J. J. Lukkien, R. A. van Santen, and A. P. Jansen, Modelling the butterfly: ($\sqrt{3}$ x $\sqrt{7}$) ordering on fcc(111) surfaces, *Surf Sci*, Vol. 527, (2004): p. 247.

111. Herrero, E., J. Mostany, J. M. Feliu, and J. Lipkowski, Thermodynamic studies of anion adsorption at the Pt(111) electrode surface in sulfuric acid solutions, *J Electroanal Chem*, Vol. 534, (2002): p. 79.

112. Garcia, G., and M. T. M. Koper, Stripping voltammetry of carbon monoxide oxidation on stepped platinum single-crystal electrodes in alkaline solution, *Phys Chem Chem Phys*, Vol. 10, (2008): p. 3802.

113. Funtikov, A. M., U. Stimming, and R. Vogel, Anion adsorption from sulfuric acid solutions on Pt(111) single crystal electrodes, *J Electroanal Chem*, Vol. 428, (1997): p. 147.

114. Garcia-Araez, N., V. Climent, P. Rodriguez, and J. M. Feliu, Thermodynamic analysis of (bi)sulphate adsorption on a Pt(111) electrode as a function of pH, *Electrochim Acta*, Vol. 53, (2008): p. 6793.

115. Braunschweig, B., and W. Daum, Superstructures and order–disorder transition of sulfate adlayers on Pt(111) in sulfuric acid solution, *Langmuir*, Vol. 25, (2009): p. 11112.

116. Wilms, M., P. Broekmann, C. Stuhlmann, and K. Wandelt, In-situ STM investigation of adsorbate structures on Cu(111) in sulfuric acid electrolyte, *Surf Sci*, Vol. 416, (1998): p. 121.

117. Magnussen, O. M., J. Hagebock, J. Hotlos, and R. J. Behm, In situ scanning tunnelling microscopy observations of a disorder–order phase transition in hydrogensulfate adlayers on Au(111), *Faraday Disc*, Vol. 94, (1992): p. 329.

118. Edens, G. J., X. Gao, and M. J. Weaver, The adsorption of sulfate on gold(111) in acidic aqueous media: Adlayer structural interferences from infrared spectroscopy and scanning tunneling microscopy, *J Electroanal Chem*, Vol. 375, (1994): p. 357.

119. Wan, L.-J., M. Hara, J. Inukai, and K. Itaya, In situ scanning tunneling microscopy of well-defined Ir(111) surface: High-resolution imaging of adsorbed sulfate, *J Phys Chem B*, Vol. 103, (1999): p. 6978.

120. Wan, L.-J., S.-L. Yau, and K. Itaya, Atomic structure of adsorbed sulfate on Rh(111) in sulfuric acid solution, *J Phys Chem*, Vol. 99, (1995): p. 9507.

121. Kim, Y.-G., J. B. Soriaga, G. Vigh, and M. P. Soriaga, Atom-resolved EC-STM studies of anion adsorption at well-defined surfaces: Pd(111) in sulfuric acid solution, *J Coll Interf Sci*, Vol. 227, (2000): p. 505.

122. Lebedeva, N. P., M. T. M. Koper, E. Herrero, J. M. Feliu, and R. A. van Santen, CO oxidation on stepped Pt[n(111)x(111)] electrodes, *J Electroanal Chem*, Vol. 487, (2000): p. 37.

123. Saravanan, C., M. T. M. Koper, N. M. Markovic, M. Head-Gordon, and P. N. Ross, Modeling base voltammetry and CO electrooxidation at the Pt(111)-electrolyte interface: Monte Carlo simulations including anion adsorption, *Phys Chem Chem Phys*, Vol. 4, (2002): p. 2660.

124. Markovic, N., and P. N. Ross, The effect of specific adsorption of ions and underpotential deposition of copper on the electro-oxidation of methanol on platinum single-crystal surfaces, *J Electroanal Chem*, Vol. 330, (1992): p. 499.

125. Nishihara, C., and H. Nozoye, Note on the hydrogen adsorption-desorption voltammogram on Pt(111) in sulfuric acid solution, *J Electroanal Chem*, Vol. 379, (1994): p. 527.

126. Koper, M. T. M., and J. J. Lukkien, Modeling the butterfly: The voltammetry of ($\sqrt{3}$ x $\sqrt{3}$)R30° and p(2x2) overlayers on (111) electrodes, *J Electroanal Chem*, Vol. 485, (2000): p. 161.

127. Koper, M. T. M., and J. J. Lukkien, Modeling the butterfly: Influence of lateral interactions and adorption geometry on the voltammetry at (111) and (100) electrodes, *J Electroanal Chem*, Vol. 498, (2002): p. 105.

128. Clavilier, J., and D. Armand, Electrochemical induction of changes in the distribu-
     tion of the hydrogen adsorption states on Pt(100) and Pt(111) surfaces in contact
     with sulphuric acid solution, *J Electroanal Chem*, Vol. 199, (1986): p. 187.
129. Clavilier, J., K. E. Achi, and A. Rodes, In situ characterization of the Pt(s)-[n(111) x
     (111)] electrode surfaces using electrosorbed hydrogen for probing terrace and step
     sites, *J Electroanal Chem*, Vol. 272, (1989): p. 253.
130. Magnussen, O. M., Ordered anion adlayers on metal electrode surfaces, *Chem Rev*,
     Vol. 102, (2002): p. 679.
131. Markovic, N. M., T. J. Schmidt, B. N. Grgur, H. A. Gasteiger, R. J. Behm, and
     P. N. Ross, Effect of temperature on suface processes at the Pt(111)-liquid interface:
     Hydrogen adsorption, oxide formation, and CO oxidation, *J Phys Chem B*, Vol. 103,
     (1999): p. 8568.
132. Berna, A., J. M. Feliu, L. Gancs, and S. Mukerjee, Voltammetric characterization of
     Pt single crystal electrodes with basal orientations in trifluoromethanesulphonic acid,
     Electrochem Commun, Vol. 10, (2008): p. 1695.
133. Fukucla, T., and A. Aramata, The kinetic study of specific adsorption of phosphate
     species on Pt(111) in acidic solutions, *J Electrochem Soc*, Vol. 440, (1997): p. 153.
134. Nikolic, B. Z., and R. R. Adzic, The electrosorption valence of adsorbed phosphoric
     acid anions on the Pt(111) surface, *J Serb Chem Soc*, Vol. 62, (1997): p. 515.
135. Mostany, J., P. Martínez, V. Climent, E. Herrero, and J. M. Feliu, Thermodynamic
     studies of phosphate adsorption on Pt(111) electrode surfaces in perchloric acid solu-
     tions, *Electrochim Acta*, Vol. 54, (2009): p. 5836.
136. Markovic, N. M., B. N. Grgur, C. A. Lucas, and P. N. Ross, Surface electrochemistry
     of CO on Pt(110)-(1x2) and Pt(110)-(1x1) surfaces, *Surf Sci*, Vol. 384, (1997):
     p. L805.
137. Lopez-Cudero, A., A. Cuesta, and C. Gutiérrez, Potential control of the CO adsorption
     site on Pt(100) electrodes, *Electrochem Comm*, Vol. 6, (2004): p. 395.
138. Gomez, R., and J. Clavilier, Electrochemical behaviour of platinum surfaces contain-
     ing (110) sites and the problem of the third oxidation peak, *J Electroanal Chem*, Vol.
     354, (1993): p. 189.
139. Armand, D., and J. Clavilier, Electrochemical behaviour of the (110) orientation of a
     platinum surface in acid medium: the role of anions, *J Electroanal Chem*, Vol. 263,
     (1989): p. 109.
140. Armand, D., and J. Clavilier, Quantitative analysis of the distribution of the hydrogen
     adsorption states at platinum surfaces. Part II. Application to Pt(110), stepped and
     polyoriented platinum surfaces in sulphuric acid medium, *J Electroanal Chem*, Vol.
     233, (1987): p. 251.
141. Feliu, J. M., A. Rodes, J. M. Orts, and J. Clavilier, The problem of surface order of
     Pt single crystals in electrochemistry, *Polish J Chem*, Vol. 68, (1994): p. 1575.
142. Tidswell, I. M., N. M. Markovic, and P. N. Ross, Potential dependent surface
     relaxation of the Pt(001)/electrolyte interface, *Phys Rev Lett*, Vol. 71, (1993): p. 1601.
143. Al-Akl, A., G. A. Attard, R. Price, and B. Timothy, Voltammetric and UHV charac-
     terisation of the (1x1) and reconstructed hex-R0.7° phases of Pt{100}, *J Electroanal
     Chem*, Vol. 467, (1999): p. 60.
144. Al-Akl, A., G. Attard, R. Price, and B. Timothy, Electrochemical and UHV char-
     acterisation of stepped Pt{100} electrode surfaces, *Phys Chem Chem Phys*, Vol. 3,
     (2001): p. 3261.
145. Lopez-Cudero, A., A. Cuesta, and C. Gutierrez, Potential dependence of
     the saturation CO coverage of Pt electrodes: The origin of the pre-peak in
     CO-stripping voltammograms. Part 2: Pt(100), *J Electroanal Chem*, Vol. 586,
     (2006): p. 204.

146. Villegas, I., and M. J. Weaver, Nature of the atomic-scale restructuring of Pt(100) electrode surfaces as evidenced by in-situ scanning tunneling microscopy, *J Electroanal Chem*, Vol. 373, (1994): p. 245.

147. Armand, D., and J. Clavilier, Quantitative analysis of the distribution of the hydrogen adsorption states at platinum surfaces. Part I. Application to Pt(100) in sulphuric acid medium, *J Electroanal Chem*, Vol. 225, (1987): p. 205.

148. Clavilier, J., J. M. Feliu, A. Fernandez-Vega, and A. Aldaz, Electrochemical behaviour of irreversibly adsorbed bismuth on Pt(100) with different degrees of crystalline surface order, *J Electroanal Chem*, Vol. 269, (1989): p. 175.

149. Rodes, A., M. A. Zamakhchari, K. El Achi, and J. Clavilier, Electrochemical behaviour of Pt(100) in various acidic media. I. On a new voltammetric profile of Pt(100) in perchloric acid and effects of surface defects, *J Electroanal Chem*, Vol. 305, (1991): p. 115.

150. Domke, K., E. Herrero, A. Rodes, and J. M. Feliu, Determination of the potentials of zero total charge of Pt(100) stepped surfaces in the [01 $\bar{1}$] zone. Effect of the step density and anion adsorption, *J Electroanal Chem*, Vol. 552, (2003): p. 115.

151. Love, B., K. Seto, and J. Lipkowski, Adsorption hydrogen on stepped Pt single crystal surfaces, *J Electroanal Chem*, Vol. 199, (1986): p. 219.

152. Markovic, N. M., N. S. Marinkovic, and R. R. Adzic, Electrosorption of hydrogen and sulfuric acid anions on single crystal platinum stepped surfaces. Part I. The [110] zone., *J Electroanal Chem*, Vol. 241, (1988): p. 309.

153. Motoo, S., and N. Furuya, Effect of terraces and steps in the electrocatalysis of formic acid oxidation on platinum, *Ber Bunsenges Phys Chem*, Vol. 91, (1987): p. 457.

154. Somorjai, G. A., Modern concepts in surface science and heterogeneous catalysis, *J Phys Chem*, Vol. 94, (1990): p. 1013.

155. Somorjai, G. A., The puzzles of surface science and recent attempts to explain them, *Surf Sci*, Vol. 242, (1991): p. 481.

156. Somorjai, G. A., *Chemistry in Two Dimensions: Surfaces*, Cornell: Ithaca, New York, 1981.

157. Lang, B., R. W. Joyner, and G. A. Somorjai, Low energy electron diffraction studies of high index crystal surfaces of platinum, *Surf Sci*, Vol. 30, (1972): p. 440.

158. Lang, B., R. W. Joyner, and G. A. Somorjai, Low energy electron diffraction studies of chemisorbed gases on stepped surfaces of platinum, *Surf Sci*, Vol. 30, (1972): p. 454.

159. Blakely, D. W., and G. A. Somorjai, The stability and structure of high Miller index platinum crystal surfaces in vacuum and in the presence of adsorbed carbon and oxygen, *Surf Sci*, Vol. 65, (1977): p. 419.

160. Davis, S. M., and G. A. Somorjai, The effect of surface oxygen on hydrocarbon reactions catalyzed by platinum crystal surfaces with variable kink concentrations, *Surf Sci*, Vol. 91, (1980): p. 73.

161. Garcia-Araez, N., V. Climent, E. Herrero, and J. M. Feliu, On the electrochemical behavior of the Pt(100) vicinal surfaces in bromide solutions, *Surf Sci*, Vol. 560, (2004): p. 269.

162. Frumkin, A. N., and O. A. Petrii, Potentials of zero total and zero free charge of platinum group metals, *Electrochim Acta*, Vol. 20, (1975): p. 347.

163. Hurwitz, H. D., Direct determination of the surface excess of specifically adsorbed ions on mercury, *J Electroanal Chem*, Vol. 10, (1965): p. 35.

164. Savich, W., G.-S. Sun, J. Lipkowski, and A. Wieckowski, Determination of the sum of Gibbs excesses of sulfate and bisulfate adsorbed at the Pt(111) electrode surface using chronocoulometry and thermodynamics of the perfectly polarized electrode, *J Electroanal Chem*, Vol. 388, (1995): p. 233.

165. Climent, V., R. Gomez, and J. Feliu, Effect of increasing amount of steps on the potential of zero total charge of Pt(111) electrodes, *Electrochim Acta*, Vol. 45, (1999): p. 629.

166. Clavilier, J., R. Albalat, R. Gomez, J. M. Orts, J. M. Feliu, and A. Aldaz, Study of the charge displacement at constant potential during CO adsorption on Pt(110) and Pt(111) electrodes in contact with a perchloric acid solution, *J Electroanal Chem*, Vol. 330, (1992): p. 489.

167. Clavilier, J., R. Albalat, R. Gomez, J. M. Orts, and J. M. Feliu, Displacement of adsorbed iodine on platinum single-crystal electrodes by irreversible adsorption of CO at controlled potential, *J Electroanal Chem*, Vol. 360, (1993): p. 325.

168. Climent, V., R. Gomez, J. M. Orts, A. Rodes, A. Aldaz, and J. M. Feliu, Electrochemistry, spectroscopy, and scanning tunneling microscopy images of small single-crystal electrodes, in: *Interfacial Electrochemistry: Theory, Experiment, and Applications*, A. Wieckowski (ed.), Dekker: New York (1999), p. 463.

169. Herrero, E., J. M. Feliu, A. Wieckowski, and J. Clavilier, The unusual adsorption states of Pt(111) electrodes studied by an iodine displacement method: Comparison with Au(111) electrodes, *Surf Sci*, Vol. 325, (1995): p. 131.

170. Bergelin, M., E. Herrero, J. M. Feliu, and M. Wasberg, Oxidation of CO adlayers on Pt(111) at low potentials: An impinging jet study in H_2SO_4 electrolyte with mathematical modeling of the current transients, *J Electroanal Chem*, Vol. 467, (1999): p. 74.

171. Gomez, R., J. M. Feliu, A. Aldaz, and M. J. Weaver, Validity of double-layer charge-corrected voltammetry for assaying carbon monoxide coverages on ordered transition metals: Comparisons with adlayer structures in electrochemical and ultrahigh vacuum environments, *Surf Sci*, Vol. 410, (1998): p. 48.

172. Weaver, M. J., Potentials of zero charge for platinum(111)-aqueous interfaces: A combined assessment from in-situ and ultrahigh-vacuum measurements, *Langmuir*, Vol. 14, (1998): p. 3932.

173. Garcia-Araez, N., V. Climent, and J. Feliu, Potential-dependent water orientation on well-defined platinum surfaces as inferred from laser-pulsed experiments, *J Phys Chem C*, Vol. 113, (2009): p. 9290.

174. Lecoeur, J., J. Andro, and R. Parsons, The behaviour of water at stepped surfaces of single crystal gold electrodes, *Surf Sci*, Vol. 114, (1982): p. 320.

175. Besocke, K., B. Krahl-Urban, and H. Wagner, Dipole moments associated with edge atoms; A comparative study on stepped Pt, Au and W surfaces, *Surf Sci*, Vol. 68, (1977): p. 39.

176. Ross, P. N., The role of defects in the specific adsorption of anions on Pt(111), *J Chim Phys*, Vol. 88, (1991): p. 1353.

177. Smoluchowski, R., Anisotropy of the electronic work function of metals, *Phys Rev*, Vol. 60, (1941): p. 661.

178. Trasatti, S., Structuring of the solvent at metal/solution interfaces and components of the electrode potential, *J Electroanal Chem*, Vol. 150, (1983): p. 1.

179. Pirug, G., and H. P. Bonzel, Electrochemical double-layer modeling under ultrahigh vacuum conditions, in: *Interfacial Electrochemistry: Theory, Experiment, and Applications*, A. Wieckowski (ed.), Dekker: New York (1999), p. 269.

180. Villegas, I., and M. J. Weaver, Modeling electrochemical interfaces in ultrahigh vacuum: Molecular roles of solvation in double-layer phenomena, *J Phys Chem B*, Vol. 101, (1997): p. 10166.

181. Climent, V., B. A. Coles, and R. G. Compton, Coulostatic potential transients induced by laser heating of a Pt(111) single-crystal electrode in aqueous acid solutions. Rate of hydrogen adsorption and potential of maximum entropy, *J Phys Chem B*, Vol. 106, (2002): p. 5988.

182. Climent, V., B. A. Coles, R. G. Compton, and J. M. Feliu, Coulostatic potential transients induced by laser heating of platinum stepped electrodes: Influence of steps on the entropy of double layer formation, *J Electroanal Chem*, Vol. 561, (2004): p. 157.
183. Climent, V., N. Garcia-Araez, R. G. Compton, and J. M. Feliu, Effect of deposited bismuth on the potential of maximum entropy of Pt(111) single-crystal electrodes, *J Phys Chem B*, Vol. 110, (2006): p. 21092.
184. Garcia-Araez, N., V. Climent, and J. M. Feliu, Evidence of water reorientation on model electrocatalytic surfaces from nanosecond-laser-pulsed experiments, *J Am Chem Soc*, Vol. 130, (2008): p. 3824.
185. Garcia-Araez, N., V. Climent, and J. Feliu, Potential-dependent water orientation on Pt(111) stepped surfaces from laser-pulsed experiments, *Electrochim Acta*, Vol. 54, (2009): p. 966.
186. Climent, V., N. García-Araez, E. Herrero, and J. Feliu, Potential of zero total charge of platinum single crystals: A local approach to stepped surfaces vicinal to Pt(111), *Russ J Elec*, Vol. 42, (2006): p. 1145.
187. Inkaew, P., W. Zhou, and C. Korzeniewski, CO monolayer oxidation at Pt(100) probed by potential step measurements in comparison to Pt(111) and Pt nanoparticle catalyst, *J Electroanal Chem*, Vol. 614, (2008): p. 93.
188. Inkaew, P., and C. Korzeniewski, Kinetic studies of adsorbed CO electrochemical oxidation on Pt(335) at full and sub-saturation coverages, *Phys Chem Chem Phys*, Vol. 10, (2008): p. 3655.
189. Rodes, A., R. Gomez, J. M. Perez, J. M. Feliu, and A. Aldaz, On the voltammetric and spectroscopic characterization of nitric oxide adlayers formed from nitrous acid on Pt(h,k,l) and Rh(h,k,l) electrodes, *Electrochim Acta*, Vol. 41, (1996): p. 729.
190. Rodes, A., V. Climent, J. M. Orts, J. M. Pérez, and A. Aldaz, Nitric oxide adsorption at Pt(100) electrode surfaces, *Electrochim Acta*, Vol. 44, (1998): p. 1077.
191. Orts, J. M., R. Gomez, J. M. Feliu, A. Aldaz, and J. Clavilier, Nature of Br adlayers on Pt(111) single-crystal surfaces. Voltammetric, charge displacement, and ex situ STM experiments, *J Phys Chem*, Vol. 100, (1996): p. 2334.
192. Orts, J. M., R. Gomez, J. M. Feliu, A. Aldaz, and J. Clavilier, Voltammetry, charge displacement experiments, and scanning tunneling microscopy of the Pt(100)-Br system, *Langmuir*, Vol. 13, (1997): p. 3016.
193. Climent, V., A. Rodes, J. M. Orts, A. Aldaz, and J. M. Feliu, Urea adsorption on Pt(111) electrodes, *J Electroanal Chem*, Vol. 461, (1999): p. 65.
194. Climent, V., A. Rodes, J. M. Orts, J. M. Feliu, J. M. Perez, and A. Aldaz, On the electrochemical and in-situ Fourier transform infrared spectroscopy characterization of urea adlayers at Pt(100) electrodes, *Langmuir*, Vol. 13, (1997): p. 2380.
195. Climent, V., A. Rodes, R. Albalat, J. Claret, J. M. Feliu, and A. Aldaz, Urea adsorption on platinum single crystal stepped surfaces, *Langmuir*, Vol. 17, (2001): p. 8260.
196. Mostany, J., E. Herrero, J. M. Feliu, and J. Lipkowski, Thermodynamic studies of anion adsorption at stepped platinum(*hkl*) electrode surfaces in sulfuric acid solutions, *J Phys Chem B*, Vol. 106, (2002): p. 12787.
197. Mostany, J., E. Herrero, J. M. Feliu, and J. Lipkowski, Determination of the Gibbs excess of H and OH adsorbed at a Pt(111) electrode surface using a thermodynamic method, *J Electroanal Chem*, Vol. 558, (2003): p. 19.
198. Garcia-Araez, N., V. Climent, E. Herrero, J. Feliu, and J. Lipkowski, Thermodynamic studies of chloride adsorption at the Pt(111) electrode surface from 0.1 M HClO$_4$ solution, *J Electroanal Chem*, Vol. 576, (2005): p. 33.
199. Garcia-Araez, N., V. Climent, E. Herrero, J. Feliu, and J. Lipkowski, Determination of the Gibbs excess of H adsorbed at a Pt(111) electrode surface in the presence of co-adsorbed chloride, *J Electroanal Chem*, Vol. 582, (2005): p. 76.

200. Garcia-Araez, N., V. Climent, E. Herrero, J. Feliu, and J. Lipkowski, Thermodynamic studies of bromide adsorption at the Pt(111) electrode surface perchloric acid solutions: Comparison with other anions, *J Electroanal Chem*, Vol. 591, (2006): p. 149.

201. Shi, Z., and J. Lipkowski, Investigations of SO_4^{2-} adsorption at the Au(111) electrode by chronocoulometry and radiochemistry, *J Electroanal Chem*, Vol. 366, (1994): p. 317.

202. Trasatti, S., and R. Parsons, Interphases in systems of conducting phases, *J Electroanal Chem*, Vol. 205, (1986): p. 359.

203. Parsons, R., The specific adsorption of ions at the metal-electrolyte interphase, *Trans Faraday Soc*, Vol. 51, (1955): p. 1518.

204. Climent, V., B. A. Coles, and R. G. Compton, Laser-induced potential transients on a Au(111) single-crystal electrode. Determination of the potential of maximum entropy of double-layer formation, *J Phys Chem B*, Vol. 106, (2002): p. 5258.

205. Harrison, J. A., J. E. B. Randles, and D. J. Schiffrin, The entropy of formation of the mercury-aqueous solution interface and the structure of the inner layer, *J Electroanal Chem*, Vol. 48, (1973): p. 359.

206. Hamelin, A., T. Vitanov, E. Sevastyanov, and A. Popov, The electrochemical double layer on sp metal single crystals: The current status of data, *J Electroanal Chem*, Vol. 145, (1983): p. 225.

207. Kolb, D. M., Physical and electrochemical properties of metal monolayers on metallic substrates, in: *Advances in Electrochemistry and Electrochemical Engineering*, H. Gerischer, and C. W. Tobias (eds.), Wiley: New York (1978), Vol. 11, p. 125.

208. Herrero, E., L. J. Buller, and H. D. Abruna, Underpotential deposition at single crystal surfaces of Au, Pt, Ag and other materials, *Chem Rev*, Vol. 101, (2001): p. 1897.

209. Climent, V., N. Garcia-Araez, and J. M. Feliu, Clues for the molecular level understanding of electrocatalysis on single-crystal platinum surfaces modified by p-block adatoms, in: *Fuel Cell Catalysis: A Surface Science Approach*, M. T. M. Koper (ed.), Wiley: Hoboken, New Jersey (2009), p. 209.

210. Clavilier, J., J. M. Feliu, and A. Aldaz, An irreversible structure sensitive adsorption step in bismuth underpotential deposition at platinum electrodes, *J Electroanal Chem*, Vol. 243, (1988): p. 419.

211. Clavilier, J., J. M. Orts, J. M. Feliu, and A. Aldaz, Study of the conditions for irreversible adsorption of lead at Pt(h,k.l) electrodes, *J Electroanal Chem*, Vol. 293, (1990): p. 197.

212. Clavilier, J., A. Fernandez-Vega, J. M. Feliu, and A. Aldaz, Electrochemical behaviour of the Pt(111)-As system in acidic medium: Adsorbed hydrogen and hydrogen reaction, *J Electroanal Chem*, Vol. 294, (1990): p. 193.

213. Solla-Gullon, J., P. Rodriguez, E. Herrero, A. Aldaz, and J. M. Feliu, Surface characterization of platinum electrodes, *Phys Chem Chem Phys*, Vol. 10, (2008): p. 1359.

214. Rodríguez, P., E. Herrero, J. Solla-Gullon, F. J. Vidal-Iglesias, A. Aldaz, and J. M. Feliu, Specific surface reactions for identification of platinum surface domains. Surface characterization and electrocatalytic tests, *Electrochim Acta*, Vol. 50, (2005): p. 4308.

215. Solla-Gullon, J., F. J. Vidal-Iglesias, P. Rodriguez, E. Herrero, J. M. Feliu, J. Clavilier, and A. Aldaz, In situ surface characterization of preferentially oriented platinum nanoparticles by using electrochemical structure sensitive adsorption reactions, *J Phys Chem B*, Vol. 108, (2004): p. 13573.

216. Rodriguez, P., J. Solla-Gullon, F. J. Vidal-Iglesias, E. Herrero, A. Aldaz, and J. M. Feliu, Determination of (111) ordered domains on platinum electrodes by irreversible adsorption of bismuth, *Anal Chem*, Vol. 77, (2005): p. 5317.

217. Clavilier, J., A. Fernandez-Vega, J. M. Feliu, and A. Aldaz, Heterogeneous electrocatalysis on well defined platinum surfaces modified by controlled amounts of irreversibly adsorbed adatoms. Part I. Formic acid oxidation on the Pt(111)-Bi system, *J Electroanal Chem*, Vol. 258, (1989): p. 89.

218. Clavilier, J., A. Fernandez-Vega, J. M. Feliu, and A. Aldaz, Heterogeneous electrocatalysis on well-defined Pt surfaces modified by controlled amounts of irreversible adsorbed adatoms. Part III. Formic acid oxidation on Pt(100)-Bi system, *J Electroanal Chem*, Vol. 261, (1989): p. 113.

219. Chang, S. C., A. Hamelin, and M. J. Weaver, New perspectives in electrochemical processes on single-crystal surfaces from real-time FTIR spectroscopy, *J Chim Phys*, Vol. 86, (1991): p. 1615.

220. Gomez, R., A. Fernandez-Vega, J. M. Feliu, and A. Aldaz, Hydrogen evolution on Pt single crystal surfaces. Effects of irreversibly adsorbed bismuth and antimony on hydrogen adsorption and evolution on Pt(100), *J Phys Chem*, Vol. 97, (1993): p. 4769.

221. Herrero, E., A. Fernandez-Vega, J. M. Feliu, and A. Aldaz, Poison formation reaction from formic acid and methanol on Pt(111) electrodes modified by irreversibly adsorbed Bi and As, *J Electroanal Chem*, Vol. 350, (1993): p. 73.

222. Herrero, E., J. M. Feliu, and A. Aldaz, Poison formation reaction from formic-acid on Pt(100) electrodes modified by irreversibly adsorbed bismuth and antimony, *J Electroanal Chem*, Vol. 368, (1994): p. 101.

223. Herrero, E., J. M. Feliu, and A. Aldaz, CO adsorption and oxidation on Pt(111) electrodes modified by irreversibly adsorbed bismuth in sulfuric acid medium, *J Catal*, Vol. 152, (1995): p. 264.

224. Schmidt, T. J., R. J. Behm, B. N. Grgur, N. M. Markovic, and P. N. Ross, Jr., Formic acid oxidation on pure and Bi-modified Pt(111): Temperature effects, *Langmuir*, Vol. 16, (2000): p. 8159.

225. Schmidt, T. J., V. R. Stamenkovic, C. A. Lucas, N. M. Markovic, and P. N. Ross, Jr., Surface processes and electrocatalysis on the Pt(hkl)/Bi-solution interface, *Phys Chem Chem Phys*, Vol. 3, (2001): p. 3879.

226. Schmidt, T. J., B. N. Grgur, R. J. Behm, N. M. Markovic, and P. N. Ross, Jr., Bi adsorption on Pt(111) in perchloric acid solution: A rotating ring–disk electrode and XPS study, *Phys Chem Chem Phys*, Vol. 2, (2000): p. 4379.

227. Rodríguez, P., E. Herrero, A. Aldaz, and J. M. Feliu, Tellurium adatoms as an in-situ surface probe of (111) two-dimensional domains at platinum surfaces, *Langmuir*, Vol. 22, (2006): p. 10329.

228. Gomez, R., M. J. Llorca, J. M. Feliu, and A. Aldaz, The behaviour of germanium adatoms irreversibly adsorbed on platinum single crystals, *J Electroanal Chem*, Vol. 340, (1992): p. 349.

229. Rodriguez, P., E. Herrero, J. Solla-Gullon, F. J. Vidal-Iglesias, A. Aldaz, and J. M. Feliu, Electrochemical characterization of irreversibly adsorbed germanium on platinum stepped surfaces vicinal to Pt(100), *Electrochim Acta*, Vol. 50, (2005): p. 3111.

230. Bockris, J. O., and A. K. N. Reddy, *Modern Electrochemistry*, Plenum: New York, 1973.

231. Appleby, A. J., Electrocatalysis, in: *Comprehensive Treatise of Electrochemistry*, B. E. Conway, J. O. M. Bockris, E. Yeager, S. U. M. Khan, and R. E. White (eds.), Plenum: New York (1983), Vol. 7, p. 173.

232. Appleby, A. J., and F. R. Foulkes, *Fuel Cell Handbook*, Van Nostrand Reinhold: New York, 1989.

233. Appleby, A. J., Electrocatalysis, in: *Modern Aspects of Electrochemistry*, B. E. Conway, and J. O. M. Bockris (eds.), Plenum: New York (1974), Vol. 9, p. 369.

234. Markovic, N. M., and P. N. Ross, Jr., Electrocatalysis at well-defined surfaces: Kinetics of oxygen reduction and hydrogen oxidation/evolution on Pt(hkl) electrodes, in: *Interfacial Electrochemistry: Theory, Experiment and Applications*, A. Wieckowski (ed.), Dekker: New York (1999), p. 821.

235. Parsons, R., and T. Vandernoot, The oxidation of small organic molecules. A survey of recent fuel cell related research, *J Electroanal Chem*, Vol. 257, (1988): p. 9.

236. Beden, B., J.-M. Leger, and C. Lamy, Electrocatalytic oxidation of oxygenated aliphatic organic compounds at noble metal electrodes, in: *Modern Aspects of Electrochemistry*, J. O. M. Bockris, B. E. Conway, and R. E. White (eds.), Plenum: New York (1992), Vol. 22, p. 97.

237. de Vooys, A. C. A., M. T. M. Koper, R. A. van Santen, and J. A. R. van Veen, The role of adsorbates in the electrochemical oxidation of ammonia on noble and transition metal electrodes, *J Electroanal Chem*, Vol. 506, (2001): p. 127.

238. Beltramo, G. L., and M. T. M. Koper, Nitric oxide reduction and oxidation on stepped Pt[n(111)x(111)] electrodes, *Langmuir*, Vol. 19, (2003): p. 8907.

239. Rosca, V., and M. T. M. Koper, Mechanism of electrocatalytic reduction of nitric oxide on Pt(100), *J Phys Chem B*, Vol. 109, (2005): p. 16750.

240. Rosca, V., M. Duca, M. T. de Groot, and M. T. M. Koper, Nitrogen cycle electrocatalysts, *Chem Rev*, Vol. 109, (2009): p. 2209.

241. Adzic, R. R., Recent advances in the kinetics of oxygen reduction, in: *Electrocatalysis*, J. Lipkowski, and P. N. Ross (eds.), Wiley-VCH: New York (1998), p. 197.

242. Sun, S.-G., Studying electrocatalytic oxidation of small organic molecules with in-situ infrared spectroscopy, in: *Electrocatalysis*, J. Lipkowski, and P. N. Ross (eds.), Wiley-VCH: New York (1998), p. 243.

243. Koper, M. T. M., S. C. S. Lai, and E. Herrero, Mechanisms of the oxidation of carbon monoxide and small organic molecules at metal electrodes, in: *Fuel Cell Catalysis: A Surface Science Approach*, M. T. M. Koper (ed.), Wiley: Hoboken, New Jersey (2009), p. 159.

244. Xu, Y., M. Shao, M. Mavrikakis, and R. R. Adzic, Recent developments in the electrocatalysis of the O_2 reduction reaction, in: *Fuel Cell Catalysis: A Surface Science Approach*, M. T. M. Koper (ed.), Wiley: Hoboken, New Jersey (2009), p. 271.

245. Markovic, N. M., T. S. Sarraf, H. A. Gasteiger, and P. N. Ross, Hydrogen electrochemistry on platinum low-index single-crystal surfaces in alkaline solution, *J Chem Soc, Faraday Trans*, Vol. 92, (1996): p. 3719.

246. Markovic, N. M., B. N. Grgur, and P. N. Ross, Temperature-dependent hydrogen electrochemistry on platinum low-index single-crystal surfaces in acid solutions, *J Phys Chem B*, Vol. 101, (1997): p. 5405.

247. Schmidt, T. J., P. N. Ross, Jr., and N. M. Markovic, Temperature dependent surface electrochemistry on Pt single crystals in alkaline electrolytes. Part 2. The hydrogen evolution/oxidation reaction, *J Electroanal Chem*, Vol. 524–525, (2002): p. 252.

248. Barber, J., S. Morin, and B. E. Conway, Specificity of the kinetics of H_2 evolution to the structure of single-crystal Pt surfaces and the relation between opd and upd H, *J Electroanal Chem*, Vol. 446, (1998): p. 125.

249. Barber, J. H., and B. E. Conway, Structural specificity of the kinetics of the hydrogen evolution reaction on the low-index surfaces of Pt single-crystal electrodes in 0.5 M NaOH, *J Electroanal Chem*, Vol. 461, (1999): p. 80.

250. Seto, K., A. Iannelli, B. Love, and J. Lipkowski, The influence of surface crystallography on the rate of hydrogen evolution at Pt electrodes, *J Electroanal Chem*, Vol. 226, (1987): p. 351.

251. Kita, H., S. Ye, and Y. Gao, Mass transfer effect in hydrogen evolution reaction on Pt single-crystal electrodes in acid solution, *J Electroanal Chem*, Vol. 334, (1992): p. 351.
252. Gottesfeld, S., Electrocatalysis of oxygen reduction in polymer electrolyte fuel cells: A brief history and a critical examination of present theory and diagnostics, in: *Fuel Cell Catalysis: A Surface Science Approach*, M. T. M. Koper (ed.), Wiley: Hoboken, New Jersey (2009), p. 1.
253. Gottesfeld, S., and T. A. Zawodzinski, Polymer electrolyte fuel cells, in: *Advances in Electrochemical Science and Engineering*, R. C. Alkire, H. Gerischer, D. M. Kolb, and C. W. Tobias (eds.), Wiley-VCH: New York (1997), Vol. 5, p. 195.
254. Gasteiger, H. A., and N. M. Markovic, Just a dream – or future reality?, *Science*, Vol. 324, (2009): p. 48.
255. Ross, P. N., Jr., Structure sensitivity in the electrocatalytic properties of Pt. II. Oxygen reduction on low index single crystals and the role of steps, *J Electrochem Soc*, Vol. 126, (1979): p. 78.
256. Markovic, N. M., R. R. Adzic, B. D. Cahan, and E. B. Yeager, Structural effects in electrocatalysis: Oxygen reduction on platinum low index single-crystal surfaces in perchloric acid solutions, *J Electroanal Chem*, Vol. 377, (1994): p. 249.
257. Kita, H., H. W. Lei, and Y. Z. Gao, Oxygen reduction on platinum single-crystal electrodes in acidic solutions, *J Electroanal Chem*, Vol. 379, (1994): p. 407.
258. El Kadiri, F., R. Faure, and R. Durand, Electrochemical reduction of molecular oxygen on platinum single crystals, *J Electroanal Chem*, Vol. 301, (1991): p. 177.
259. Markovic, N. M., H. A. Gasteiger, and P. N. Ross, Jr., Oxygen reduction on platinum low-index single-crystal surfaces in alkaline solution: Rotating ring disk-Pt(hkl) studies, *J Phys Chem*, Vol. 100, (1996): p. 6715.
260. Wang, J. X., N. M. Markovic, and R. R. Adzic, Kinetic analysis of oxygen reduction on Pt(111) in acid solutions: Intrinsic kinetic parameters and anion adsorption effects, *J Phys Chem B*, Vol. 108, (2004): p. 4127.
261. Zou, S., I. Villegas, C. Stuhlmann, and M. J. Weaver, Nanoscale phenomena in surface electrochemistry: Some insights from scanning tunneling microscopy and infrared spectroscopy, *Electrochim Acta*, Vol. 43, (1998): p. 2811.
262. Korzeniewski, C., Vibrational coupling effects on infrared spectra of adsorbates on electrodes, in: *Interfacial Electrochemistry: Theory, Experiment and Applications*, A. Wieckowski (ed.), Dekker: New York (1999), p. 345.
263. Korzeniewski, C., Infrared spectroelectrochemistry, in: *Handbook of Vibrational Spectroscopy*, J. M. Chalmers, and P. R. Griffiths (eds.), Wiley: New York (2002), Vol. 4, p. 2699.
264. Chang, S. C., and M. J. Weaver, Coverage-dependent dipole coupling for carbon monoxide adsorbed at ordered platinum(111)-aqueous interfaces: Structural and electrochemical implications, *J Chem Phys*, Vol. 92, (1990): p. 4582.
265. Chang, S. C., J. D. Roth. and M. J. Weaver, Formation and dissipation of adlayer domains during electrooxidation of carbon monoxide on ordered monocrystalline and disordered platinum and rhodium surfaces as probed by FTIR spectroscopy, *Surf Sci*, Vol. 244, (1991): p. 113.
266. Severson, M. W., and M. J. Weaver, Nanoscale island formation during oxidation of carbon monoxide adlayers at ordered electrochemical interfaces: A dipole-coupling analysis of coverage-dependent infrared spectra, *Langmuir*, Vol. 14, (1998): p. 5603.
267. Beden, B., C. Lamy, N. R. de Tacconi, and A. J. Arvia, The electrooxidation of CO: A test reaction in electrocatalysis, *Electrochim Acta*, Vol. 35, (1990): p. 691.
268. Beden, B., S. Bilmes, C. Lamy, and J. M. Leger, Electrosorption of carbon monoxide on platinum single crystals in perchloric acid medium, *J Electroanal Chem*, Vol. 149, (1983): p. 295.

269. Palaikis, L., D. Zurawski, M. Hourani, and A. Wieckowski, Surface electrochemistry of carbon monoxide adsorbed from electrolytic solutions at single crystal surfaces of Pt(111) and Pt(100), *Surf Sci*, Vol. 199, (1988): p. 183.

270. de Becdelievre, A. M., J. de Becdelievre, and J. Clavilier, Electrochemical oxidation of adsorbed carbon monoxide on platinum spherical single crystals, *J Electroanal Chem*, Vol. 294, (1990): p. 97.

271. Zurawski, D., M. Wasberg, and A. Wieckowski, Low-energy electron diffraction and voltammetry of carbon monoxide electrosorbed on Pt(111), *J Phys Chem*, Vol. 94, (1990): p. 2076.

272. Kitamura, F., M. Takeda, M. Takahashi, and M. Ito, CO adsorption on Pt(111) and Pt(100) single crystal surfaces in aqueous solutions studied by infrared reflection-absorption spectroscopy, *Chem Phys Lett*, Vol. 142, (1987): p. 318.

273. Kitamura, F., M. Takahashi, and M. Ito, Adsorption site interconversion induced by electrode potential of CO on the Pt(100) single-crystal electrode, *J Phys Chem*, Vol. 92, (1988): p. 3320.

274. Chang, S.-C., L.-W. H. Leung, and M. J. Weaver, Comparisons between coverage-dependent infrared frequencies for carbon monoxide adsorbed on ordered Pt(111), Pt(100) and Pt(110) in electrochemical and ultrahigh-vacuum environments, *J Phys Chem*, Vol. 93, (1989): p. 5341.

275. Henderson, M. A., A. Szabo, and J. T. Yates, Jr., The structure of CO on the Pt(112) stepped surface—a sensitive view of bonding configurations using electron stimulated desorption, *J Chem Phys*, Vol. 91, (1989): p. 7245.

276. Severson, M. W., C. Stuhlmann, I. Villegas, and M. J. Weaver, Dipole-dipole coupling effects upon infrared spectroscopy of compresssed electrochemical adlayers: Application to the Pt(111)/CO system, *J Chem Phys*, Vol. 103, (1995): p. 9832.

277. Roth, J. D., S.-C. Chang, and M. J. Weaver, IR spectroscopy of carbon monoxide electrosorption on platinum over wide potential ranges. Delineation of site occupancy changes and nonlinear band frequency-potential shifts, *J Electroanal Chem*, Vol. 288, (1990): p. 285.

278. Roth, J. D., and M. J. Weaver, Role of the double-layer cation on the potential-dependent stretching frequencies and binding geometries of carbon monoxide at platinum-nonaqueous interfaces, *Langmuir*, Vol. 8, (1992): p. 1451.

279. Wasileski, S. A., M. T. M. Koper, and M. J. Weaver, Metal electrode–chemisorbate bonding: General influence of surface bond polarization on field-dependent binding energetics and vibrational frequencies *J Chem Phys*, Vol. 115, (2001): p. 8193.

280. Kim, C. S., C. Korzeniewski, and W. J. Tornquist, Site specific co-adsorption at Pt(335) as probed by infrared spectroscopy: Structural alterations in the CO adlayer under aqueous electrochemical conditions, *J Chem Phys*, Vol. 100, (1994): p. 628.

281. Kim, C. S., W. J. Tornquist, and C. Korzeniewski, Site-dependent vibrational coupling of CO adsorbates on well-defined step and terrace sites of monocrystalline platinum: Mixed isotope studies at Pt(335) and Pt(111) in the aqueous electrochemical environment, *J Chem Phys*, Vol. 101, (1994): p. 9113.

282. Kim, C. S., and C. Korzeniewski, Vibrational coupling as a probe of adsorption at different structural sites on a stepped single-crystal electrode, *Anal Chem*, Vol. 69, (1997): p. 2349.

283. Lebedeva, N. P., A. Rodes, J. M. Feliu, M. T. M. Koper, and R. A. van Santen, Role of crystalline defects in electrocatalysis: CO adsorption and oxidation on stepped platinum electrodes as studied by in situ infrared spectroscopy, *J Phys Chem B*, Vol. 106, (2002): p. 9863.

284. Rodes, A., R. Gomez, J. M. Feliu, and M. J. Weaver, Sensitivity of compressed carbon monoxide adlayers on Platinum (111) electrodes to long-range substrate structure: Influence of monoatomic steps, *Langmuir*, Vol. 16, (2000): p. 811.

285. Tolmachev, Y. V., A. Menzel, A. V. Tkachuk, Y. S. Chu, and H. D. You, In situ surface X-ray scattering observation of long-range ordered ($\sqrt{19}$ x $\sqrt{19}$)R23.4 degrees-13CO structure on Pt(111) in aqueous electrolytes, *Electrochemical and Solid State Letters*, Vol. 7, (2004): p. E23.

286. Lebedeva, N. P., M. T. M. Koper, J. M. Feliu, and R. A. van Santen, Mechanism and kinetics of the electrochemical CO adlayer oxidation on Pt(111), *J Electroanal Chem*, Vol. 524–525, (2002): p. 242.

287. Garcia, G., and M. T. M. Koper, Dual reactivity of step-bound carbon monoxide during oxidation on a stepped platinum electrode in alkaline media, *J Am Chem Soc*, Vol. 131, (2009): p. 5384.

288. Housmans, T. H. M., C. G. M. Hermse, and M. T. M. Koper, CO oxidation on stepped single crystal electrodes: A dynamic Monte Carlo study, *J Electroanal Chem*, Vol. 607, (2007): p. 69.

289. Love, B., and J. Lipkowski, Effect of surface crystallography on electrocatalytic oxidation of carbon monoxide on platinum electrodes, in: *Electrochemical Surface Science. Molecular Phenomena at Electrode Surfaces*, M. P. Soriaga (ed.), American Chemical Society: Washington, DC (1988), ACS Symposium Series 378, p. 484.

290. Petukhov, A. V., W. Akemann, K. A. Friedrich, and U. Stimming, Kinetics of electrooxidation of a CO monolayer at the platinum/electrolyte interface, *Surf Sci*, Vol. 402–404, (1998): p. 182.

291. Vidal-Iglesias, F. J., J. Solla-Gullon, J. M. Campina, E. Herrero, A. Aldaz, and J. M. Feliu, CO monolayer oxidation on stepped Pt(s)-[(n−1)(100) × (110)] surfaces, *Electrochim Acta*, Vol. 54, (2009): p. 4459.

292. Andreaus, B., F. Maillard, J. Kocylo, E. R. Savinova, and M. Eikerling, Kinetic modeling of CO_{ad} monolayer oxidation on carbon-supported platinum nanoparticles, *J Phys Chem B*, Vol. 110, (2006): p. 21028

293. Maillard, F., E. R. Savinova, and U. Stimming, CO monolayer oxidation on Pt nanoparticles: Further insights into the particle size effects, *J Electroanal Chem*, Vol. 599, (2007): p. 221.

294. Arenz, M., K. J. J. Mayrhofer, V. Stamenkovic, B. B. Blizanac, T. Tomoyuki, P. N. Ross, and N. M. Markovic, The effect of the particle size on the kinetics of CO electrooxidation on high surface area Pt catalysts, *J Am Chem Soc*, Vol. 127, (2005): p. 6819.

295. Friedrich, K. A., F. Henglein, U. Stimming, and W. Unkauf, Size dependence of the CO monolayer oxidation on nanosized Pt particles supported on gold, *Electrochim Acta*, Vol. 45, (2000): p. 3283.

296. Cherstiouk, O. V., P. A. Simonov, and E. R. Savinova, Model approach to evaluate particle size effects in electrocatalysis: Preparation and properties of Pt nanoparticles on GC and HOPG, *Electrochim Acta*, Vol. 48, (2003): p. 3851.

297. Cherstiouk, O. V., P. A. Simonov, V. I. Zaikovskii, and E. R. Savinova, CO monolayer oxidation at Pt nanoparticles supported on glassy carbon electrodes, *J Electroanal Chem*, Vol. 554–555, (2003): p. 241.

298. Wieckowski, A., M. Rubel, and C. Gutierrez, Reactive sites in bulk carbon monoxide electro-oxidation on oxide-free platinum(111), *J Electroanal Chem*, Vol. 382, (1995): p. 97.

299. Kita, H., H. Naohara, T. Nakato, S. Taguchi, and A. Aramata, Effects of adsorbed CO on hydrogen ionization and CO oxidation reactions at Pt single-crystal electrodes in acidic solution, *J Electroanal Chem*, Vol. 386, (1995): p. 197.

300. Markovic, N. M., C. A. Lucas, B. N. Grgur, and P. N. Ross, Surface electrochemistry of CO and H_2/CO mixtures at Pt(100) interfaces: Electrode kinetics and interfacial structures, *J Phys Chem B*, Vol. 103, (1999): p. 9616.

301. Lopez-Cudero, A., A. Cuesta, and C. Gutierrez, Potential dependence of the saturation CO coverage of Pt electrodes: The origin of the pre-peak in CO-stripping voltammograms. Part 1: Pt(111), *J Electroanal Chem*, Vol. 579, (2005): p. 1.

302. Gasteiger, H. A., N. M. Markovic, and P. N. Ross, Jr., H_2 and CO electrooxidation on well-characterized Pt, Ru, and Pt-Ru. 1. Rotating disk electrode studies of the pure gases including temperature effects, *J Phys Chem*, Vol. 99, (1995): p. 8290.

303. Koper, M. T. M., T. J. Schmidt, N. M. Markovic, and P. N. Ross, Potential oscillations and S-shaped polarization curve in the continuous electro-oxidation of CO on platinum single-crystal electrodes, *J Phys Chem B*, Vol. 105, (2001): p. 8381.

304. Angelucci, C. A., F. C. Nart, E. Herrero, and J. M. Feliu, Anion re-adsorption and displacement at platinum single crystal electrodes in CO-containing solutions, *Electrochem Comm*, Vol. 9, (2007): p. 1113.

305. Feliu, J. M., and E. Herrero, Formic acid oxidation, in: *Handbook of Fuel Cells: Fundamentals, Technology and Applications*, W. Vielstich, A. Lamm, and H. A. Gasteiger (eds.), Wiley-VCH: New York (2003), Vol. 2, p. 625.

306. Jusys, Z., and R. J. Behm, Methanol oxidation on carbon-supported Pt fuel cell catalyst – A kinetic and mechanistic study by differential electrochemical mass spectrometry, *J Phys Chem B*, Vol. 105, (2001): p. 10874.

307. Jusys, Z., J. Kaiser, and R. J. Behm, Methanol electrooxidation over Pt/C fuel cell catalysts: Dependence of product yields on catalyst loading, *Langmuir*, Vol. 19, (2003): p. 6759.

308. Housmans, T. H. M., A. H. Wonders, and M. T. M. Koper, Structure sensitivity of methanol electrooxidation pathways on platinum: An on-line electrochemical mass spectrometry study, *J Phys Chem B*, Vol. 110, (2006): p. 10021.

309. Narayan, S. R., and T. I. Valdez, High-energy portable fuel cell power sources, *The Electrochemical Society Interface*, Vol. 17(4), (2008): p. 40.

310. Leger, J.-M., C. Coutanceau, and C. Lamy, Electrocatalysis for the direct alcohol fuel cell, in: *Fuel Cell Catalysis: A Surface Science Approach*, M. T. M. Koper (ed.), Wiley: Hoboken, New Jersey (2009), p. 343.

311. Franaszczuk, K., E. Herrero, P. Zelenay, A. Wieckowski, J. Wang, and R. I. Masel, A comparison of electrochemical and gas-phase decomposition of methanol on platinum surfaces, *J Phys Chem*, Vol. 96, (1992): p. 8509.

312. Yu, X., and P. G. Pickup, Recent advances in direct formic acid fuel cells (DFAFC), *J Power Sources*, Vol. 182, (2008): p. 124.

313. Rice, C., S. Ha, R. I. Masel, P. Waszczuk, A. Wieckowski, and T. Barnard, Direct formic acid fuel cells, *J Power Sources*, Vol. 111, (2002): p. 83.

314. Lai, S. C. S., and M. T. M. Koper, Electro-oxidation of ethanol and acetaldehyde on platinum single-crystal electrodes, *Faraday Discuss*, Vol. 140, (2008): p. 399.

315. Lai, S. C. S., S. E. F. Kleyn, V. Rosca, and M. T. M. Koper, Mechanism of the dissociation and electrooxidation of ethanol and acetaldehyde on platinum as studied by SERS, *J Phys Chem C*, Vol. 112, (2008): p. 19080.

316. Wang, Q., G. Q. Sun, L. H. Jiang, Q. Xin, S. G. Sun, Y. X. Jiang, S. P. Chen, Z. Jusys, and R. J. Behm, Adsorption and oxidation of ethanol on colloid-based Pt/C, PtRu/C and Pt_3Sn/C catalysts: In situ FTIR spectroscopy and on-line DEMS studies, *Phys Chem Chem Phys*, Vol. 9, (2007): p. 2686.

317. Del Colle, V., A. Berna, G. Tremiliosi-Filho, E. Herrero, and J. M. Feliu, Ethanol electrooxidation onto stepped surfaces modified by Ru deposition: Electrochemical and spectroscopic studies, *Phys Chem Chem Phys*, Vol. 10, (2008): p. 3766.

318. Colmati, F., G. Tremiliosi-Filho, E. R. Gonzalez, A. Berna, E. Herrero, and J. M. Feliu, Surface structure effects on the electrochemical oxidation of ethanol on platinum single crystal electrodes *Faraday Disc*, Vol. 140, (2008): p. 379.

319. Adzic, R. R., J. Zhang, K. Sasaki, M. B. Vukmirovic, M. Shao, J. X. Wang, A. U. Nilekar, M. Mavrikakis, J. A. Valerio, and F. Uribe, Platinum monolayer fuel cell electrocatalysts, *Top Catal*, Vol. 46, (2007): p. 249.
320. Stamenkovic, V., B. S. Mun, M. Arenz, K. J. J. Mayrhofer, C. A. Lucas, G. Wang, P. N. Ross, and N. M. Markovic, Trends in electrocatalysis on extended and nanoscale Pt-bimetallic alloy surfaces, *Nature Mater*, Vol. 6, (2007): p. 241.
321. Stamenkovic, V., B. S. Mun, K. J. J. Mayrhofer, P. N. Ross, N. M. Markovic, J. Rossmeisl, J. Greeley, and J. K. Norskov, Changing the activity of electrocatalysts for oxygen reduction by tuning the surface electronic structure, *Angew Chem Int Ed*, Vol. 45, (2006): p. 2897.
322. Zhang, J., M. B. Vukmirovic, K. Sasaki, A. U. Nilekar, M. Mavrikakis, and R. R. Adzic, Mixed-metal Pt monolayer electrocatalysts for enhanced oxygen reduction kinetics, *J Am Chem Soc*, Vol. 127, (2005): p. 12480.
323. Zhang, J., M. B. Vukmirovic, Y. Xu, M. Mavrikakis, and R. R. Adzic, Controlling the catalytic activity of platinum-monolayer electrocatalysts for oxygen reduction with different substrates, *Angew Chem Int Ed*, Vol. 44, (2005): p. 2132.
324. Bligaard, T., and J. K. Norskov, Ligand effects in heterogeneous catalysis and electrochemistry, *Electrochim Acta*, Vol. 52, (2007): p. 5512.
325. Kibler, L. A., A. M. El-Aziz, R. Hoyer, and D. M. Kolb, Tuning reaction rates by lateral strain in a palladium monolayer, *Angew Chem Int Ed*, Vol. 44, (2005): p. 2080.
326. Shibata, M., O. Takahashi, and S. Motoo, Electrocatalysis by ad-atoms: Part XXII. S-hole control by ad-atoms on HCOOH oxidation, *J Electroanal Chem*, Vol. 249, (1988): p. 253.
327. Angerstein-Kozlowska, H., B. MacDougall, and B. E. Conway, Origin of activation effects of acetonitrile and mercury in electrocatalytic oxidation of formic acid, *J Electrochem Soc*, Vol. 120, (1973): p. 756.
328. Watanabe, M., M. Horiuchi, and S. Motoo, Electrocatalysis by ad-atoms: Part XXIII. Design of platinum ad-electrodes for formic acid fuel cells with ad-atoms of the IVth and the Vth groups, *J Electroanal Chem*, Vol. 250, (1988): p. 117.
329. Shibata, M., N. Furuya, M. Watanabe, and S. Motoo, Electrocatalysis by ad-atoms: Part XXIV. Effect of arrangement of Bi ad-atoms on formic acid oxidation, *J Electroanal Chem*, Vol. 263, (1989): p. 97.
330. Gasteiger, H. A., N. Markovic, P. N. Ross, Jr., and E. J. Cairns, Methanol electrooxidation on well-characterized Pt-Ru alloys, *J Phys Chem*, Vol. 97, (1993): p. 12020.
331. Watanabe, M., and S. Motoo, Electrocatalysis by ad-atoms. Part III. Enhancement of the oxidation of carbon monoxide on platinum by ruthenium ad-atoms, *J Electroanal Chem*, Vol. 60, (1975): p. 275.
332. Watanabe, M., and S. Motoo, Electrocatalysis by ad-atoms: Part XVI. Enhancement of carbon monoxide oxidation on platinum electrode in acid solution by the VIth group ad-atoms, *J Electroanal Chem*, Vol. 194, (1985): p. 275.
333. Shibata, M., and S. Motoo, Electrocatalysis by ad-atoms: Part XV. Enhancement of CO oxidation on platinum by the electronegativity of ad-atoms, *J Electroanal Chem*, Vol. 194, (1985): p. 261.
334. Gasteiger, H. A., N. Markovic, P. N. Ross, Jr., and E. J. Cairns, CO electrooxidation on well-characterized Pt-Ru alloys, *J Phys Chem*, Vol. 98, (1994): p. 617.
335. Adzic, R. R., A. V. Tripkovic, and N. M. Markovic, Structural effects in electrocatalysis. Oxidation of formic acid and oxygen reduction on single-crystal electrodes and the effects of foreign metal adatoms, *J Electroanal Chem*, Vol. 150, (1983): p. 79.
336. Fernandez-Vega, A., J. M. Feliu, A. Aldaz, and J. Clavilier, Heterogeneous electrocatalysis on well-defined Pt surfaces modified by controlled amounts of irreversible adsorbed adatoms. Part II. Formic acid oxidation on Pt(100)-Sb system, *J Electroanal Chem*, Vol. 258, (1989): p. 101.

337. Fernandez-Vega, A., J. M. Feliu, A. Aldaz, and J. Clavilier, Heterogeneous electro-catalysis on well-defined Pt surfaces modified by controlled amounts of irreversible adsorbed adatoms. Part IV. Formic acid oxidation on Pt(111)-As system, *J Electroanal Chem*, Vol. 305, (1991): p. 229.

338. Chang, S. C., and M. J. Weaver, Influence of coadsorbed bismuth and copper on carbon monoxide adlayer structures at ordered low-index platinum-aqueous interfaces, *Surf Sci*, Vol. 241, (1991): p. 11.

339. Chang, S. C., Y. Ho, and M. J. Weaver, Applications of real-time infrared spectroscopy to electrocatalysis at bimetallic surfaces. I. Electrooxidation of formic acid and methanol on bismuth-modified platinum (111) and platinum (100), *Surf Sci*, Vol. 265, (1992): p. 81.

340. Grozovski, V., V. Climent, E. Herrero, and J. M. Feliu, Intrinsic activity and poisoning rate for HCOOH oxidation at Pt(100) and vicinal surfaces containing monoatomic (111) steps, *Chem Phys Chem*, Vol. 10, (2009): p. 1922.

341. Smith, S. P. E., and H. D. Abruna, Structural effects on the oxidation of HCOOH by bismuth modified Pt(111) elecrodes with (110) monatomic steps, *J Electroanal Chem*, Vol. 467, (1999): p. 43.

342. Smith, S. P. E., K. F. Ben-Dor, and H. D. Abruna, Structural effects on the oxidation of HCOOH by bismuth-modified Pt(111) electrodes with (100) monatomic steps, *Langmuir*, Vol. 15, (1999): p. 7325.

343. Smith, S. P. E., K. F. Ben-Dor, and H. D. Abruna, Poison formation upon the dissociative adsorption of formic acid on bismuth-modified stepped platinum electrodes, *Langmuir*, Vol. 16, (2000): p. 787.

344. Herrero, E., A. Fernandez-Vega, J. M. Feliu, and A. Aldaz, Poison formation reaction from formic acid and methanol on Pt(111) electrodes modified by irreversibly adsorbed Bi and As, *J Electroanal Chem*, Vol. 350, (1993): p. 73.

345. Zhu, M. Y., G. Q. Sun, and Q. Xin, Effect of alloying degree in PtSn catalyst on the catalytic behavior for ethanol electro-oxidation, *Electrochim Acta*, Vol. 54, (2009): p. 1511.

346. Simoes, F. C., D. M. dos Anjos, F. Vigier, J.-M. Leger, F. Hahn, C. Coutanceau, E. R. Gonzalez, et al., Electroactivity of tin modified platinum electrodes for ethanol electrooxidation, *J Power Sources*, Vol. 167, (2007): p. 1.

347. Kim, J. H., S. M. Choi, S. H. Nam, M. H. Seo, S. H. Choi, and W. B. Kim, Influence of Sn content on PtSn/C catalysts for electrooxidation of C-1-C-3 alcohols: Synthesis, characterization, and electrocatalytic activity, *Appl Cat B: Environ*, Vol. 82, (2008): p. 89.

348. Friedrich, K. A., K.-P. Geyzers, F. Henglein, A. Marmann, U. Stimming, W. Unkauf, and R. Vogel, CO adsorption and oxidation on nanostructured electrode surfaces studied by STM and IR spectroscopy, in: *Electrode Processes VI*, A. Wieckowski, and K. Itaya (eds.), The Electrochemical Society: Pennington, New Jersey (1996), PV 96–98, p. 119.

349. Friedrich, K. A., K.-P. Geyzers, U. Linke, U. Stimming, and J. Stumper, CO adsorption and oxidation on a Pt(111) electrode modified by ruthenium deposition: An IR spectroscopic study, *J Electroanal Chem*, Vol. 402, (1996): p. 123.

350. Chrzanowski, W., and A. Wieckowski, Ultrathin films of ruthenium on low index platinum single crystal surfaces: An electrochemical study, *Langmiur*, Vol. 13, (1997): p. 5974.

351. Chrzanowski, W., and A. Wieckowski, Surface structure effects in platinum/ruthenium methanol oxidation electrocatalysis, *Langmuir*, Vol. 14, (1998): p. 1967.

352. Chrzanowski, W., and A. Wieckowski, Methanol oxidation catalysis on well-defined platinum/ruthenium electrodes: Ultrahigh vacuum surface science and electrochemistry approach, in: *Interfacial Electrochemistry*, A. Wieckowski (ed.), Dekker: New York (1999), p. 937.

353. Arenz, M., T. J. Schmidt, K. Wandelt, P. N. Ross, and N. M. Markovic, The oxygen reduction reaction on thin palladium films supported on a Pt(111) electrode, *J Phys Chem B*, Vol. 107, (2003): p. 9813.
354. Vukmirovic, M. B., J. Zhang, K. Sasaki, A. U. Nilekar, F. Uribe, M. Mavrikakis, and A. R. R., Platinum monolayer electrocatalysts for oxygen reduction, *Electrochim Acta*, Vol. 52, (2007): p. 2257.
355. Crown, A., I. R. Moraes, and A. Wieckowski, Examination of Pt(111)/Ru and Pt(111)/Os surfaces: STM imaging and methanol oxidation activity, *J Electroanal Chem*, Vol. 500, (2001): p. 333.
356. Crown, A., C. Johnston, and A. Wieckowski, Growth of ruthenium islands on Pt(hkl) electrodes obtained via repetitive spontaneous deposition, *Surf Sci*, Vol. 506, (2002): p. L268.
357. Rhee, C. K., M. Wakisaka, Y. V. Tolmachev, C. M. Johnston, R. Haasch, K. Attenkofer, G. Q. Lu, H. You, and A. Wieckowski, Osmium nanoislands spontaneously deposited on a Pt(111) electrode: An XPS, STM and GIF-XAS study, *J Electroanal Chem*, Vol. 554–555, (2003): p. 367.
358. Climent, V., N. M. Markovic, and P. N. Ross, Kinetics of oxygen reduction on an epitaxial film of palladium on Pt(111), *J Phys Chem B*, Vol. 104, (2000): p. 3116.
359. Gasteiger, H. A., N. Markovic, P. N. Ross, Jr., and E. J. Cairns, Temperature-dependent methanol electro-oxidation on well-characterized Pt-Ru alloys, *J Electrochem Soc*, Vol. 141, (1994): p. 1795.
360. Gasteiger, H. A., N. M. Markovic, and P. N. Ross, Jr., H_2 and CO electrooxidation on well-characterized Pt, Ru, and Pt-Ru. 2. Rotating disk electrode studies of CO/H_2 Mixtures at 62°C, *J Phys Chem*, Vol. 99, (1995): p. 16757.
361. Stamenkovic, V., M. Arenz, B. B. Blizanac, K. J. J. Mayrhofer, P. N. Ross, and N. M. Markovic, In situ CO oxidation on well characterized Pt_3Sn(hkl) surfaces: A selective review, *Surf Sci*, Vol. 576, (2005): p. 145.
362. Stamenkovic, V., T. J. Schmidt, P. N. Ross, and N. M. Markovic, Surface composition effects in electrocatalysis: Kinetics of oxygen reduction on well-defined Pt_3Ni and Pt_3Co alloy surfaces, *J Phys Chem B*, Vol. 106, (2002): p. 11970.
363. Casado-Rivera, E., D. J. Volpe, L. Alden, C. Lind, C. Downie, T. Vazquez-Alvarez, A. C. D. Angelo, F. J. DiSalvo, and H. D. Abruna, Electrocatalytic activity of ordered intermetallic phases for fuel cell applications, *J Am Chem Soc*, Vol. 126, (2004): p. 4043.
364. Kardash, D., C. Korzeniewski, and N. Markovic, Effects of thermal activation on the oxidation pathways of methanol at Pt-Ru alloy electrodes, *J Electroanal Chem*, Vol. 500, (2001): p. 518.
365. Stamenkovic, V. R., B. S. Mun, K. J. J. Mayrhofer, P. N. Ross, and N. M. Markovic, Effect of surface composition on electronic structure, stability, and electrocatalytic properties of Pt-transition metal alloys: Pt-skin versus Pt-skeleton surfaces, *J Am Chem Soc*, Vol. 128, (2006): p. 8813.
366. Gauthier, Y., Pt-metal alloy surfaces: Systematic trends, *Surf Rev Lett*, Vol. 3, (2001): p. 1663.
367. Gauthier, Y., R. Baudoing-Savois, J. M. Bugnard, U. Bardi, and A. Atrei, Influence of the transition metal and of order on the composition profile of $Pt_{80}M_{20}$(111) (M = Ni, Co, Fe) alloy surfaces: LEED study of $Pt_{80}Co_{20}$(111), *Surf Sci*, Vol. 276, (1992): p. 1.
368. Zhang, J., Y. Mo, M. B. Vukmirovic, R. Klie, K. Sasaki, and R. R. Adzic, Platinum monolayer electrocatalysts for O_2 reduction: Pt monolayer on Pd(111) and on carbon-supported Pd nanoparticles, *J Phys Chem B*, Vol. 108, (2004): p. 10955.
369. Stamenkovic, V., T. J. Schmidt, P. N. Ross, and N. M. Markovic, Surface segregation effects in electrocatalysis: Kinetics of oxygen reduction reaction on polycrystalline Pt_3Ni alloy surfaces, *J Electroanal Chem*, Vol. 554–555, (2003): p. 191.

370. Solla-Gullon, J., F. J. Vidal-Iglesias, E. Herrero, J. M. Feliu, and A. Aldaz, CO monolayer oxidation on semi-spherical and preferentially oriented (100) and (111) platinum nanoparticles, *Electrochem Comm*, Vol. 8, (2006): p. 189.

371. Solla-Gullon, J., V. Montiel, A. Aldaz, and J. Clavilier, Electrochemical characterisation of platinum nanoparticles prepared by microemulsion: How to clean them without loss of crystalline surface structure, *J Electroanal Chem*, Vol. 491, (2000): p. 69.

372. Solla-Gullon, J., V. Montiel, A. Aldaz, and J. Clavilier, Electrochemical and electrocatalytic behaviour of platinum-palladium nanoparticle alloys, *Electrochem Commun*, Vol. 4, (2002): p. 716.

373. Solla-Gullon, J., A. Rodes, V. Montiel, A. Aldaz, and J. Clavilier, Electrochemical characterisation of platinum/palladium nanoparticles prepared in a water-in-oil microemulsion, *J Electroanal Chem*, Vol. 554–555, (2003): p. 273.

374. Solla-Gullon, J., V. Montiel, A. Aldaz, and J. Clavilier, Synthesis and electrochemical decontamination of platinum-palladium nanoparticles prepared by water-in-oil microemulsion, *J Electrochem Soc*, Vol. 150, (2003): p. E104.

375. Solla-Gullon, J., F. J. Vidal-Iglesias, A. Lopez-Cudero, E. Garnier, J. M. Feliu, and A. Aldaz, Shape-dependent electrocatalysis: methanol and formic acid electrooxidation on preferentially oriented Pt nanoparticles, *Phys Chem Chem Phys*, Vol. 10, (2008): p. 3689.

376. Lopez-Cudero, A., F. J. Vidal-Iglesias, J. Solla-Gullon, E. Herrero, A. Aldaz, and J. M. Feliu, Formic acid electrooxidation on Bi-modified polyoriented and preferential (111) Pt nanoparticles, *Phys Chem Chem Phys*, Vol. 11, (2009): p. 416.

377. Paulus, U. A., A. Wokaun, G. G. Scherer, T. J. Schmidt, V. Stamenkovic, V. Radmilovic, N. M. Markovic, and P. N. Ross, Oxygen reduction on carbon-supported Pt-Ni and Pt-Co alloy catalysts, *J Phys Chem B*, Vol. 106, (2002): p. 4181.

378. Ahmadi, T. S., Z. L. Wang, T. C. Green, A. Henglein, and M. A. El-Sayed, Shape-controlled synthesis of colloidal platinum nanoparticles, *Science*, Vol. 272, (1996): p. 1924.

379. El-Sayed, M. A., Some interesting properties of metals confined in time and nanometer space of different shapes, *Acc Chem Res*, Vol. 34, (2001): p. 257.

380. Friedrich, K. A., F. Henglein, U. Stimming, and W. Unkauf, Investigation of Pt particles on gold substrates by IR spectroscopy – Particle structure and catalytic activity, *Colloids Surf A: Physiochem Eng Asp*, Vol. 134, (1998): p. 193.

381. Schmidt, T. J., H. A. Gasteiger, G. D. Stab, P. M. Urban, D. M. Kolb, and R. J. Behm, Characterization of high-surface-area electrocatalysts using a rotating disk electrode configuration, *J Electrochem Soc*, Vol. 145, (1998): p. 2354.

382. Mayrhofer, K. J. J., D. Strmcnik, B. B. Blizanac, V. Stamenkovic, M. Arenz, and N. M. Markovic, Measurement of oxygen reduction activities via the rotating disc electrode method: From Pt model surfaces to carbon-supported high surface area catalysts, *Electrochim Acta*, Vol. 53, (2003): p. 3181.

383. Gasteiger, H. A., S. S. Kocha, B. Sompalli, and F. T. Wagner, Activity benchmarks and requirements for Pt, Pt-alloy, and non-Pt oxygen reduction catalysts for PEMFCs, *Appl Cat B: Environ*, Vol. 56, (2005): p. 10.

384. Nicholas, J. F., *An Atlas of Models of Crystal Surfaces*, Gordon and Breach: New York, 1965.

385. Somorjai, G. A., *Introduction to Surface Chemistry and Catalysis*, Wiley: New York, 1994.

386. Wood, E. A., *Crystal Orientation Manual*, Columbia University Press: New York, 1963.

APPENDIX: SURFACE CRYSTALLOGRAPHY

A. General Considerations

Pt crystallizes in the fcc structure. Figure A.1 includes a model of an fcc unit cell. The corner atoms form a cube shape, and an atom is present in the center of each cube face.

The surfaces of crystals are typically denoted by Miller indices, a universal notation that references surface planes to the unit cell dimensions along the crystal axes. In the cubic system, there are three orthogonal axes [3,384]. With the scale on the axes graduated in multiples of the crystal unit cell (Figure A.1), the Miller indices of a plane can be determined from the following steps [3]: (i) find the intercepts of the plane on the axes; (ii) take the reciprocal of the intercepts; (iii) remove common factors to reduce the numbers to the three smallest integer values; and (iv) enclose the values in parentheses. The Miller indices of an arbitrary surface plane on a cubic crystal are denoted "(hkl)." A bar above a number indicates a negative intercept.

As depicted in Figure A.1, low Miller index planes intercept one or more axis at one unit cell length and are oriented parallel to the remaining axes. High Miller index planes intercept a minimum of two crystal axes, and at least one of the intercepts occurs at a fraction of the unit cell length. High Miller index surfaces

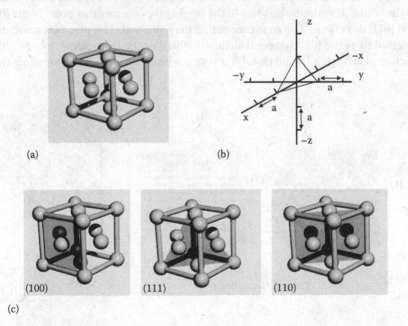

(a)   (b)

(100)   (111)   (110)

(c)

FIGURE A.1   (a) Space filling model showing the unit cell of an fcc crystal. (b) The axis system with the (111) plane included for the cubic system. (c) The low index planes and their Miller indices for the cubic system.

consist of regularly spaced steps and terraces, and the steps may have kinks [156,385]. Figures A.2 and A.3 show the spatial arrangement of atoms in the low index planes and a few high index planes of Pt, respectively.

Indices enclosed by square brackets, as in [hkl], give the direction of a line extending from the origin of the crystal axes to the coordinate h,k,l. In the cubic system, the direction $[h_1k_1l_1]$ is perpendicular to the plane $(h_1k_1l_1)$. The relationship is illustrated in Figure A.4a for the low index planes. These directional vectors also are included on the surfaces in Figure A.2.

Figure A.4a is also useful for understanding the concepts of crystallographic zones, zone axes, and stereographic projections. The figure shows a cubic crystal at the center of a sphere. For each low index surface plane on the crystal, a perpendicular line extends from the origin to a point on the sphere surface. The points, called "poles," are labeled by the surface plane's Miller indices. The lines that connect the poles map to crystallographic zones. For example, along the zone between the (100) and (111) poles are points that correspond to poles of high Miller index surfaces. The surfaces contain steps and terraces, each of which displays either the (100) or (111) structure, depending upon the position along the zone. Details of these surfaces are discussed further below. Planes in a zone have a common axis. A zone axis intercepts the origin of the crystal axes and runs parallel to the line formed by the intersection of the planes. The line formed where the (100) and (111) planes meet can be observed on the crystal in Figure A.4a. Translating this line to the origin gives vectors that point in the $[0\bar{1}1]$ and $[01\bar{1}]$ directions. The zone connecting the (100) and (111) planes is sometimes referred to as the $[0\bar{1}1]$ (or equivalently the $[01\bar{1}]$) zone. In general, for two intersecting planes, $(h_1k_1l_1)$ and $(h_2k_2l_2)$, a vector with direction [u,v,w] extending from

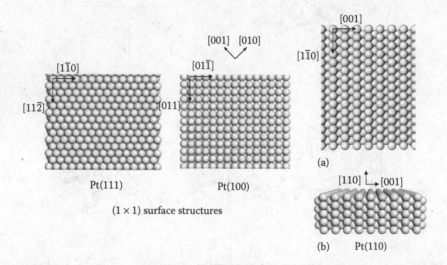

Pt(111)   Pt(100)

(1 × 1) surface structures

(a)

(b)   Pt(110)

FIGURE A.2 Drawings showing the packing arrangement of atoms in the low Miller index surface planes of an fcc crystal. The surfaces are in their unreconstructed (1 × 1) forms. For Pt(110), views from the top (a) and side (b) are included.

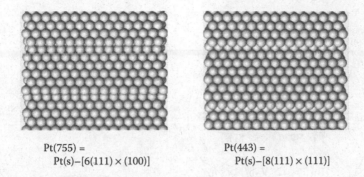

Pt(755) =
Pt(s)–[6(111) × (100)]

Pt(443) =
Pt(s)–[8(111) × (111)]

FIGURE A.3 Drawings showing the packing arrangement of atoms in a few high Miller index surface planes of an fcc crystal.

the origin and running parallel to the line of intersection can be determined from the products of the Miller indices, as follows: $u = k_1l_2 - l_1k_2$; $v = l_1h_2 - h_1l_2$; and $h_1k_2 - k_1h_2$ [386].

Instead of the spherical projection shown in Figure A.4a, crystal faces are more often depicted by projection into a plane (Figures A.4b and c). The method of projection is shown in Figure A.4b. From the $(00\bar{1})$ pole at the bottom of the sphere, dashed lines are drawn to the poles on the upper half of the sphere. Where the dashed lines cross the center plane of the sphere, points are indicated. These points comprise the stereographic projection of the poles on the upper half of the sphere. The plane is shown in two dimensions in Figure A.4c.

Due to the high symmetry of the cubic system, it is often sufficient to represent crystal surface planes of interest on a triangle connecting the low Miller index surfaces. A unit projected stereographic triangle for the cubic system is shown in Figure A.5. The (100), (111), and (110) poles at the vertexes are connected by zones on which are marked poles for high Miller index planes.

B. High Miller Index Surfaces

To more easily relate the Miller indices to structure, a notation was devised that specifies the crystallographic orientation of the steps and terraces on stepped high Miller index surfaces. The surface depicted in Figure A.3a has six atom wide, (111) oriented terraces, which are separated by single atom steps. Additionally, the orientation of the plane formed by the face of the step is (100). The Miller index designation for the surface is (755), whereas the step-terrace notation is $6(111) × (100)$ (or $6(111) - (100)$, see Figure A.5). The naming system was devised by Somorjai and coworkers [156–158] and includes specification of the number of atoms in a terrace (n), the Miller index for the terrace ($h_tk_tl_t$), and the Miller index for the step ($h_sk_sl_s$). In describing a stepped Pt surface, the general notation is Pt(s)-[$n(h_tk_tl_t) × (h_sk_sl_s)$].

Figure A.5 shows that in progressing along the zone from the (111) to the (100) pole, a series of high index planes consisting of (111) oriented terraces separated

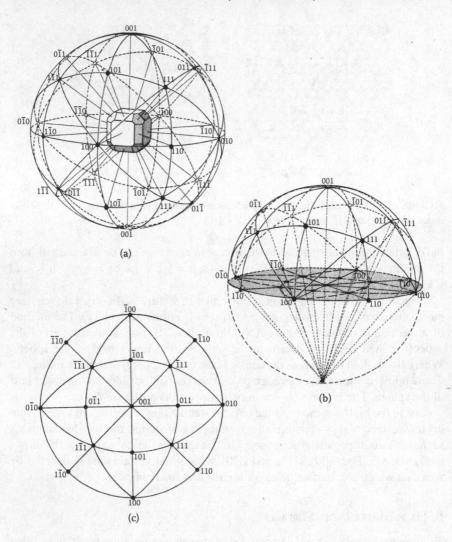

FIGURE A.4 Development of the stereographic projection. (Adapted from Wood, *Crystal Orientation Manual*, Columbia University Press: New York, 1963. With permission.) (a) An fcc crystal with low Miller index planes exposed is positioned at the center of a sphere. The poles formed by the intersection of the sphere surface with vectors that extend from the origin of the crystal axes and run normal to the crystal surface planes are indicated. (b) Development of the (001) stereographic projection of poles into a plane. (c) A (001) stereographic projection for an fcc system of poles and zone circles.

by single atom high, (100) steps is encountered up to the (311) pole. The (311) surface has two equivalent step-terrace notations, and moving from (311) toward the (100) pole, the step and terrace orientations reverse. The (311) pole is called the "turning point" of the zone. The related trend connecting the (100) and (110) planes is apparent through the (100)–(110) zone. It is notable that a (110) plane of

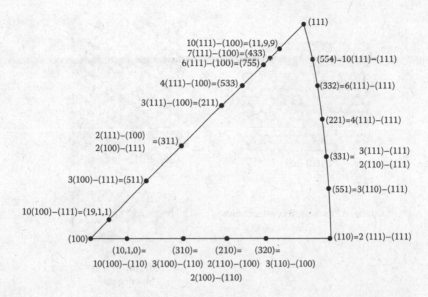

FIGURE A.5 Unit stereographic triangle for the fcc structure with the (100) face at the center of the projection. (Adapted from Hamelin et al., in: *Modern Aspects of Electrochemistry*, R. E. White, J. O. M. Bockris, and B. E. Conway (eds.), Plenum Press: New York (1985), Vol. 16, p. 1. With permission.)

an fcc crystal has a structure equivalent to a stepped surface comprising two atom wide, (111) oriented terraces separated by single atom, (111) oriented steps [3]. Therefore, surfaces in the zone between the (111) and (110) poles are often designated as $n(111) \times (111)$ surfaces [3,28].

Progressing along a zone line is equivalent to rotating a crystal about its zone axis. If the crystal in Figure A.4a is viewed from a fixed position behind the (100) pole marked on the sphere and rotated about the [0$\bar{1}$1] zone axis, then the rotation will bring high index planes in the (100)–(111) zone into alignment with the viewing position. The high index plane brought into view will depend upon the angle of rotation. Relationships between the structure of stepped high index surfaces

TABLE A.1
Relationships Between the Step-terrace Notation and the Miller Indices for Stepped Surfaces in the Cubic System

Step-terrace notation	Miller Index
$n(111)–(111)$	$(n, n, n-2)$
$n(111)–(100)$	$(n+1, n-1, n-1)$
$n(100)–(111)$	$(2n-1, 1, 1)$
$n(100)–(110)$	$(n, 1, 0)$
$n(110)–(111)$	$(2n-1, 2n-1, 1)$
$n(110)–(100)$	$(n, n-1, 0)$

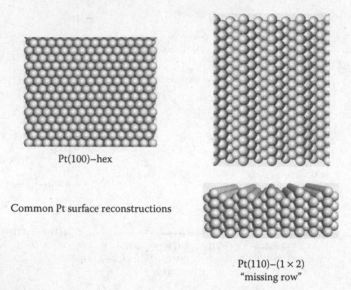

Pt(100)–hex

Common Pt surface reconstructions

Pt(110)–(1 × 2)
"missing row"

FIGURE A.6 Drawings showing the packing arrangement of atoms in the common reconstructed structures for Pt(100) and Pt(110) electrodes. Pt(111) electrodes retain their (1 × 1) structure.

and the angle of cubic crystal rotation relative to the low index terrace plane are given in Ref. [3]. Table A.1 shows the transformations between the step-terrace notation and the Miller indices for stepped surfaces in the cubic system.

C. Surface Reconstruction and Relaxation

The atoms near the surface of metals often shift position relative to the atoms in the bulk crystal lattice to lower the surface free energy in response to forces from the surrounding environment. Displacement of atoms normal to the surface by a fraction of the bond length is called *relaxation*. Surface *reconstruction* is the result of small lateral displacements. Models for Pt low index planes in unreconstructed and their common reconstructed forms are shown in Figures A.2 and A.6, respectively. The notation (1 × 1) is used in referring to a surface that has the same structure as its plane in the bulk crystal. The notation used for reconstructed surfaces usually carries information about the unit cell of the top layer of atoms with respect to an unreconstructed underlying layer.

D. Surface Atom Density Determinations

The number of surface atoms per unit area ($N_{(hkl)}$) is referred to as the surface atom density. Surface atom densities for different planes of Pt single crystals can be determined from geometric relationships relative to the surface unit cell. Figure A.7 shows unit cells for the (111) and (553) planes on a Pt(553) surface.

TABLE A.2

Surface Atom Densities and Related Parameters for High Miller Index Surface Planes

Terrace Row/n	Pt(s)-[n(111) x (111)]			Pt(s)-[n(111) x (100)]			Pt(s)-[n(100) x (111)]		
	(h k l)	β/°	atoms/cm²	(h k l)	β/°	atoms/cm²	(h,k,l)	β/°	atoms/cm²
2	(110)	35.3	9.22×10^{14}	(311)	29.5	1.57×10^{15}	(311)	25.2	1.57×10^{15}
3	(331)	22.0	1.20×10^{15}	(211)	19.5	1.60×10^{15}	(511)	15.8	1.50×10^{15}
4	(221)	15.8	1.30×10^{15}	(533)	14.4	1.59×10^{15}	(711)	11.4	1.46×10^{15}
5	(552)	12.3	1.36×10^{15}	(322)	11.4	1.58×10^{15}	(911)	8.9	1.43×10^{15}
6	(332)	10.0	1.39×10^{15}	(755)	9.4	1.57×10^{15}	(11 1 1)	7.3	1.41×10^{15}
7	(775)	8.5	1.41×10^{15}	(433)	8.0	1.56×10^{15}	(13 1 1)	6.2	1.40×10^{15}
8	(443)	7.3	1.42×10^{15}	(977)	7.0	1.56×10^{15}	(15 1 1)	5.4	1.38×10^{15}
9	(997)	6.4	1.44×10^{15}	(544)	6.2	1.55×10^{15}	(17 1 1)	4.8	1.38×10^{15}
10	(554)	5.8	1.44×10^{15}	(11 9 9)	5.6	1.55×10^{15}	(19 1 1)	4.3	1.37×10^{15}
∞	(111)	0.0	1.50×10^{15}	(111)	0.0	1.50×10^{15}	(100)	0.0	1.30×10^{15}

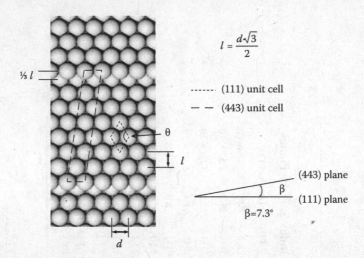

$$l = \frac{d\sqrt{3}}{2}$$

------ (111) unit cell

— — (443) unit cell

(443) plane

β

(111) plane

$\beta = 7.3°$

FIGURE A.7 Drawing showing a step and terrace on the unreconstructed Pt(443) $\equiv$ Pt(s)-[8(111) $\times$ (111)] surface plane. Unit cells for the (111) (dotted line) and (443) (dashed line) planes are indicated. The inset to the right shows that the (443) plane lies at an angle, β, of 7.3° from the (111) plane. The distance between nearest neighbor Pt atoms on the (111) plane is indicated by d (2.77×10^{-8} cm), and l represents the distance between rows of Pt atoms on the (111) plane. The distances d and l are related through the angle θ indicated in the unit cell ($\theta = 120°$). Projected onto the (111) plane, the center-to-center distance between the row of step atoms and the row of atoms on the plane immediately below is $l/3$.

For Pt(111) (and Pt(100)), the surface atom density is given by the reciprocal of the unit cell area. With reference to Figure A.7, the Pt(111) unit cell area is $d \times l$, where d is the nearest neighbor distance between Pt atoms (2.77×10^{-8} cm) and l is the distance between rows of Pt atoms ($d\sin(\theta/2)$). Surface atom densities for Pt(111) and Pt(100) are included in Table A.2.

For stepped surfaces, surface atom densities can be decomposed into the sum of contributions of step and terrace atoms. The area of the unit cell for the high Miller index surface plane is readily determined from its projection onto the terrace plane. For the Pt(s)-[n(111) $\times$ 111] surface in Figure A.7, the projected area (S) is given by:

$$S = d^2 \frac{\sqrt{3}}{2}\left(n - \frac{2}{3}\right) \tag{A.1}$$

Therefore, the unit cell area is $S/\cos \beta$. Since the unit cell contains one step, the number of step atoms per unit surface area is $\cos \beta/S$. When the step plane is defined as the (110) plane extending from the center of the row of step atoms through the center of the closest row of atoms fully exposed on the terrace (see Figure A.7), then the terrace contribution to the atomic site density is (($n - 2$)

$\cos \beta)/S$. Taking the sum of the step and terrace atom contributions and substituting for S from Equation A.1 gives:

$$N_{n(111)-(111)} = \frac{2(n-1)\cos\beta}{d^2\sqrt{3}\times(n-2/3)}$$ (A.2)

as the number of surface atoms per unit area on the Pt(s)-[$n(111) \times 111$] plane [129,165]. The approach has been applied to other stepped surfaces, including Pt(s)-[$n(111) \times 100$] [42] and Pt(s)-[$n(100) \times 111$] [150]. The general expressions for the surface atom densities of these planes are:

$$N_{n(111)-(100)} = \frac{2\cos\beta}{d^2\sqrt{3}}\left(1+\frac{1}{3(n-1/3)}\right)$$ (A.3)

and

$$N_{n(100)-(111)} = \frac{\cos\beta}{d^2}\left(1+\frac{1}{2(n-1/2)}\right)$$ (A.4)

respectively. Table A.2 reports Miller indices, angles relative to the terrace plane, and surface atom coverages for high Miller index planes from zones encompassing many commonly employed stepped Pt electrodes.

3 Application of SECM to Corrosion Studies

Mark B. Jensen and Dennis E. Tallman

CONTENTS

3.1   INTRODUCTION

3.1.1   Definition of Corrosion

The corrosion of a material may be defined as the irreversible reaction of the material with its environment, usually resulting in degradation of the material and/or its properties. For pure metals and metal alloys, the reactions are redox reactions involving the oxidation of one or more of the metals comprising the alloy, driven by reduction of one or more oxidants, such as dioxygen, hydrogen ion, and/or water. These oxidants are typical of those encountered in aqueous (or environmental) corrosion under near-ambient conditions, the most common corrosion environment, and the principal subject of this chapter. Corrosion of metals has also been called "extractive metallurgy in reverse" [1], reflecting the fact that considerable energy must be expended to extract and purify a metal such as Al or Fe from its ore (i.e., its oxide forms). Thus, corrosion is the thermodynamically favored process by which the metal reverts back to its oxide form(s). As a result, it is virtually impossible to completely stop such a thermodynamically favorable process. Therefore, corrosion control strategies are designed to slow the corrosion rate by retarding the rate of metal oxidation, the rate of oxidant reduction, the rate of ion movement between sites of oxidation and reduction (such ion movement is required for maintaining overall charge balance), or some combination of these approaches.

3.1.2 Costs of Corrosion

The major motivation for performing corrosion research is the very high cost to society associated with the corrosion of engineering metals. As long as we continue to use active (i.e., corrodible) metals for the construction of automobiles, airplanes, helicopters, ships, bridges, buildings, industrial reactors, and other objects, there will be corrosion problems and associated costs that impact virtually all sectors of society, including utilities, transportation, infrastructure, production and manufacturing, and government (e.g., Department of Defense). The annual direct cost of corrosion and its control for a developed country is estimated to be approximately 3%−4% of the country's gross domestic product; for the United States that amounts to over $300 billion per year [2]. These direct costs include the costs of design, manufacturing, and construction that incorporate strategies to mitigate corrosion, as well as management costs (e.g., maintenance costs such as material replacement, stripping and/or applying coatings, cathodic protection of gas and liquid transmission pipelines, etc.). There are additional indirect costs (mostly associated with loss of productivity or product due to outages, delays, failures, and litigation), which are estimated to be nearly as large as the direct costs. Catastrophic failures attributed to corrosion such as airplane crashes [3] and bridge collapses [4] have led to loss of life. A goal of corrosion research is to reduce these costs.

3.1.3 Types of Corrosion

Corrosion occurs in various forms, including uniform corrosion, pitting corrosion, crevice corrosion, galvanic corrosion, and a few others (such as hydrogen damage, intergranular corrosion, dealloying and erosion corrosion) [1]. Uniform corrosion occurs for metals that are compositionally and metallurgically uniform and for which a homogeneous corrosive environment has equal access to all areas of the metal surface. This form of corrosion is characterized by numerous small rapidly interchanging anodic and cathodic sites across the metal surface that lead to a predictable and uniform dissolution of the metal. Uniform corrosion is conveniently studied by a number of electrochemical techniques such as electrochemical impedance spectroscopy (EIS), electrochemical noise methods (ENM), and potentiodynamic techniques, such as the polarization resistance method and the Tafel extrapolation method [1,5]. These electrochemical techniques provide a surface-averaged response that is appropriate for uniform corrosion. The scanning probe techniques discussed in this chapter provide little additional information about uniform corrosion and, thus, find limited application to this form of corrosion.

Localized forms of corrosion (such as pitting, crevice, and galvanic corrosion) arise when the metal surface is not compositionally uniform or when there is not a uniform exposure of the metal surface to the corrosive environment. These forms of corrosion lead to nonuniform current density distribution across the metal surface as well as nonuniform distributions of species concentration (such as H^+

and O_2). For these localized forms of corrosion, the scanning probe techniques provide valuable spatial and temporal information not available from the surface-averaging techniques mentioned above. A brief review of scanning probe techniques as applied to localized corrosion studies has recently appeared [6].

3.2　SCANNING ELECTROCHEMICAL PROBE TECHNIQUES—AN OVERVIEW

In this chapter, we focus on scanning probe techniques that provide a probe response (signal) that originates from some electrochemical phenomenon and that have been applied to corrosion research. This section provides a brief description of each of these techniques. The reader is referred to the cited references and to other chapters in this volume for greater detail on the theory and instrumentation. Although each of these techniques may be considered a form of scanning electrochemical microscopy (SECM), we will reserve the use of the acronym SECM to refer to the technique in which a faradaic current at a microelectrode is measured as it is held or moved in a solution in the vicinity of a substrate surface. The term *substrate* will be used to refer to the sample being examined by the scanning probe technique.

3.2.1　Scanning Electrochemical Microscopy

The basic SECM experiment involves the movement of a very small amperometric microelectrode (usually a disk microelectrode of a few μm or smaller radius, referred to as the tip or the probe) near the surface of a substrate immersed in an electrolyte containing at least one redox active species (a mediator) [7]. The two most common modes of operation are the *feedback* mode and the *generator-collector* mode. In the feedback mode, the potential of the tip is maintained constant by a potentiostat and the substrate is at the open circuit potential (OCP), which is dictated by the concentration(s) of redox species in the electrolyte (according to the Nernst equation). Only the tip current from a reaction such as $O \rightarrow R$ (where O is an appropriately chosen mediator) is measured. The tip current is perturbed by the presence of the substrate. When the tip is sufficiently far from the substrate, a steady-state current is observed, given by the expression:

$$i_{T,\infty} = 4nFDca \tag{3.1}$$

where a is the electrode radius, c is the mediator concentration, D is the diffusion coefficient of the mediator, n the number of electrons transferred at the tip, and F is the Faraday. When the tip is moved toward the surface of an insulating substrate, the tip current decreases from $i_{T,\infty}$ because the insulating nature of the substrate blocks diffusion of O to the tip from the bulk solution (known as negative feedback). If the substrate is conductive, the species R generated by the tip can be oxidized back to O (or recycled, since the substrate is poised at a sufficiently positive potential by the bulk redox species), producing an additional flux of O to the tip and, hence, an increase in tip current from $i_{T,\infty}$ (known as positive feedback).

For the generator–collector mode of operation, the tip and substrate potentials are in most cases independently controlled by a bipotentiostat such that a species generated at the tip (e.g., R from the reaction $O \rightarrow R$) is collected (detected) at the substrate by forcing the reverse reaction ($R \rightarrow O$) at the substrate. This mode is known as the tip generation/substrate collection (TG/SC) mode. Conversely, the substrate may generate R (from the reaction $O \rightarrow R$), which is then collected (detected) at the tip, resulting in the complementary mode known as the substrate generation/tip collection (SG/TC) mode. In these generator–collector modes, both tip current and substrate current can be measured. When the tip is rastered in the x-y plane above the substrate, the tip current variation represents changes in topography and/or changes in the reactivity of the substrate. These two effects can be separated and, thus, SECM permits assessment of the kinetics of heterogeneous electron transfer occurring at the substrate surface as well as the kinetics of homogeneous chemical reactions coupled to electron transfer [7].

Additional experiments performed with SECM include current–distance measurements, whereby the steady-state tip current is measured as the tip is slowly moved from bulk solution toward the substrate in steps (at a fixed x-y position above the substrate), yielding a probe-approach curve (PAC). Theory can be used to fit the experimentally observed PACs, permitting the extraction of rate constants for irreversible substrate kinetics [7]. The scanning tip can also be used to detect variations in the concentrations of dissolved species near the substrate surface (such as dioxygen or hydrogen ion, particularly important species in corrosion processes) or to locally modify near-surface solution conditions (e.g., pH) by means of a tip redox reaction.

3.2.2 ALTERNATING CURRENT SECM (AC-SECM)

There has been an increase in the use of AC-SECM for corrosion research in recent years. The main difference between SECM and AC-SECM is that an alternating potential (usually sinusoidal rather than, or perhaps in addition to, a constant potential) is applied to the tip in AC-SECM, and the experiment can often be performed without addition of a mediator [8]. In most cases, the substrate is not connected to a potentiostat and, thus, remains at its OCP. A lock-in amplifier is used to determine both the magnitude of the tip current and its phase angle with respect to the applied AC potential. In this regard, the technique bears some resemblance to impedance spectroscopy and enjoys the signal-to-noise advantage generally associated with frequency selective amplification. The AC current at the tip displays a dependence on the tip-to-substrate distance and can be used for precise positioning of the tip. This distance dependence arises from the influence of the substrate on the local electrolyte conductivity between the tip and the counter electrode in bulk solution. Thus, AC-SECM can be utilized for topographic imaging (of insulating or electrochemically homogeneous substrates) as well as for providing information about the local electrochemical properties of the substrate. The ability to perform experiments in the absence of an added mediator is an advantage for systems that might be sensitive to any disturbance caused by such an addition (e.g., the Nernst potential imposed by addition of dissolved redox species). The frequency of the applied AC tip potential may also be varied and has an influence on the electrochemical contrast of AC-SECM. Scanning the frequency at each x-y tip position leads to a four-dimensional data set (4D AC-SECM), which can be visualized by means of a movie in which the individual AC-SECM images are sequentially displayed with increasing perturbation frequency [8].

3.2.3 POTENTIOMETRIC SECM

Potentiometric SECM (sometimes referred to as the scanning ion electrode technique, SIET) makes use of ion-selective microelectrodes that provide a potential in response to a particular ion concentration (or more specifically, ion activity), as dictated by the Nernst equation. Such electrodes can be used to map temporal and spatial variations in ion concentration near the surface of a corroding substrate. These measurements must be made rather slowly (approximately 0.5 to 1 s per point) due to the mechanical disturbance of the concentration gradient by the electrode movement. Otherwise, potentiometric measurements have the advantage over amperometric measurements of not perturbing the target ion concentration. The most common measurement made by potentiometric SECM is pH, a particularly important measurement in corrosion studies.

A variety of methods have been used to fabricate potentiometric microelectrodes for potentiometric SECM [9]. In some cases, the electrodes are liquid ion exchange (LIX) electrodes fabricated in drawn capillaries (or micropipettes) for measuring ion concentrations down to picomolar levels [10]. These electrodes can be fabricated with micron or submicron tip diameters and have been used for several decades by life scientists to monitor ion activities in living organisms [11]. Ion exchange liquids for H^+ and for various metal ions are commercially available, making this approach quite versatile. However, such electrodes are very fragile and are easily damaged if brought into contact with the substrate surface during a potentiometric SECM experiment. LIX electrodes typically have time constants on the order of tenths of a second.

Solid-state microelectrodes tend to be more robust and can be fabricated by a number of approaches. For example, pH microelectrodes for use in corrosion studies have been fabricated based on the reversible Ir/IrO_2 couple [12]:

$$IrO_2 + 4H^+ + 4e^- \leftrightarrow Ir + 2H_2O$$

These iridium oxide electrodes can be prepared by sputtering as well as by thermal or electrochemical oxidation of iridium wire. Electrochemical oxidation of an iridium substrate is the most common method of preparation, by potential cycling of iridium in sulfuric acid between potentials of −0.25 and +1.25 V versus SCE.

A combination electrode for chloride ion having a micrometer-sized tip and containing both a liquid membrane reference electrode and a solid-state Ag/AgCl sensing electrode has been described [13]. This electrode was used to map in situ chloride ion distribution in several localized corrosion systems, providing information on the distribution of chloride ions near the metal/electrolyte interface. Other so-called "electrodes of the second kind" have been described, mostly for chloride ion mapping [9].

A general approach to fabricating solid-state ion selective microelectrodes has been described, whereby a conducting electroactive polymer, which is both an electronic and an ionic conductor (e.g., polypyrrole, polythiophene, or polyaniline), is used to mediate charge exchange between an ion selective membrane (an ion conductor) and a metal substrate (an electronic conductor) [14]. These electrodes are reported to be robust and mechanically flexible while exhibiting good potential stability with no redox sensitivity. While applications to potentiometric SECM have been described [14], these electrodes are yet to find use in corrosion studies.

3.2.4 Electrochemical Microcell and Scanning Capillary Microscopy

Another approach for the study of local electrochemistry involves the use of an electrochemical microcell, a glass microcapillary filled with electrolyte and containing counter and/or reference electrodes. In these experiments, the substrate is not immersed in electrolyte, but rather a very small volume of electrolyte (a droplet) bridges the gap between capillary tip and substrate. The capillary is either stationary (for small area measurements) or scanned above the substrate for spatially resolved measurements [15]. For stationary experiments, the tip is typically a few microns in diameter and is sealed to the substrate surface (e.g., with silicone rubber), permitting polarization of microscopic surface areas and localized electrochemical measurement. This technique has found application in corrosion studies, especially for probing the effect of microheterogeneities on localized corrosion [16–18]. Positionable microcells have been described for making spatially resolved measurements, often within droplets of electrolyte contacting both the substrate and the capillary [19–20]. For example, measurements in subnanoliter solution volumes have been performed by scanning a theta glass capillary containing a carbon fiber microelectrode in one channel and a Ag/AgCl reference electrode in the other channel to different locations within an electrolyte droplet [20]. In another approach, the capillary tip–substrate distance is modulated as in tapping mode atomic force microscopy (AFM), with amplitude and phase of the modulation providing feedback for control of tip–substrate distance [21]. In some cases, electrolyte flows through the drop by means of a multibarreled capillary, continuously refreshing the solution in the droplet cell [22–23]. These capillary techniques (sometimes referred to as scanning capillary microscopy or the scanning droplet cell [SDC]) permit all common potentiostatic and galvanostatic measurements, such as cyclic voltammetry, chronoamperometry, impedance spectroscopy, and electrochemical noise measurements [19]. A recent review of the many electrochemical configurations using single-barreled and multibarreled capillary cells has been provided by Lohrengel [24].

3.2.5 SCANNING REFERENCE ELECTRODE TECHNIQUE (SRET) AND SCANNING VIBRATING ELECTRODE TECHNIQUE (SVET)

Localized corrosion on a metal surface immersed in an electrolyte engenders a flow of ionic current in the electrolyte between local anodic and cathodic sites on the metal surface (due to the requirement for charge balance). This ionic current leads to a potential field and associated potential gradients within the electrolyte, the relationship between the local current density and the potential gradient being given by:

$$\bar{\iota}\left(x,y,z\right) = \frac{\nabla \bar{V}\left(x,y,z\right)}{\rho} \tag{3.2}$$

where $\bar{\iota}(x, y, z)$ is the current density vector having magnitude (A/cm^2) and direction, ρ is the solution resistivity (Ω cm), and $\nabla \bar{V}(x, y, z)$ is the potential gradient (V/cm) at location (x, y, z). By scanning a small reference electrode near the substrate surface in two dimensions and measuring its potential versus a remote reference electrode, a map of the potential variation (but not the potential gradient) near the surface can be obtained. One of the earliest examples of this technique was reported in 1940 by Evans [25] for calculating corrosion rates on a water pipeline. Isaacs [26] further developed this approach for monitoring pit propagation on stainless steels (SS), using a drawn glass capillary leading from a reference electrode, and first referred to this approach as SRET.

To measure the local current density, the local potential gradient $\nabla \bar{V}(x, y, z)$ must be determined (Equation 3.2). There are two approaches to measuring the local potential gradient. One approach uses a twin-electrode assembly consisting of two microreference (or pseudoreference) electrodes separated from one another along the direction normal to the sample surface (the z direction) [27]. The second approach (known as SVET) uses a single microelectrode that is vibrated either in one direction (the z direction) or in two directions (the x and z directions, each at different vibration frequencies) [28–29]. Vibration in two directions permits measurement of the components of current density both normal and parallel to the substrate surface. This second approach (SVET) possesses a signal-to-noise advantage over the single or dual electrode SRET measurements since lock-in amplifiers are used to measure the voltage amplitudes at the frequencies of vibration. Thus, SVET with its improved sensitivity and higher spatial resolution has largely replaced SRET.

Originally developed to study ion transport in biology, SVET has found wide application in corrosion research as pioneered by Isaacs and coworkers [30]. The current density vector can be determined from Equation 3.2, provided the electrolyte resistivity is known, typically obtained from a calibration experiment involving the injection of a known current density from a point source. The vibrating probe is scanned just above the surface of the substrate (which is normally at the OCP), producing a map of DC current density distribution.

3.2.6 LOCAL ELECTROCHEMICAL IMPEDANCE SPECTROSCOPY AND MAPPING (LEIS/LEIM)

LEIS and LEIM have proven useful in corrosion research, especially for the study of coated metals. These techniques usually employ a five-electrode arrangement, with reference and counter electrodes used to apply an AC potential perturbation to the substrate (or working electrode, as in global EIS). The local AC current flow through the electrolyte just above the substrate surface is then measured, typically using the twin-microelectrode approach [31]. The use of a single vibrating microelectrode has also been described for measuring the local current [32–33]. The LEIS experiment involves varying the frequency of the AC perturbation at a fixed position of the twin-electrode above the sample surface. In the LEIM experiment, the twin-electrode probe is scanned in the x-y plane above the sample surface while holding the AC perturbation frequency constant. The use of higher frequencies of the AC perturbation lowers the impedance of dielectric-type samples, such as coated metals. Thus, LEIS and LEIM can be used for both bare and coated metals. This approach has been used to probe intrinsic and extrinsic defects in organic coatings applied to metals [34].

It is important to distinguish between LEIS/LEIM (a five-electrode technique with the AC perturbation applied to the substrate) and AC-SECM (a three-electrode technique with the AC perturbation applied to the probe). Additionally, it should also be noted that LEIS/LEIM as usually implemented does not provide the true local impedance since it is computed from the local current using the global potential, not the local potential. To obtain true local impedance, the local potential must also be measured (or controlled), and some efforts in this direction have been made [35].

3.2.7 OTHER SCANNING PROBE TECHNIQUES

We conclude this section by mentioning other scanning probe techniques that have found extensive application in corrosion research, but for which the probe response does not originate directly from an electrochemical phenomenon. These include scanning Kelvin probe (SKP), scanning tunneling microscopy (STM), AFM and its various elaborations, and near-field scanning optical microscopy (NSOM). Certain of these techniques have been integrated with SECM and applied to corrosion studies, and a few examples of such combination techniques are discussed later in this chapter.

3.3 BASICS OF CORROSION

This section provides a very brief introduction to the principles of corrosion, emphasizing the nomenclature that will be encountered throughout the remainder of this chapter. More detailed discussions of corrosion and its control can be found elsewhere [1,36]. The majority of the scanning probe studies applied to corrosion and discussed later in this chapter are on Fe, Al, and their alloys, and we use these metals as examples in our discussion in this section.

The aqueous corrosion of a metal requires the presence of an electrolyte at the metal surface. Ions move (or migrate) through this electrolyte to maintain charge balance as oxidation and reduction reactions occur on the metal surface. Oxidation of the metal may lead to soluble ionic species of the metal and/or to insoluble forms, as exemplified by the following reactions:

$$Fe \rightarrow Fe^{2+} + 2e^-$$

$$Fe + 3H_2O \rightarrow Fe(OH)_3 + 3H^+ + 3e^-$$

$$2Al + 3H_2O \rightarrow Al_2O_3 + 6H^+ + 6e^-$$

The specific forms of the oxidized metal produced are influenced by the interfacial potential, pH, and other electrolyte constituents. Such oxidation reactions often lead to a local decrease in pH at the sites of oxidation (i.e., at *local anodes*).

The dominant forms of a metal as a function of potential (E) and pH can be computed from Nernst equations and other thermodynamic expressions (acid dissociation constants, solubility product constants, etc.) and displayed in the form of a potential/pH diagram, also known as a *Pourbaix* diagram [37,1]. Such diagrams are useful for predicting or understanding the nature of the corrosion products formed under specified conditions.

The most common oxidants (for environmental corrosion) are O_2, H^+ and H_2O and the corresponding reduction reactions may be written as:

$$O_2 + 2H_2O + 4e^- \rightarrow 4OH^-$$

$$2H^+ + 2e^- \rightarrow H_2$$

$$2H_2O + 2e^- \rightarrow H_2 + 2OH^-$$

Note that each of these reduction reactions engenders a local increase in pH at the sites of reduction (i.e., at *local cathodes*). Depending on environmental conditions, only one of these reduction reactions may be important. A local anode and cathode along with the electrolyte constitute a *corrosion cell*. Electrons flow from anode to cathode through the metal substrate while cations migrate from anode to cathode (anions in the opposite direction) through the electrolyte to maintain an overall charge balance.

3.3.1 MIXED POTENTIAL THEORY

Because two (or more) redox reactions take place simultaneously on the metal surface, the interfacial potential of the metal at open circuit (i.e., the OCP) cannot be predicted from thermodynamics alone. Even though there is no net current flow to/from the metal under OCP conditions, there is an internal current flow between the local anodes and cathodes. As a result, the interfacial potential of the metal under open-circuit conditions is not an equilibrium potential, but rather a time-dependent potential called a *mixed potential*, which is dictated by both the thermodynamics and the kinetics of the two (or more) redox systems. This OCP is in fact the *corrosion potential* (denoted E_{corr}), and the total internal oxidation current density (due to metal oxidation) is called the *corrosion current density* (denoted i_{corr}) and is equal to the total internal reduction current density (the net current being zero at the OCP). E_{corr} can be measured directly using a suitable reference electrode and a high impedance voltmeter. However, i_{corr} cannot be measured directly using global electrochemical techniques, but it can be determined indirectly by a number of methods [5]. We distinguish here between the global (or average) corrosion current density discussed above, defined as the total oxidation current divided by the area of the substrate exposed to electrolyte, and the local corrosion current density, which can vary widely across the substrate surface, especially for metals undergoing localized corrosion. The local i_{corr} on a substrate can be directly measured using the SVET (as discussed above). The global corrosion current density is related to the average corrosion rate (r, mass loss per unit area per unit time) by the following equation [1]:

$$r = \frac{i_{corr}a}{nF} \tag{3.3}$$

where a is the atomic mass of the metal and n the number of electrons transferred.

If the potential of the metal substrate is polarized away from its OCP, the polarization being $E - E_{corr}$, the net current density in the absence of ohmic and concentration polarization (i.e., only activation or charge-transfer kinetics controls the current) is given by the equation [1,38]:

$$i = i_{corr}\left\{\exp\left[\frac{2.3\left(E - E_{corr}\right)}{\beta_a}\right] - \exp\left[\frac{2.3\left(E - E_{corr}\right)}{\beta_c}\right]\right\} \tag{3.4}$$

where

$$\beta_a = \frac{2.3RT}{\alpha_a n_a F} \quad \text{and} \quad \beta_c = -\frac{2.3RT}{\alpha_c n_c F} \tag{3.5}$$

are known as the *Tafel coefficients*, with α_a, α_c representing the *transfer coefficients* for the anodic and cathodic reactions, respectively, and n_a, n_c representing the number of electrons transferred in the anodic and cathodic reactions, respectively. At 25°C for a one-electron transfer and for $\alpha = 0.5$, each Tafel coefficient is predicted to have a value of 0.12 V. Experimentally observed values of these Tafel coefficients for oxidation and reduction reactions generally fall in the range 0.05 to 0.2 V [1].

Equation 3.4 is very similar in form to the Butler–Volmer equation, which describes the current for a single redox reaction occurring at an electrode, in which case the first and second terms of the equation represent the forward (oxidation) and reverse (reduction) directions of that reaction, respectively, and i_{corr} is replaced by the exchange current density [39]. However, for a mixed redox system (such as a corrosion reaction), the two terms in Equation 3.4 represent the two different redox reactions, the first term representing the metal oxidation current (defined as positive) and the second term representing the oxidant reduction current (defined as negative). For sufficiently positive polarization ($\eta_a = E - E_{corr} > 0$), the second term in Equation 3.4 becomes negligible (note β_c is negative), the net current (i) is an anodic current (i_a), and the equation reduces to the following:

$$\eta_a = E - E_{corr} = \beta_a \log \frac{i_a}{i_{corr}} \tag{3.6}$$

Similarly, at sufficiently negative polarization ($\eta_c = E - E_{corr} < 0$), the net current is cathodic (i_c) and Equation 3.3 reduces to:

$$\eta_c = E - E_{corr} = \beta_c \log \frac{-i_c}{i_{corr}} \tag{3.7}$$

Equations 3.6 and 3.7 are called Tafel equations and predict a linear relationship between potential or polarization and the logarithm of current.

The origin of the corrosion potential can be understood from *mixed potential theory* [1]. Figure 3.1 illustrates a potential versus log current plot for a mixed redox system consisting of metal oxidation and oxygen reduction reactions (such plots are referred to as Evans diagrams). There are two sets of curves on this diagram. The top set originating from the point labeled i_{01}, E_1 is for the O_2/OH^- redox system and consists of an anodic branch (the dashed line) and a cathodic branch (the solid line). The two linear branches are just the Tafel plots for this redox system and their positions on the diagram are determined by the exchange current density for the O_2/OH^- reaction (i_{01}) on the metal surface, the equilibrium potential (E_1) for this reaction (as predicted by the Nernst equation), and the Tafel coefficients for the anodic and cathodic reactions. Only the cathodic branch of slope β_c is of interest here. Also shown for this O_2/OH^- redox system are two examples of the limiting current behavior when O_2 concentration polarization occurs (the dotted lines), one example for a high O_2 concentration (resulting in $i_{lim(1)}$) and one for a low O_2 concentration (resulting in a lower $i_{lim(2)}$). The bottom set of curves originating from the point labeled i_{02}, E_2 is for the M^{++}/M redox system and again

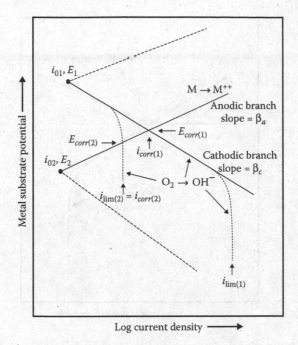

FIGURE 3.1 Potential versus log current plot (Evans diagram) for a mixed redox system consisting of metal oxidation and oxygen reduction reactions (see text for a description of the curves).

consists of an anodic branch (solid line) and a cathodic branch (dashed line). Only the anodic branch is of interest here and its position on the diagram is determined by the exchange current density for the M^{++}/M reaction (i_{02}) at the metal surface, the equilibrium potential (E_2) for this reaction, and the Tafel coefficient β_a.

If only the M^{++}/M redox couple were present, the metal would be at the equilibrium potential E_2. Similarly, if only the O_2/OH^- redox couple were present, the metal would be at the equilibrium potential E_1. When both are present, the metal assumes a (nonequilibrium) potential somewhere between E_1 and E_2 where the anodic and cathodic currents are equal. For the high oxygen concentration case of Figure 3.1, that point would be the intersection of the two solid curves at $E_{corr(1)}$ and $i_{corr(1)}$ and concentration polarization is not a factor (i.e., activation polarization controls the corrosion rate). For the low oxygen concentration case, the intersection would be at $E_{corr(2)}$ and $i_{corr(2)}$ (which equals $i_{lim(2)}$) and the mass transport of O_2 to the metal surface (concentration polarization) controls the corrosion rate.

A polarization experiment for the high oxygen concentration case of Figure 3.1 would yield a result similar to that shown schematically by the solid curve in Figure 3.2. The solid curve (up to the onset of concentration polarization) is described by Equation 3.4. The two linear regions of the curve (extrapolated with dashed lines) are the Tafel regions where either anodic or cathodic activation limits the current and are described by Equations 3.6 and 3.7, respectively. The

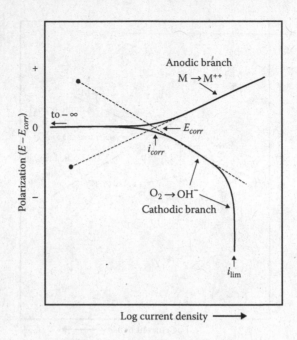

FIGURE 3.2 Schematic polarization curve (solid line) for the high oxygen concentration case of Figure 3.1. E_{corr} and i_{corr} in this figure correspond to $E_{corr(I)}$ and $i_{corr(I)}$, respectively, in Figure 3.1.

net current goes to zero (i.e., to a large negative value on the log scale) at the OCP (E_{corr}). Extrapolation of the linear Tafel regions permits determination of E_{corr} and i_{corr}, one of the indirect methods by which i_{corr} can be determined [5].

3.3.2 PASSIVE METALS AND LOCALIZED CORROSION

A number of structural metals, including alloys of iron, aluminum, titanium, and chromium, form passive films consisting of an oxide or hydrated oxide layer of limited ionic conductivity that adheres to the metal and provides a barrier to further metal dissolution. In some cases, a salt film precipitation may be involved. For such metals, the corrosion process is usually not uniform, but rather occurs at localized sites on the metal surface at positions where the passive layer has been compromised, leading to *pitting corrosion*. A breach in the passive layer is often the result of localized chemical damage to or mechanical disruption of the protective oxide film, promoted by such factors as acidity or alkalinity, high concentrations of chloride, low dissolved oxygen concentrations (rendering the passive film less stable), or the presence of heterogeneities (such as nonmetallic inclusions or second-phase constituents) on the metal surface.

Pitting corrosion is more difficult to detect and predict than uniform corrosion and is considered to be far more dangerous. Highly corrosive conditions can develop within the pits, leading to very low pH and high salt (e.g., chloride ion) concentration. This, in turn, can lead to rapid propagation of the pit through the metal, leading to perforation of the structure or initiation of a crack, and failure can occur unexpectedly with minimal overall metal loss.

Figure 3.3 illustrates a polarization diagram for typical thin film active-passive behavior (e.g., Cr or Ni in sulfuric acid). The dashed portion of the curve intersecting the point $E_{(M/M^{++})}$, $i_{0(M/M^{++})}$ is the cathodic branch along which metal ion reduction occurs and the solid portion of the curve represents the anodic branch along which metal oxidation occurs. As the potential is increased positive of the equilibrium potential $E_{(M/M^{++})}$, the current (corrosion rate) increases until a critical passivation potential E_{pp} or critical current density i_{crit} is reached, at which point the current drops to a lower value (i_p), signaling the onset of passivation. In this passive region, the current remains low even though a large driving force (positive potential) is applied to the metal. At a sufficiently positive potential E_t, the current again increases as the oxide film dissolves uniformly, often accompanied by oxygen evolution (the transpassive region). Some metals (e.g., Al or Mg in water) exhibit thick film passivity, which differs from that illustrated in Figure 3.3 in that little or no drop in current is observed at E_{pp}, but rather the current levels off from the linear active branch directly to a plateau current, i_p [5]. A more detailed discussion of the passivity of metals and alloys is provided elsewhere [40]. As noted earlier, the scanning probe techniques are particularly useful for the study of localized corrosion of passive metals.

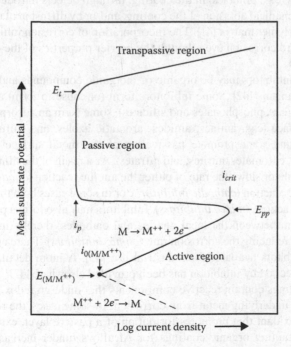

FIGURE 3.3 Schematic polarization diagram for typical thin film active–passive behavior.

3.3.3 METHODS OF CORROSION CONTROL

We conclude this section by summarizing the methods commonly used for controlling corrosion, including the use of coatings (barrier coatings as well as active coatings containing reactive components), the use of inorganic and organic inhibitors, and the use of anodic and cathodic protection. More complete discussions of these methods can be found elsewhere [1,36].

The most common approach for protecting a metal from corrosion is to coat the metal with an organic polymer that provides a barrier between the metal and the corrosive environment [41]. Usually a complete coating system is applied, consisting of two (or more) coating layers, each designed for a specific function. The first coating applied to the metal (often after some type of metal surface pretreatment or preparation) is a *primer coating* designed for good adhesion to the metal. This primer coating often contains corrosion inhibitors, but it is not a particularly good barrier coating. To provide the barrier properties, a *topcoat* is applied in one or more layers on top of the primer. The topcoat often contains additional corrosion inhibitors as well as pigments that impart aesthetic or other desirable properties to the coating (e.g., weathering resistance or improved barrier properties). The coating retards the access of water, oxygen, and ions to the metal surface and also increases the electrolyte resistance between local anodes and cathodes on the metal surface (by decreasing ion mobility). However, all barrier coating systems eventually fail due to ingress of water, oxygen, and ions, through natural defects (e.g., pinholes) in the coating, through defects introduced accidentally from mechanical abrasion of the coating, and/or by diffusion of these species through the polymer matrix [41]. The incorporation of corrosion inhibitors serves to further retard corrosion even though the barrier properties of the coating have been breached.

Corrosion inhibitors may be organic or inorganic compounds and function by various mechanisms [42]. Some inhibitors form (or cause to form) a film on the metal surface (e.g., phosphonates and silicates), some form an adsorption layer on the metal surface (e.g., amines, amides, aromatic azoles, and sulfur-containing compounds), and some promote passivation of the metal surface (*passivating inhibitors*, e.g., chromates, nitrites, and nitrates). As a result of the film or layer that forms, the inhibitor slows the rate of either the anodic reaction (*anodic inhibitors*) or the cathodic reaction (*cathodic inhibitors*), or in some cases both the anodic and the cathodic reactions (*mixed inhibitors*). Inhibitors may also form a poor ionically conducting film between the local anodes and cathodes, thereby increasing the resistance and reducing the corrosion rate (*ohmic inhibitors*). There are over 1,100 corrosion inhibitors available for industrial use [43]. A more detailed review of corrosion protection by inhibition has been provided elsewhere [42].

Active coatings contain reactive components that undergo redox reactions for protecting the underlying metal from corrosion. In some cases, the reactive component is an oxidant that promotes formation of a passive layer, examples being chromate-containing organic coatings for Al alloys (under increasing scrutiny due to the toxicity of Cr(VI) and issues related to its safe handling and disposal

[44]) and certain conjugated polymer coatings for steel [45]. In other cases, the reactive component is a reductant that protects the underlying metal by a sacrificial or cathodic protection (see below) mechanism, examples being Zn-rich organic coatings for the protection of steel [46] or Mg-rich organic coatings for the protection of Al alloys [47]. A similar sacrificial protection mechanism has been suggested for an n-doped conjugated polymer coating applied to an Al alloy [48], the first example of such protection by an all-organic coating. Metal coatings are also used for corrosion control [49], a common example being Zn coatings used on steel (galvanized steel), in which case the Zn cathodically protects the underlying steel by a sacrificial process.

Cathodic protection is frequently used to protect storage tanks, buried pipelines, offshore structures, and ship hulls and can be used for virtually all metals [50]. Using either an external DC power supply or a sacrificial anode (s.a.), the potential of the metal is forced to a value more negative than E_{corr} such that only cathodic reactions take place at the metal surface to any appreciable extent, thereby reducing the corrosion rate of the metal. When used in combination with a protective coating, the required current is small, namely that necessary to protect metals exposed at coating defects. A s.a. (made of a metal that is more active, i.e., more easily corrodible, than the metal to be protected) can be used when an electrical power source is not available. For example, s.a.'s made of aluminum and/or magnesium alloys are often used to protect steel ship hulls. The sacrificial metal may be incorporated into an organic coating as described above for Zn-rich and Mg-rich coatings.

Anodic protection is applicable only to metals that exhibit active–passive behavior (Figure 3.3) and, thus, is not as widely used as cathodic protection [50]. A DC power supply is used to force the potential of the metal to a more positive value in the passive region (Figure 3.3) where the corrosion rate is low (just i_p), thereby maintaining the metal in the passive state. Anodic protection is particularly useful under extremely corrosive environments, such as strongly alkaline or acidic environments, conditions where cathodic protection is not practical. An additional advantage of anodic protection is the low current requirement, i_p, which also provides a direct means for monitoring the corrosion rate of the system. A disadvantage of anodic protection is the risk of greatly accelerating corrosion of the metal if proper potential control is not maintained and the potential is allowed to become more positive than E_t or more negative than E_{pp} (Figure 3.3). Example applications of anodic protection include the protection of mild or SS equipment used to handle and store concentrated sulfuric acid, as well as pulp and paper mill digesters, and clarifiers and storage tanks.

3.4   APPLICATIONS OF SECM TO CORROSION STUDIES

In this section, we present a comprehensive summary of DC-SECM investigations related to corrosion on metals. We have chosen to include only those reports that utilize various DC modes of SECM to explore metals under typical corrosion conditions. While a number of related and very relevant studies involve examination of electron transfer on passive oxide films, as well as corrosion-related investigations involving AC-SECM techniques, we have chosen to high-light those in separate sections of this chapter.

3.4.1 IRON AND ITS ALLOYS

Much of the SECM work performed on iron and its alloys has taken advantage of the localized nature of the technique to investigate the inherently localized process of pitting corrosion on the metal surface. In this regard, the SECM probe has been used to both identify and generate pitting sites. The earliest demonstration of each was in a single report by Wipf, who used SECM to first detect the products of pitting corrosion on SS and then generate chloride ions with the micro-electrode to initiate pit formation on the surface [51]. For detection of corrosion products above an active pit, the sample (304 SS, $E_{substrate}$ = +0.5 V vs. Ag/AgCl) was immersed in chloride solution, and SG/TC mode was used (E_{probe} = +1.0 V) to oxidize corrosion products released from a pit. The probe was positioned 20 µm above the sample surface and rastered in the x-y plane while monitoring the probe current to create a concentration profile of the oxidizable corrosion products above a pitting site. An SECM image of an active pit with a diameter of ~120 µm was shown, and a heterogeneous distribution of corrosion products above the pit was observed, with the maximum concentration approaching 2 mM. Cyclic voltammetry performed with the microelectrode above the pit detected a broad voltammetric wave with a half-wave potential of +0.86 V—consistent with the Fe^{2+}/Fe^{3+} couple—suggesting that the oxidation of ferrous ions is responsible for the anodic current observed in the SECM image.

In the same study, an active pitting site was manually generated on the steel surface by the localized production of chloride ions with the SECM probe, which in this case was a 50-µm Au amalgam microelectrode placed approximately 100 µm above the substrate [51]. Chloride ions were produced at the probe by the reduction of trichloroacetic acid from the solution at a probe potential of −1.1 V. (The amalgam microelectrode was used to prevent hydrogen ion reduction at this potential.) After 1,400 s of chloride production, a large anodic current spike was observed at the steel substrate, with a concurrent increase in cathodic current at the probe (most likely due to reduction of Fe^{2+}). These observations were seen as consistent with the initiation, growth, and repassivation of a single corrosion pit, and optical microscopy did indeed detect a single 15 µm × 20 µm pit located in the area where the probe had been placed. In a repeat of the experiment, however, one single pit was not generated; instead, several smaller pits with diameters of 1–5 µm were detected within a 30-µm diameter region under the probe.

In a more detailed investigation, Still and Wipf used local generation of chloride ions via reduction of trichloroacetic acid to investigate the breakdown of iron passive layers under a variety of conditions in pH 6.0 phosphate/citrate buffer [52]. In these experiments, the SECM probe was first positioned close to the iron surface with both E_{probe} and $E_{substrate}$ at rest potentials. The substrate was then stepped to a potential in the passive region, and after a delay time of 0–5 s to allow for passive layer growth, the probe potential was decreased to begin generation of chloride ions. Both the substrate current and probe current were monitored for evidence of passive layer breakdown as a function of passive layer growth time (0–5 s), passive layer growth potential (−0.2 V to +0.2 V vs. MSE), and probe diameter (12.7 µm vs. 100 µm). Breakdown of the passive layer was deemed successful if two criteria were met: correlated tip and substrate current increases, and micrographic observation of a corrosion pit at the probe location. Since complete reproducibility proved difficult to obtain in these experiments, overall tendencies were noted in lieu of absolute conclusions. There was less chance of passive layer breakdown, for instance, at longer growth times and higher growth potentials, suggesting that a critical point is achieved in the passive layer thickness that results in complete protection from corrosion attack. Not surprisingly, the likelihood of passive layer breakdown was also higher when using a larger microelectrode for chloride generation, since a larger electrode is more likely to cover an area of the substrate at which the passive layer is either defective or incompletely formed. Interestingly, similar experiments performed on fully formed passive layers (1 h of film growth on polished Fe) were found to never result in breakdown if the probe was placed in the center of the substrate. The probability of breakdown increased, however, when the probe was moved toward the edges of the substrate where etching analysis revealed smaller grains and a corresponding higher density of grain boundaries, suggesting that pit formation tends to occur on areas with a higher defect density.

The addition of a chemical species such as trichloroacetic acid to the test solution for the local generation of aggressive ions may introduce undesirable experimental interferences into the system. Trichloroacetic acid, for instance, is known to inhibit pitting on 304-type SS [52]. As an alternative approach, Fushimi and coworkers used a silver/silver chloride microelectrode, which they term a liquid-phase ion gun (LPIG), as a source of chloride ions to investigate the local breakdown of passive films of iron in deaerated borate buffer solution [53–54]. The LPIG consists of a 180-µm diameter silver wire embedded in glass and polarized for 30 min in HCl solution to create a layer of AgCl on the tip surface. Cathodic polarization of the resulting electrode at −0.1 V (vs. SHE) was shown to rapidly produce a significant chloride ion concentration in the immediate solution estimated at 1 mol/L [54]. Complete removal of the AgCl layer during this step results in a bare silver electrode biased at a potential (−0.1 V) that the authors demonstrate is effective for the diffusion-limited cathodic detection of dissolved ferric ions, but gives no response to ferrous ions [53].

Passive film breakdown on iron was investigated by first positioning the LPIG 75 µm above an iron substrate passivated at +1.2 V (vs. SHE) in borate solution [53]. Chloride ions were cathodically released from the microelectrode while monitoring both the microelectrode and substrate currents. After an induction period in which the local concentration of chloride ions at the iron surface increased, the anodic current at the iron substrate rose rapidly, accompanied by a corresponding increase in the cathodic current at the Ag microelectrode due to the apparent reduction of Fe^{3+}. Optical microscopy revealed a corrosion pit with a diameter of 100 μm and depth of 5 μm. Similar experiments were carried out at passivation potentials of 0.2 V, 0.7 V, and 1.2 V, with each resulting in formation of a single corrosion pit under the microelectrode [53]. Plots of anodic charge passed at the substrate versus cathodic charge passed at the microelectrode revealed three distinct domains of activity. The first corresponds to the induction period in which chloride production at the microelectrode increases the chloride concentration at the substrate, but with little activity observed at the substrate. Not surprisingly, the length of this induction period was found to increase with the thickness of the passive layer [54]. Also, it was determined through etching analysis that the induction period (first domain) required for passive layer breakdown is dependent on the grain orientation of the underlying iron substrate, with the induction period order of (110) < (111) < (100) being consistent with the order of increasing film thickness determined in another study [54]. In the second domain a linear relationship was observed between the substrate charge and microelectrode charge, resulting from the promotion of the dissolution of the iron passive layer by the chloride ions. A positive feedback mechanism was proposed, whereby Fe^{3+} produced at the substrate is reduced to Fe^{2+} at the microelectrode, which then diffuses back to the substrate for reoxidation to Fe^{3+} [53]. The third domain is characterized by a rapid increase in the anodic substrate charge with little corresponding increase in the cathodic microelectrode charge. This domain is believed to correspond to complete dissolution of the passive film and the exposure of the bare Fe surface under the microelectrode. Corrosion of the bare metal results in an increased depth of the corrosion pit and the anodic generation of dissolved Fe^{2+} ions, which are not reduced (detected) at the potential of the microelectrode (see above).

A series of reports by Gabrielli and coworkers [55–57] have similarly focused on examinations of pitting corrosion on iron passive films using a silver/silver chloride microelectrode for pit generation similar to the LPIG described above. In this case, however, the authors have combined SECM with an EQCM (SECM/EQCM) to simultaneously measure both electrochemical activity and small mass changes in the substrate [55–56]. For a portion of this work, they have employed a twin-electrode design consisting of a 160-μm diameter silver/silver chloride microelectrode and a 60-μm diameter Pt microelectrode sealed together in a glass tube of 1 mm diameter, with the Pt microelectrode used for positioning the probe at a controlled height above the substrate surface [56]. The combined SECM/EQCM technique uses a modified substrate consisting of a quartz crystal sandwiched between two gold films for electrical connection to the oscillator circuit. A layer of iron 2–3 μm in thickness was plasma sputtered onto the upper gold electrode, cathodically pretreated to remove the air-formed oxide, and then anodically biased to form the passive layer. The close proximity of the SECM probe to the substrate was found to cause predictable frequency oscillations in the quartz crystal as a function of separation distance, and this behavior was used for precise microelectrode positioning [55]. The combined SECM/EQCM experiments are complicated by the inherent positional dependence of the mass sensitivity of the quartz electrode in the EQCM, with the maximum sensitivity occurring at the center of the electrode and the minimum on the edge. This sensitivity was assumed to follow a Gaussian distribution from the center to the edge, so the exact location of the pit was determined by SEM after the experiment and the local quartz sensitivity was applied based on calibration data and calculations [55–56].

The SECM/EQCM experiments were conducted in a manner similar to those by Fushimi described above. The probe was placed 75 μm above the iron surface and chloride generation was initiated while measuring the probe current, substrate current and, in this case, the frequency (mass) change in the substrate. Figure 3.4 shows experimental results for pit generation on iron in aerated 0.1 M NaOH [55]. The passive layer was formed at 0.08 V (vs. SSE) and chloride generation was initiated at $t = 60$ s by decreasing the probe potential to -0.75 V (Figure 3.4a). As reported by Fushimi, three distinct domains are visible, and these are separated by the vertical dotted lines in Figure 3.4. The first domain

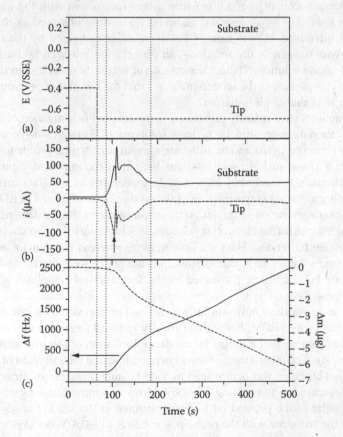

FIGURE 3.4 (a) Simultaneous potential; (b) current; and (c) frequency variation curves recorded during the formation and propagation of a single pit in 0.1 M NaOH. (From Gabrielli et al., *J Electrochem Soc*, Vol. 153, (2006): p. B68. With permission.)

shows an induction time of 30 s during which cathodic current due to chloride generation is detected at the tip, but no substrate current or change in mass is observed. It was estimated that a chloride concentration of 0.09 mol/L develops over this time [55], although the minimum amount of chloride required to initiate a single pit was found to be dependent on both substrate potential and solution pH [56]. The second domain is marked by symmetrical spikes in both the tip and substrate currents, along with a sharp linear decrease in substrate mass, from $t = 90$ s to $t = 180$ s. This domain corresponds to the dissolution of the passive layer and generation of Fe^{3+}, which is then reduced at the probe. The current oscillations within this domain were at first attributed to repassivation of the surface through precipitation of iron(III) compounds [55], but were later reported as stray electrical coupling between the probe and substrate [56]. The third domain at $t > 180$ s is marked by a steady substrate current, a small residual tip current, and a decrease in the rate of substrate mass loss. This domain results from iron dissolution, and the combined electrochemical and frequency change data indicate an electron exchange of 1.9 per iron atom—consistent with the formation of ferrous ions. The small residual cathodic tip current observed in this third domain is attributed to reduction of Fe(III) species generated by the reaction of Fe(II) with oxygen in the solution—an observation not seen by Fushimi in deaerated borate solution. The decreased rate of substrate mass loss in the third domain in comparison to the second indicates that the kinetics of pit propagation are slower than that of pit initiation.

These authors have recently reported a more detailed investigation of induced single pit formation on iron in various solutions [57]. Interestingly, a fourth domain is observed in the anodic substrate current versus time profile for borate buffer (pH 8.4) and sulfuric acid solutions long after the initiation of pit formation. This domain is marked by an abrupt disappearance of substrate current, an observation attributed to repassivation of the pit that is not observed in KOH solution. EIS characterization of the pit propagation in each solution demonstrates a strong dependence on the electrolyte solution. Insoluble corrosion products slowly form in borate buffer and block pit propagation, whereas pits remain active in strongly basic solution. In acidic solution, pits are active for a short time as chloride ions are believed to be consumed by the formation of complexes with iron corrosion products.

Luong and coworkers have similarly examined pit formation on iron in NaCl/borate buffer using SECM, but without locally induced formation of a single pit [58]. Instead, pitting was initiated by anodic polarization of the iron substrate at -0.1 V (vs. Ag/AgCl) until the substrate current indicated the initiation of pitting. The location of the pit was determined by video camera, and the substrate potential was decreased to stop pitting. The SECM tip was moved to a location 5 μm over the center of the pit, and pitting was resumed at the site via anodic polarization of the substrate with the probe potential set to -0.4 V (vs. Ag/AgCl) for cathodic detection of Fe^{3+} ejected from the pit. When pit propagation was carried out slowly at a substrate potential of -0.2 V, the tip current gradually increased for 180 s before leveling off, then quickly dropped to zero at 280 s, most likely due to repassivation of the pit. During more rapid pit propagation at a substrate potential of $+0.1$ V, a much smaller amount of Fe^{3+} was detected due to dissolution of the underlying iron substrate through production of Fe^{2+}, consistent with the results of both Fushimi and Gabrielli reported above.

Fushimi and Seo used a graphite reinforcement carbon (GRC) SECM probe (made from pencil lead) to measure the dissolution distribution of ferrous and ferric ions above a polycrystalline iron electrode in pH 2.3 sulfate solution [59]. The GRC probe was positioned 75 μm above the iron surface and biased at either $+1.2$ V or -0.2 V (SHE) to detect Fe^{2+} or Fe^{3+}, respectively, while cyclic voltammetry was performed on the substrate. In this case, both ferrous and ferric ions were detected at the probe, dependent on the substrate potential. As the substrate potential increased into the active potential region beginning at -0.3 V, ferrous species dissolving from the iron were detected. This was followed by a decrease in the Fe^{2+} current with increasing potential that was believed to result from IR drop associated with the formation of insoluble ferrous salts in the space between the probe and substrate. These salts apparently dissolved as the substrate potential moved into the passive region at $+0.06$ V where the Fe^{2+} current reappeared and then slowly dropped with increasing potential. Ferric ion was detected in the transpassive region beyond $+1.4$ V, indicating dissolution of iron as Fe^{3+}. SECM imaging was carried out to determine the distribution of anodic dissolution as a function of crystal grain orientation on the passive layer of an iron substrate etched in ethanol/HNO_3 to reveal crystal grains. Images collected in TG/SC mode with $Fe(CN)_6^{4-}$ as the mediator revealed distinct differences between crystal grains in the rate of mediator reduction—differences believed to result from variations in the thickness of the passive layer formed on different substrate grains. Subsequent line scans in SG/TC mode with detection of Fe^{2+} across several grain boundaries revealed that the active dissolution rate of ferrous ion was lower from those substrate grains with a thicker passive film than from grains with thinner passive films.

There has also been much interest not only in characterizing active pitting sites, but also in identifying sites that are precursors to pitting. Tanabe and coworkers, for example, attempted to locate pitting precursor sites by using an SECM probe to simultaneously measure local concentrations of both H^+ and Cl^- above SS samples polarized at pitting precursor potentials in NaCl solution [60–61]. Simultaneous detection of these ions was carried out by first stepping the microelectrode potential cathodically for 50 μs to reduce H^+ to H_2, followed by a 50-μs anodic step to oxidize Cl^- to Cl_2, and measuring the resulting tip current after each step. SECM images obtained in this manner on types 304 and 316L SSs successfully identified areas rich in either H^+ or Cl^-, or both (interestingly, these features are absent on more nitrogen-rich HR8C SS). The highly concentrated areas became more intense at higher polarization potentials, leading the authors to speculate that these areas correspond to pitting precursor sites. It should be noted that the resolution of the SECM images shown in these reports is remarkable, with individual features of ~100 nm diameter clearly identified. The authors report using a 20-μm diameter Pt probe sealed in a glass micropipett and electrochemically etched in KCN/NaOH, although no final diameter of the resulting tip was given [60].

Along with the detection of pitting precursor sites comes the desire to use SECM to better understand the pitting process. For example, the mechanism involved in the initial stages of pitting on SS is not clear, and SECM studies have been carried out to better understand the sequence of events that eventually lead to stable pit formation. Zhu and Williams examined spatial fluctuations in the passive current density on type 304 SS in dilute chloride solution to determine if these fluctuations play a role in pit formation [62]. Using a unique two-electrode configuration with an insulated Pt-Ir STM tip as the probe, SECM images revealed sharp anodic current spikes of ca. 40 pA in amplitude, presumably resulting from Fe^{2+} oxidation at the probe, emanating from an anodically polarized steel substrate. The lifetime of these spikes was very short—usually less than the 0.25 s required for a single line scan—and they were attributed to pitting precursor events that passivate before formation of a stable pit. In some instances, a very small (~1 pA) increase in the passive current was detected over a larger area. An example is seen in Figure 3.5, which shows such a feature growing in size over 70 min of immersion at a relatively low polarization potential of +30 mV (SCE). Upon increasing the polarization potential to +230 mV, SECM images identified nucleation of a pit on the very site where the slight increase is

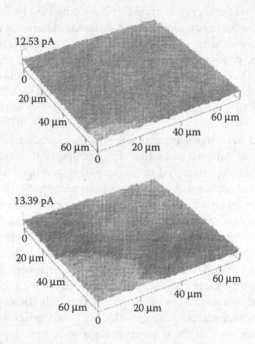

FIGURE 3.5 Development of localized minor perturbation of passive current density (+30 mV, SCE) after 5 min (top) and 70 min (bottom). (From Zhu and Williams, *J Electrochem Soc*, Vol. 144, (1997): p. L43. With permission.)

passive current had been detected, leading the authors to speculate that this was the site of a sulfide inclusion in the steel.

A subsequent study by the same authors used SECM to follow the nucleation, growth, and contraction of a pit formed at a sulfide inclusion site on an anodically polarized steel sample in chloride solution [63]. The SECM results, combined with photoelectrochemical and optical microscopy, led the authors to propose a mechanism for stable pit formation, which begins with chloride-catalyzed dissolution of an MnS inclusion. This dissolution process is accelerated by a large local increase in Cl^- concentration at the pit due to the electromigration of chloride ions to support the current during dissolution. The dissolution reactions create a local pH decrease and produce either sulfide or thiosulfate as corrosion products, both of which are ultimately converted to elemental sulfur, which subsequently forms a sulfur crust over the inclusion. This crust forms a locally occluded zone that provides the isolation necessary to maintain the extreme pH and chloride concentrations in order to prevent repassivation of the stable pit. The trigger event for the whole process is the initial rapid dissolution of the sulfide inclusion, which is believed to result from chromium depletion zones that develop around sulfide inclusions upon cooling of the alloy during its processing [64]. The dissolution of these chromium-depleted edges results in the locally extreme conditions that lead to complete dissolution of the inclusion. This model is supported by high-resolution SECM images, which show a region of high dissolution activity around the perimeter of an inclusion [64].

Evidence of metastable pitting at the open-circuit potential of SS in HCl solution was reported by González-García and coworkers [65]. SECM images were obtained in SG/TC mode with a 10-μm Pt tip biased at potentials from 0.547 to 1.077 V (SHE) to oxidize (detect) Fe^{2+} generated by the corroding substrate. Figure 3.6a shows a 100 μm × 100 μm image obtained from a 304 SS sample at its OCP in 0.1 M HCl. A number of small anodic current spikes are clearly evident, and these are attributed to Fe^{2+} cations released from metastable pitting sites. These metastable pits are random in both time and space, and the observation that no evidence of an individual pit is seen in two successive line scans leads to the conclusion that the lifetime of these pits is less than the 6 s required for a single line scan. Closer examination of Figure 3.6a reveals a number of spikes pointing in the cathodic direction as well. The image is inverted and shown in Figure 3.6b such that the upward direction is now cathodic. It can be seen that each anodically directed spike in Figure 3.6a is followed by a cathodically directed spike in Figure 3.6b. This pairing is clearly seen in the single line scan of Figure 3.6c, noting that the background is adjusted to zero. The authors argue that these apparent cathodic peaks are not produced by cathodic reactions such as the reduction of Fe^{3+}, O_2, or H_2O_2, but are instead the result of a decrease in the anodic signal arising from the passive background current. This passive current arises from the slow steady-state dissolution of the alloy and produces a small flux of Fe^{2+}, which is oxidized at the SECM tip to establish the baseline tip current. At the OCP, anodic generation of Fe^{2+} by a metastable pit will likely be accompanied by the reduction of O_2 to produce H_2O_2 in the nondeaerated solution. The H_2O_2

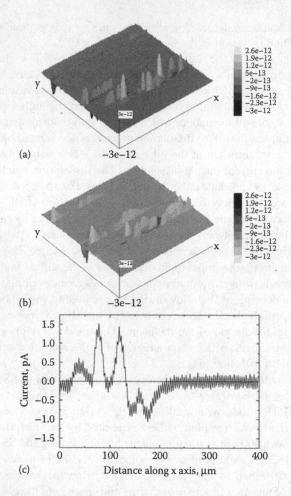

FIGURE 3.6 SECM of SS surface under immersion in 0.1 M HCl at the open-circuit potential. (a) The image as recorded, showing anodic peaks due to metastable pits. The upward direction is anodic. The x and y axes show an area 100 mm by 100 mm. The tip potential was 0.697 V (SHE). (b) The image is (a) turned upside down. In this image, the upward direction in z is cathodic. The tip potential was 0.697 V (SHE). (c) A line scan showing the positive (anodic) and negative (cathodic) wings of the tip current as it passes over a metastable pit and beyond. The tip potential was 0.997 V (SHE). (Background rescaled to zero.) (From González-Garcia et al., *Electrochem Commun*, Vol. 6, (2004): p. 637. With permission.)

produced may then react in the solution phase with the passively generated Fe^{2+}, oxidizing it to Fe^{3+} and thereby causing a localized decrease in the baseline tip current. Further support for this model comes from similar experiments carried out on an anodically polarized substrate. In this case, anodic spikes were again observed at the tip and attributed to metastable pitting, but with no accompanying cathodic features. This result is consistent with the proposed model since potentiostatic control would cause the cathodic reactions to occur at the counter electrode instead of the substrate surface.

If ferrous ions dissolving from the passive current at the OCP can be detected by the SECM tip as just described, then a corresponding depletion of oxidant should be detected as well since small microcells are established on the substrate in which metal oxidation is balanced by the reduction of O_2 in neutral/alkaline solutions or H^+ in acidic solution. This hypothesis was tested in a related study by the same group [66]. A pure iron sample mounted in an epoxy resin sleeve was immersed in 0.1 M NaCl, and SECM line scans were performed over the substrate with the Pt tip biased at either +0.60 V (Ag/AgCl) for Fe^{2+} oxidation or −0.70 V for O_2 reduction. The Fe^{2+} oxidation signal was observed to increase as the tip passed from the resin to the substrate, and a corresponding decrease was seen in the O_2 reduction signal as well, thus confirming that concentration profiles of both cathodic and anodic redox species participating in the microcell reactions can be measured by adequate selection of the tip potential.

Paik and coworkers used a redox mediator to detect sulfur-containing species over MnS inclusions in types 303 and 304 SSs during the initiation of pitting corrosion in chloride solution [67]. Previous studies had reported the generation of both thiosulfate and hydrogen sulfide from these inclusions; however, both of these species exhibited slow electron-transfer kinetics at the carbon fiber SECM tip used in this investigation. The I^-/I_3^- redox couple was therefore introduced as a mediator via the addition of iodide to the solution to take advantage of the following reactions:

$$3I^- \rightarrow I_3^- + 2e^- \text{ (SECM tip)}$$

$$I_3^- + HS^- \rightarrow 3I^- + H^+ + S \text{ (inclusion)}$$

$$I_3^- + 2S_2O_3^{2-} \rightarrow 3I^- + S_4O_6^{2-} \text{ (inclusion)}$$

Both inclusion reactions were shown to result in an enhanced tip current due to positive feedback from regeneration of I^-. SECM images were produced identifying dissolved sulfur species above inclusions with sizes ranging from 1 to 20 μm. Active dissolution sites were observed at substrate potentials well below the measured pitting potential, with dissolution rates varying widely for individual inclusions.

Similar investigations using the I^-/I_3^- redox couple to detect sulfur-containing species from type 304 SS in chloride solution were carried out by Lister and Pinhero [68]. In this work, however, galvanostatic polarization of the substrate was used instead of potentiostatic polarization to force the total corrosion rate to remain constant during the completion of an SECM scan. At low current densities, images similar to those seen by Paik were reported in which localized areas of high sulfur concentration were detected with peak tip currents ranging from 2 nA to over 80 nA above the background I^- oxidation current. However, an interesting inversion was observed in the tip current at higher substrate current densities where peaks pointing in the cathodic direction emerged as a leading edge to the anodic sulfur peaks. The current in these peaks remained anodic, but with levels dropping below the background, and the effect was found to be completely reversible when the substrate current density was cycled between low and high values. Current level shifts due to changes in pH or dissolved metal ion concentrations were ruled out as possible causes of this phenomenon. Instead, these inverted peaks are believed to result from a localized increase in the electric field at the corrosion site (*IR* drop), which shifts the potential of the microelectrode when positioned over the site. This effect was demonstrated by performing cyclic voltammetry with the tip while it was positioned over a sulfide inclusion at various substrate current densities. A linear relationship was observed between $E_{1/2}$ for iodide oxidation and the current density, indicating that the shift in potential is proportional to the current passing through an active site. Furthermore, the slope of the $E_{1/2}$ versus current plot yielded a resistance that was in agreement with the cell impedance as determined by EIS. The authors speculate that the combination of chemical and electric field measurements demonstrated in this investigation could be useful in future mechanistic determinations.

Lister and Pinhero have also used SECM in SG/TC mode to detect corrosion dynamics on type 304 SS in iodide solution [69]. The substrate was biased at 500 mV (Ag/AgCl) where cyclic voltammetry indicated the oxidation of I^-, while the carbon-fiber tip was held at a potential to reduce I_3^- generated at the substrate back to I^-. Figure 3.7 shows selected images from a sequence of 25 images demonstrating the temporal and spatial dependence of iodide oxidation over a 3.0 mm^2 area of substrate, with an acquisition time of 49.8 min per image. The initial image displays a number of sharp features, while subsequent images show these features to appear and disappear over time with decreasing frequency until no features are evident in the final image. These features are believed to demonstrate the apparently random emergence, growth, and passivation of individual pits, with reemergence of some sites observed after an initial passivation. The formation of a pit exposes the underlying metal surface where oxidation of I^- may occur (detected by the SECM), and conventional microscopy of the surfaces following the SECM measurements revealed pitting corrosion. In a separate experiment, the SECM tip was held stationary over an initially active pit for an extended time with the substrate anodically polarized. The tip current profile showed five separate cycles of pit formation/passivation occurring at this one site over a 2,000 s time period, once again demonstrating the dynamic nature of the corrosion process.

The images presented in Figure 3.7 reveal the need for better temporal resolution in the detection of corrosion processes occurring over a large spatial area. It is impossible to detect and monitor individual events occurring on a fast time scale when scanning a large area with a single tip. In an attempt to overcome this limitation, Lister and Pinhero developed an instrument for continuous monitoring of individual electrode currents from a two-dimensional (2D) array of closely spaced microelectrodes [70]. This instrument, called a microelectrode array microscope (MEAM), consists of 100 stationary microelectrodes spaced 400 μm apart in a square 10 × 10 arrangement, for an effective analysis area

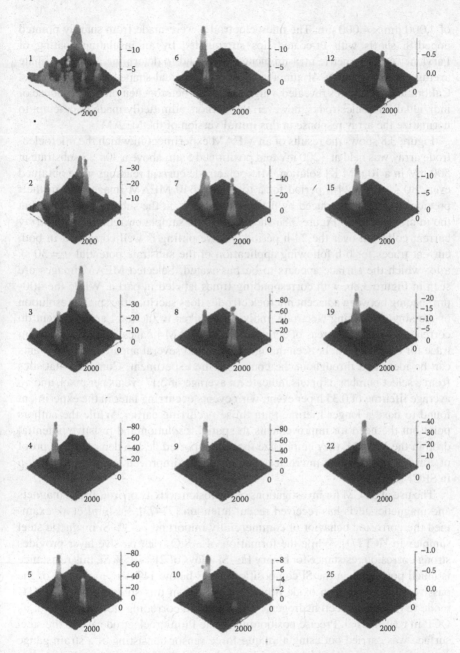

FIGURE 3.7 Sequence of SECM images of 3.0 mm² area of a fresh type SS304 sample initiated after jumping the potential from 0 to 500 mV (Ag/AgCl). The current axis (z axis) is in nanoamperes and the xy axes in micrometers. A total of 25 images were taken, the first 10 images are displayed along with selected images of the remainder (number next to plot corresponds to the scan number). (From Lister and Pinhero, *Electrochem Solid-State Lett*, Vol. 5, (2002): p. B33. With permission.)

of 4,000 μm × 4,000 μm. The microelectrodes were made from sharply pointed doped-Si shafts with Pt-coated tips surrounded by an insulating coating of Parylene. Amperometric measurements were used to determine an approximate average radius of 60 ± 41 μm for the hemispherical-shaped microelectrodes. Calibration procedures revealed a high degree of heterogeneity in the response of individual microelectrodes; however, the authors admittedly made no attempt to normalize the array response in this initial version of the MEAM.

Figure 3.8 shows the results of an MEAM experiment in which the microelectrode array was held at −200 mV and positioned 5 μm above a 304 SS substrate at 500 mV in a 10 mM KI solution. Microelectrode current readings were obtained every 10 s over a 24-h period for a total of 8,640 MEAM images. As in their previous report, I_3^- produced by the oxidation of I⁻ at the substrate is reduced at the microelectrodes. Figure 3.8a shows both the sample current and total array current collected over the 24-h period. Active pitting is well correlated in both current traces for 6 h following application of the substrate potential ($t = 30$ s), after which the surface appears to be passivated. Selected MEAM images are seen in Figure 3.8b, with corresponding times labeled in part a. While the 400-μm spacing between adjacent microelectrodes does sacrifice the spatial resolution of the single scanning electrode, individual spikes in the total array current do correlate to localized areas of activity in the MEAM images. The entire current trace for one pixel can be seen in Figure 3.8c, and several apparent pitting events can be identified throughout the course of the experiment. Compiled statistics from a select number of pixels indicate an average of 2.17 events per pixel, and an average lifetime of 0.35 h per event, with events occurring later in the experiment found to have a longer lifetime than those occurring earlier. While the authors point out the need for improvements in spatial resolution and resistive potential drop in the MEAM, they emphasize that the reported design demonstrates proof of concept in the use of microelectrode arrays for improved temporal resolution in SECM imaging.

The use of SECM for investigations of corrosion activity on iron-based magnets and magnetic steels has received recent attention [71–73]. Fushimi et al. examined the corrosion behavior of commercially important Fe-3% Si magnetic steel samples in HCl [71]. While the formation of a SiO_2-rich passive layer provides strong corrosion resistance for binary Fe–Si alloys of 21%–25% Si, this resistance is much poorer when the Si composition drops below 14% due to control of the passive layer by an iron oxide film. The authors in this study used the SG/TC mode to oxidize (detect) hydrogen evolved from a corroding steel substrate at its OCP in 0.01 M HCl. Precise positioning of the Pt microelectrode above the steel surface was carried out using a unique force sensor consisting of a strain gauge on a solid lever block, which shows a steep rise in the force curve at the moment of contact between the probe and sample surface. SECM images acquired in the initial stages of HCl immersion showed H_2 evolution occurring over the entire surface, but with distinct regions of varying activity. Postexperimental analysis of the surface by electron backscattering deflection (EBSD) revealed these regions to be associated with particular grain directions, with the highest activity coming

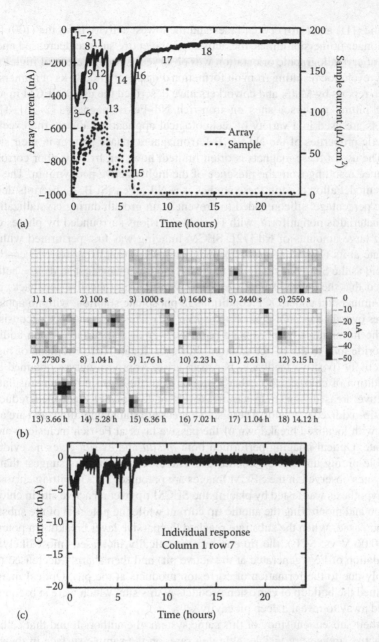

FIGURE 3.8 MEAM experiment on 304 SS in 10 mM KI held at 0.50 V (Ag/AgCl) and analyzed for 24 h. (a) shows the sample current and the total array current during the experiment. The total array current has numerical tags refering to the MEAM images in (b), extracted from the data file at important times during the experiment. The sample potential was stepped from 0 to 0.50 V at t = 25 s. One microelectrode current versus time plot for the entire experiment is also shown (c). (From Lister and Pinhero, *Anal Chem*, Vol. 77, (2005): p. 2601. With permission.)

from the (111) and (110) crystal faces and the lowest activity from the (100) face. With longer immersion times, the activity of the entire surface decreased and no effect of crystallographic orientation was observed, although areas of high localized H_2 evolution resulting from pit formation did appear after 80 ks of immersion.

Two reports by Malik and coworkers have described the use of SECM in identifying pitting precursor sites on iron-rich Nd–Fe–B magnets [72–73]. These magnets are used in a variety of technological applications due to the excellent magnetic properties of the $Nd_2Fe_{14}B$ ferromagnetic phase present in their structure. The use of these magnets is often limited, however, by their poor corrosion resistance resulting from the presence of the highly active neodymium. The particular model alloy used in these studies was $Nd_{13.5}Fe_{79.5}Si_1B_6$ (subscripts denote atomic percentage; silicon added to prevent grain growth during crystallization). This material is nonuniform, with Fe-rich inclusions surrounded by phases containing large amounts of Nd [72]. SECM imaging was first performed with the substrate at its OCP in neutral phosphate buffer solution using ferrocenecarboxylic acid as the SECM mediator in feedback mode. In their first study, the authors reported that the entire sample surface was insulating (negative feedback) after 5 h of immersion under OCP conditions in phosphate solution, with no apparent changes over an exposure time of up to 75 h [72]. This observation was consistent with the formation of a phosphate passive layer on the surface. Upon addition of chloride ions to the buffer solution, however, locally active regions of higher conductivity (positive feedback) began to arise. Approach curves obtained after the addition of chloride show that while most of the surface remains insulating, the active areas demonstrate various levels of kinetically controlled reduction of the tip-oxidized mediator. These active areas are stable with time, and are associated with localized breakdown of the passive layer at Fe-rich inclusions on the substrate. Optical imaging performed both *in situ* and *ex situ* gives no evidence of stable pitting under OCP conditions. Therefore, the authors suggest that the active sites observed in the SECM images are precursor sites to pitting corrosion. This hypothesis was tested by placing the SECM tip over an active site in chloride solution and monitoring the anodic tip current while the potential of the substrate was increased. When the substrate reached the passive layer breakdown potential ($E_b = 0.06$ V vs. SCE), the tip current dramatically increased (most likely due to oxidation of Fe^{2+} generated at the active pit) and then sharply decreased (presumably due to the formation of corrosion products at the pit). Optical imaging confirmed the buildup of corrosion products at this site, which were subsequently washed away to reveal a deep pit cavity.

In their subsequent study of this same system, the authors found that cathodically active areas were visible, although rare, on the sample surface in the early stages of immersion in the neutral phosphate buffer [73]. In fact, the entire surface showed noninsulating levels of kinetically controlled reduction of the tip-oxidized mediator. As immersion time increased beyond 7 h, however, the phosphate passive layer became fully formed and the entire surface was insulating—consistent with their previous report. Once again these originally active sites were attributed to Fe-rich inclusions based on separate OCP versus time experiments that showed the rate of formation of passive phosphate films to be much slower on pure iron than on the bulk material of the magnet. As in the previous study, these Fe-rich inclusions proved to be pitting precursors, as increasing the substrate potential after the addition of chloride generated pits at these sites. The authors suggest that chloride attack and subsequent pit formation on the Nd–Fe–B surface occurs preferentially at Fe-rich sites due to the relative stability of complexes of Fe–Cl over those of Nd–Cl.

In a final application, SECM has proven useful for the detection of corrosion processes occurring on model galvanic cells [74–75]. Simões and coworkers used SECM and SVET to detect corrosion reactions occurring at iron and zinc electrodes coupled together in chloride solution [74]. Small 1 mm × 1 mm samples of iron and zinc embedded in epoxy resin at a separation distance of 1 mm were electrically connected at the nonsolution side and immersed in naturally aerated 0.1 M NaCl. SECM line scan measurements in SG/TC mode were first taken separately on the unconnected electrodes, using a 10-μm Pt probe at potentials of +0.60 V (Ag/AgCl) to detect Fe^{2+} and −0.70 V to detect O_2. Dissolved Fe^{2+} ions were detected above the iron substrate at its OCP, but only after corrosion was induced by making a scratch on the iron surface. While the Fe^{2+} concentration peaked over the scratch, a comparatively broad minimum was observed in the O_2 signal, indicating that anodic dissolution of iron localized over the scratch was balanced by larger areas of cathodic O_2 consumption. A sharper minimum in the O_2 current was observed above the Zn substrate at its OCP, indicative of more localized cathodic activity on zinc than on iron. When the samples were connected to form the galvanic couple, a large drop in the O_2 current was observed above the iron surface, as would be expected given its cathodic role in the cell. SVET measurements clearly showed cathodic activity at the iron electrode and anodic activity at the sacrificial zinc electrode.

The authors demonstrated that Zn^{2+} could be detected with the SECM by cathodic polarization of the Pt probe at −1.15 V and anodic polarization of the Zn substrate at −0.83 V. This scheme, however, necessitated removal of the deposited zinc from the probe after each run, thereby preventing a systematic SECM investigation of Zn oxidation. An alternative approach to Zn^{2+} detection was employed by Tada and coworkers who examined the spatial distribution of Zn^{2+} during corrosion of a Zn/steel galvanic couple using a 0.5-mm diameter zinc disk electrode polarized at −1.15 V (Ag/AgCl) as an amperometric sensor for Zn^{2+} detection [75]. The sample consisted of a 3-μm thick layer of zinc electrodeposited at the center of a coupon of mild steel, resulting in a 1:12 area ratio of zinc to steel. Immersion of the sample in 0.01 M NaCl led to corrosion of the Zn layer, with complete disappearance observed at 270 min. During corrosion the spatial concentration of Zn^{2+} was measured by scanning the sensor electrode over the sample in the horizontal and vertical directions. As expected, the Zn^{2+} concentration over the zinc layer increased in the vertical direction with increasing immersion time. The horizontal profile, however, showed much less change with time, with an apparent barrier to Zn^{2+} diffusion existing approximately 5 mm from the edge of the zinc layer. This observation was explained as the result of a pH gradient that affects the solubility of zinc corrosion products. Hydrolysis of Zn^{2+} above the zinc layer proceeds according to the reaction:

$$Zn^{2+} + H_2O \rightarrow ZnOH^+ + H^+$$

This reaction leads to a locally low pH that prevents precipitation. On the steel, however, the generation of OH^- through reduction of O_2 produces a high pH that precipitates zinc as $Zn(OH)_2$. Visual analysis of the sample after complete dissolution of the zinc revealed the presence of corrosion products on the majority of the steel surface, but no corrosion products in an area including the location of the zinc and extending 5–8 mm on either side.

3.4.2 ALUMINUM AND ITS ALLOYS

Much less has been published regarding SECM investigations of corrosion processes on aluminum than on iron [51,76–83]. The earliest work on aluminum is a brief report by Wipf who used the previously mentioned amalgamated gold microelectrode to generate localized chloride ions next to a pure aluminum substrate through the reduction of trichloroacetic acid at the microelectrode tip [51]. A wildly fluctuating substrate current was observed approximately 100 s after introduction of chloride ions, while large pits and bubbles began to form on the Al surface. The author notes, however, that while the pit formation appears linked to the chloride ions generated at the microelectrode tip, it is unclear whether the pits were truly triggered by the chloride ions since the corrosion rate of aluminum is high at pH of 2.4 used in the experiment.

In a more recent study from the same laboratory, the localized cathodic activity of an aluminum-based metal-matrix composite (MMC) was investigated [77]. These materials are attractive in a number of industrial applications due to their low density and enhanced mechanical properties. The particular substrate used in this investigation was an Al/Si(13.5%)/Mg(9%) alloy matrix with reinforcing silicon carbide particles (SiC_p). There is apparent controversy in the literature regarding the corrosion behavior of this material, including conflicting reports as to the location of the cathodic reaction sites. Díaz-Ballote and coworkers, therefore, used SECM imaging and approach curves to identify the cathodically active areas of this substrate at its OCP [77]. SECM images were acquired using ferrocenemethanol as the mediator in feedback mode in both corrosive NaCl and less aggressive borate buffer solutions at the same pH of 6.8. Figure 3.9 shows both optical (a) and SECM (b) images of the SiC_p/Al composite in 0.1 M NaCl. The area labeled 1 in the optical image is a SiC particle, while area 2 is the aluminum alloy matrix. It can be clearly seen that the areas of highest tip feedback current in the SECM image correspond to SiC particles in the optical image, indicating that these particles reduce the tip-oxidized mediator. Similar images were obtained in borate buffer. Separate imaging experiments carried out in SG/TC mode using oxygen as the mediator confirm the cathodic behavior of the SiC particles, as

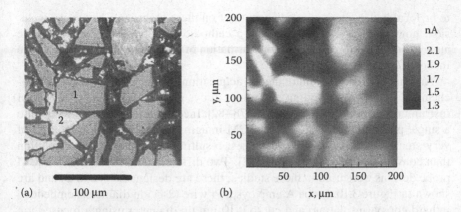

FIGURE 3.9 (a) Optical and (b) SECM images of the SiC/Al composite. For the SECM image, the substrate-tip separation was less than 5 μm using a 10-μm diameter Au tip biased at 600 mV vs. Ag/AgCl. The SECM image was recorded in an aqueous solution of 0.1 M NaCl and 2 mM ferrocenemethanol (pH 6.8). The substrate was at OCP (−650 mV vs. Ag/AgCl). (From Díaz-Ballote et al., *J Electrochem Soc*, Vol. 151, (2004): p. B299. With permission.)

the cathodic tip current resulting from oxygen reduction was shown to decrease over the SiC sites, indicating that oxygen is consumed as these sites act as local cathodes for oxygen reduction. As can be seen in Figure 3.9, not all SiC particles showed cathodic activity, so a series of probe approach curves were measured across the substrate to examine the electron-transfer kinetics as a function of position. The results demonstrated that electron-transfer kinetics varied greatly, even across a single active SiC site, and the authors speculate that these variations result from slow passivation of the active sites caused by corrosion products from the Al matrix.

There has been much interest in developing the methods for simultaneously acquiring both electrochemical and topographical information from a substrate [84–86,78], and two groups have applied these efforts toward the studies of corrosion on aluminum alloys. Büchler and coworkers applied a shear-force feedback technique used in NSOM to control the tip-to-sample distance of the SECM scanning probe and thereby obtain both topographical and electrochemical images [76]. An insulated tungsten wire with an exposed tip for SECM measurements was mounted onto one arm of a tuning fork attached to a drive piezo, which causes the tuning fork to vibrate at its resonant frequency [84]. As the probe approaches the substrate surface its movement is damped by viscous drag, and this sensing of the surface is used as feedback to control the tip-to-sample distance. Scanning the probe in the x-y plane therefore produces a topographic image of the surface. This instrument was used to investigate the cathodic polarization behavior of Al 2024 using ferricyanide as the SECM mediator in SG/TC mode [76]. Topographical images readily detected intermetallic s-phase (Al-Cu-Mg) inclusions, and SECM measurements showed the cathodic current to be localized at sites either adjacent to or on these inclusions. Not every inclusion, however, was found to support a cathodic reaction—an observation the authors suggest may result from the formation of an insulating oxide film at the inclusion/electrolyte interface.

Another approach to simultaneous acquisition of topographical and electrochemical images of corroding aluminum alloys is a combination AFM/SECM instrument developed by Davoodi [87,78–82]. Incorporation of the AFM tip into a single probe for both AFM and SECM imaging allows for precise control of very small SECM tip-sample distances, resulting in higher lateral resolution than conventional SECM imaging [78]. Two different dual-mode AFM/SECM probe designs were used in these studies; these are designated A and B, and are shown in Figure 3.10. Probe A employs a Pt wire (2–5 μm diameter) embedded in hard epoxy (insulating) and cut to a 10-μm tip diameter using a focused ion beam (FIB) to expose the Pt disk used as the SECM microelectrode. A small tip is then built out of epoxy next to the microelectrode with FIB, and this tip is used for AFM imaging. The entire probe is mounted at the end of an L-shaped cantilever, the top of which is flattened and coated with gold for laser reflection

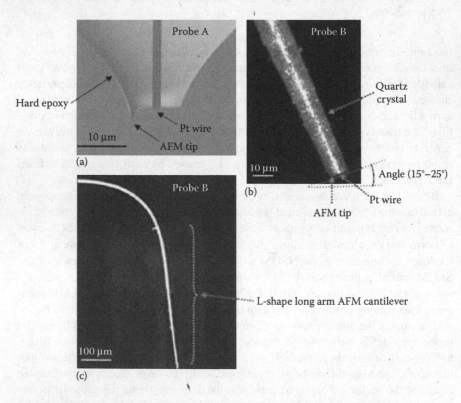

FIGURE 3.10 Dual-mode AFM/SECM probes. (a) SEM micrograph of probe A, (b) confocal laser scanning micrograph of probe B, and (c) long arm of probe B. (From Davoodi, et al., *J Electrochem Soc*, Vol. 155, (2008): p. C474. With permission.)

in AFM feedback; this top portion stays above the solution during imaging. Alternatively, the design in probe B was developed to increase SECM resolution and increase probe stability and flexibility [87]. Probe B consists of a Pt wire embedded in quartz and polished to expose the Pt surface with an inclination angle of 15°–25°. The end of the quartz tip is used as the AFM probe, and the exposed Pt disk serves as the SECM microelectrode. This probe was also connected to an L-shaped cantilever coated with gold (Figure 3.10c). With both designs, AFM and SECM images were obtained by first performing a normal AFM line scan in the x-direction to obtain a topographic (height) profile, and then raising the probe to a desired height and repeating the line scan in SECM mode for the current profile. The probe was then moved a short distance in the y-direction and the process was repeated [87]. Testing each probe design with well-characterized calibration samples showed that probe A produces a better AFM topography image, while a higher resolution SECM current image is produced by probe B. AFM images produced with either dual-mode probe, however, are of lower quality than those produced with a traditional AFM instrument [87].

Initial corrosion-related studies with the AFM/SECM compared concurrent topographic and electrochemical current images obtained on aluminum alloys in chloride/iodide solution at various anodization potentials [78–80]. At small anodization potentials, dissolution sites could be identified with the SECM in the initial stages of corrosion before any change in topography was evident with the AFM. At higher anodization potentials, features became evident in images obtained with both techniques. Figure 3.11 shows AFM (a) and SECM (b) images obtained with probe design A for AA3003 in chloride/iodide solution at a potential in the passive region of the alloy [78]. The SECM tip potential (+750 mV vs. Ag/AgCl) was set to oxidize the iodide to tri-iodide ($3I^- \rightarrow I_3^- + 2e^-$), thereby establishing a diffusion-limited anodic current. An increase in this current (lighter areas in Figure 3.11b) is attributed to dissolution of the aluminum substrate. While the description of this mechanism is somewhat unclear, it is presumed that corrosion of the surface creates holes in the passive oxide layer, resulting in exposure of the underlying bare aluminum to the solution. Tri-iodide generated at the SECM microelectrode then reacts directly with the exposed aluminum and is reduced back to iodide ($2Al + 3I_3^- \rightarrow 2Al^{3+} + 9I^-$, $E° = 2.212$ V), which diffuses back to the microelectrode where it is reoxidized generating positive feedback. Figure 3.11 identifies two active regions on the alloy with both AFM and SECM. The higher areas in the AFM image (sites A and C) are attributed to deposition of aluminum oxyhydroxide corrosion products, and the high current areas (sites B and D) to active corrosion processes occurring at these sites. Interestingly, the electrochemical activity at site B appears to be between the deposition areas in the ring-like structure of site A, indicating possible dissolution of Al at the boundary of a large intermetallic particle (IMP). Electrochemical activity at site D, on the other hand, occurs at the same location as the deposition in site C, and is likely a result of pitting corrosion in the aluminum matrix [78].

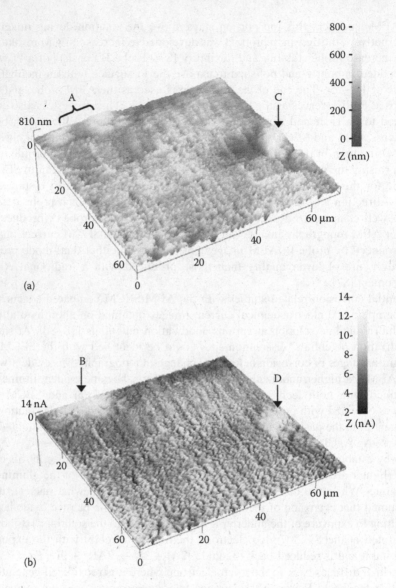

FIGURE 3.11 AFM/SECM images of AA3003 in 10 mM NaCl + 5 mM KI at 100 mV anodic polarization, and tip at +750 mV vs. Ag/AgCl. (a) Topography and (b) electrochemical activity map. (From Davoodi et al., *Electrochem Solid-State Lett*, Vol. 8, (2005): p. B21. With permission.)

The combined AFM/SECM was used in a more detailed investigation of the mechanism of localized corrosion of the Al–Mn alloy EN-AW 3003 [81–82]. Both cathodic and anodic activities were observed and correlated with topographic AFM images. Figure 3.12 shows concurrent AFM and SECM images, obtained with probe B design (Figure 3.10), for a cathodically polarized alloy

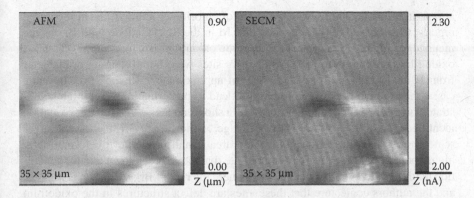

FIGURE 3.12 Concurrent AFM and SECM images of EN AW-3003 alloy exposed to 10 mM NaCl + 2 mM $K_4Fe(CN)_6$ as a mediator, with 200 mV cathodic polarization. (From Davoodi et al., *J Electrochem Soc*, Vol. 155, (2008): p. C138. With permission.)

sample in chloride solution with ferrocyanide as the SECM mediator, presumably in TG/SC mode [82]. The size and shape of the raised regions in the AFM image are consistent with large IMPs, and the corresponding SECM image shows that some of these particles show high cathodic activity. The difference in cathodic activity among IMPs is likely due to differences in composition [81]. As the polarization potential was increased to the passive region of the substrate, anodically active sites were identified, although AFM images still indicated a smooth surface [82]. A further increase in the polarization potential to the region of passivity breakdown resulted in both an increase in the local anodic dissolution current and topography changes due to early stages of pit formation, although little correlation was found between the areas of high SECM current and the location of pits identified by AFM. It is interesting to note that high-resolution SECM images of large IMPs under these conditions give evidence that anodic dissolution occurs at the edges of these particles—an observation consistent with accompanying SKP force microscopy measurements, which show a Volta potential maximum (most noble) on the IMPs and a minimum (most active) at the boundary between the IMP and the alloy matrix [81].

3.4.3 OTHER METALS

One of the earliest reports on the use of SECM in corrosion studies was by Gilbert and coworkers, who introduced SECM to the biomaterials research community by mapping the oxygen reduction reaction and detecting substrate dissolution products on Ag–Hg dental amalgam crystals and Co–Cr–Mo alloy beads used in hip prostheses [88]. Much of the early published work that followed was related to SECM investigations of corrosion-related processes on titanium substrates [89–96]. These studies focused primarily on the identification of precursor sites for pitting corrosion on native and anodically grown TiO_2 passive films, particularly in bromide solutions since bromide is known to cause pitting on titanium at potentials significantly lower than that of chloride or iodide [90]. In the initial reports, Casillas and coworkers used SECM in the SG/TC mode with Br^- as the mediator to identify active sites for bromide oxidation (Br_2 formation) on a 50 Å oxide film on titanium foil [89–90]. Active sites were found ranging in diameter from 10 to 100 µm, with a site density of approximately 30 sites/cm^2. In many cases, these active sites were observed to lead to pit formation when higher substrate potentials were applied. Figure 3.13a shows an active site for Br^- oxidation identified in a 400 × 400 $µm^2$ SECM image, while Figure 3.13b shows a larger scale photograph that includes the same area (small box) after pit formation at higher potential. These images demonstrate that oxide breakdown and pit generation tend to preferentially occur at sites where the rate of Br^- oxidation is highest, and the authors conjecture that these sites are defect structures in the oxide film that exhibit significantly higher conductivity than the average of the surface.

The localized behavior of bromide oxidation on TiO_2 demonstrated by the SECM results, taken in conjunction with microscopy and voltammetry data, suggest a mechanism for pit formation in which Ti–Br bonds form at the oxide/solution interface, leading to dissolution of the film and the growth of pits [90]. In this model, pit formation will occur first on the areas where the oxide film is thinnest (defect), and these are the same areas that would show the highest electrochemical activity (i.e., the highest rate of bromide oxidation) prior to pit formation. A subsequent SECM investigation looked for evidence that chemisorption of bromide ions at sites where the oxide layer is relatively thin may alter the electronic structure of these locations and thereby result in their high activity toward bromide oxidation [91]. However, it was found that sites that are highly active for oxidation of bromide are active for ferrocyanide oxidation as well, leading to the conclusion that the localized activity of Ti/TiO_2 reflects inherent properties of the oxide layer, rather than a specific interaction with the bromide ion. Garfias-Mesias and coworkers moved from the titanium foil samples used in the previously described investigations to an SECM examination of bromide and ferrocyanide oxidation on polycrystalline titanium plates [96]. They also found that both mediators are oxidized at the same localized sites, and that these locations were usually found to be precursor sites for pitting corrosion at higher potentials in bromide solution. Unlike the previous studies, however, SEM/EDX analysis showed these active sites to be associated with particles containing Al and Si, and not simply defects in the oxide film.

A thorough kinetic study of bromide oxidation on polycrystalline titanium was conducted by Basame and White [95]. The surfaces of five electrodes (0.079 cm^2 area) were first scanned in their entirety by SECM at low resolution to identify all active sites, resulting in an average site density of ~180 sites/cm^2. This value was six times larger than that found on unpolished Ti foil electrodes—an observation attributed to the polishing process on the Ti disk resulting in less uniform film thickness. The current generated at an individual active site (i_{site}) is given by Equation 3.8:

$$i_{site} = 4nFDC_s a \qquad (3.8)$$

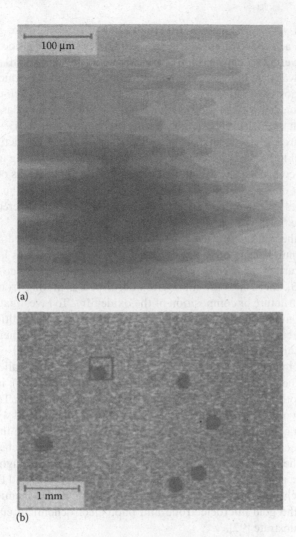

FIGURE 3.13   (a) SECM image of a precursor site for corrosion pitting obtained at 1.5 V in a 1 M KBr/0.05 M H$_2$SO$_4$ solution. (b) Optical photograph showing the position of oxide breakdown after the potential was increased to 3.0 V. The box outlined in part (b) corresponds to the area imaged by SECM in part (a). (From Casillas et al., *J Electrochem Soc*, Vol. 140, (1993): p. L142. With permission.)

where a is the site radius, and D and C_s are the diffusivity and surface concentration, respectively, of electrogenerated Br$_2$. The concentration profile at the center of the site ($r = 0$) at a distance z above the surface is given by Equation 3.9:

$$C(r = 0, z) = \frac{2C_s}{\pi} \tan^{-1}\left(\frac{a}{z}\right) \tag{3.9}$$

Probe current versus distance (z direction) curves were therefore generated at $r = 0$ for each active site on three separate electrodes, and the probe current values were converted to concentration to generate a plot of $C(r = 0, z)$ versus z. Equation 3.9 was then fit to this plot to give values for C_s and a, which were then used in Equation 3.8 to give i_{site}. When the total current from each of the active sites was summed, it was determined that 60%–70% of the total measured current passes through the active sites, yet these sites account for only 0.01%–0.1% of the total surface area. The current densities widely varied among the active sites, and all were found to be far below the mass-transport limit, indicating a wide range of kinetically controlled electron transfer rates. Interestingly, it was observed that higher current densities tend to come from smaller active sites—an observation that the authors speculate may arise from smaller sites being covered by thinner layers of oxide.

While all the work described above points to a high rate of redox activity at a very small number of active sites on the TiO$_2$ surface, little is known about the actual nature of these sites. While one report did find evidence of chemical impurities [96], these sites are most often described as arising from variations in either the structure or composition of the oxide film. To investigate the nature of these sites more closely, Smyrl and coworkers combined traditional SECM with photoelectrochemical microscopy (PEM), which uses either a scanned laser spot or an optical fiber to stimulate photoelectrochemical processes on a semiconductor surface [92–94]. The combined technique, called scanning photoelectrochemical and electrochemical microscopy (SPECM), is capable of performing both SECM and PEM measurements simultaneously. The probe for SPECM, shown in Figure 3.14, consists of an optical fiber coated with gold and encased in an outer polymer film. The fiber is cleaved to expose the gold ring, which then becomes the microelectrode for performing electrochemical reactions, while the substrate surface is illuminated with light through the optical fiber core. The probe is rastered over the substrate, and SECM and PEM images are acquired either sequentially or concurrently, with electrochemical currents measured at the gold microelectrode and photoelectrochemical currents measured at the substrate [92].

The SPECM was used to examine pitting precursor sites on Ti/TiO$_2$ in bromide solution [92–94]. Results are presented in Figure 3.15 showing images of both electrochemical and photoelectrochemical currents acquired under a variety of conditions [94]. Figure 3.15a was acquired in traditional SECM fashion (no illumination) with a carbon fiber microelectrode to identify an active site for bromide oxidation; the potential of the substrate was +1.0 V to oxidize Br$^-$ to Br$_2$, and the potential of the probe was +0.6 V to reduce Br$_2$ back to Br$^-$ (SG/TC mode). The carbon fiber was then replaced with the SPECM electrode, and the same area was scanned again in SECM SG/TC mode using the Au ring microelectrode, with the resulting image shown in Figure 3.15b. The same active spot is again visible; however, it is slightly shifted due to uncertainty in repositioning the probe. The image appears elongated as well, most likely a result of stirring caused by the larger optical fiber probe.

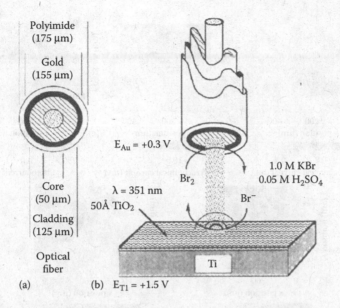

Polyimide
(175 μm)

Gold
(155 μm)

$E_{Au} = +0.3\text{ V}$

Core
(50 μm)

$\lambda = 351$ nm
50Å TiO$_2$

Br_2

1.0 M KBr
0.05 M H$_2$SO$_4$

Br$^-$

Cladding
(125 μm)

Optical
fiber

Ti

(a)

(b) $E_{Ti} = +1.5\text{ V}$

FIGURE 3.14 (a) Idealized diagram of the optical microelectrode probe used to perform SPECM, top view. (b) Side view of SPECM experiment illustrating typical experimental conditions and the major reactions that occur. (From James et al., *J Electrochem Soc*, Vol. 143, (1996): p. 3853. With permission.)

Figure 3.15c shows an image of the substrate current (photocurrent) that results when the rastering optical fiber illuminates the sample with 351 nm light from an argon ion laser. In this case, the potential of the substrate has been lowered to +0.5 V, so no electrochemical oxidation of Br$^-$ is possible at the substrate surface. The observed photocurrent (anodic) is therefore attributed to the promotion of electrons from the valence band to the conduction band within the semiconducting oxide layer, and the subsequent oxidation of bromide by the promoted electron. Of particular note in Figure 3.15c is the observation that the photocurrent is lowest at the site that is most electrochemically active for bromide oxidation. Figures 3.15d and e show SECM and PEM currents obtained concurrently, with both showing the position of the active site. The images are quite similar to those acquired sequentially (b and c). Once again, the photocurrent observed at the substrate was lowest at the active site. While SECM alone cannot determine if the activity of these sites is due to defects in the oxide or to a reduced thickness of the oxide, the addition of the PEM allows for more insight. A lower photocurrent on the active sites is consistent with a thinner oxide layer at these locations, whereas defects in the oxide would likely be more highly doped and result in a higher photocurrent [93]. The combination of the SECM and PEM, therefore, provides complementary information that could not be obtained with either technique individually.

Moving away from titanium, SECM has been applied to corrosion-related studies of nickel-based alloys. Paik and Alkire used iodide as a mediator in chloride

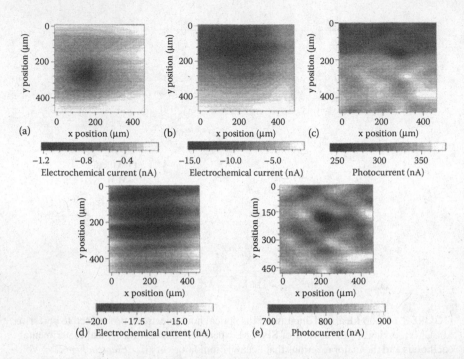

FIGURE 3.15 Gray scale images generated by performing SECM, PEM, and SPECM over a pitting precursor site on titanium in 0.05 M H_2SO_4 + 1.0 M KBr. The carbon fiber was biased at +0.6 V (SCE) and the Au ring at +0.3 V (SCE). (a) SECM via carbon fiber, Ti biased at +1.0 V (SCE). (b) SECM via optical fiber, Ti biased at +1.0 V. (c) PEM via optical fiber, Ti biased at +0.5 (SCE). (d) SECM image obtained during SPECM via optical fiber, Ti biased at +1.0 V (SCE). (e) PEM image obtained during SPECM via optical fiber, Ti biased at +1.0 V (SCE). (From James et al., *J Electrochem Soc*, Vol. 143, (1996): p. 3853. With permission.)

solution to detect dissolving sulfur-containing species near sulfide inclusions on Ni200 [97]. Line scans showed sulfide inclusion dissolution to begin well below the pitting potential and continue for tens of minutes after the pitting potential had been surpassed.

Lister and coworkers used SECM to examine the corrosion behavior of a NiCrMoGd alloy in iodide solution [98]. This alloy is of interest due to its potential as a neutron-absorbing material for application in high-level nuclear waste storage. NiCrMo alloys are quite corrosion-resistant, and gadolinium is added (2.4% in this case) because of its neutron-absorbing ability. Gadolinium, however, is highly reactive, and its addition to the base material decreases its corrosion resistance. A two-phase alloy forms, with a Ni_5Gd (gadolinide) secondary phase, and it is this phase that is known to dissolve rapidly at potentials above the OCP. These secondary phase particles are 2–5 μm in diameter and are generally distributed into regions of lower density (2.6×10^5 particles/cm²) and higher density (5.6×10^5 particles/cm²) [98]. Figure 3.16 shows a series of

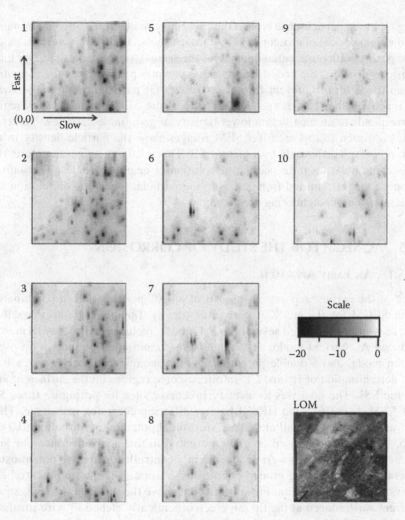

FIGURE 3.16 Consecutively acquired SECM micrographs. The sample was biased at 600 mV about 5 min prior to beginning acquisition. Images are 3 mm × 3 mm and the z-scale 0 to −20 nA. LOM images (four stitched images) were taken at 100×, representative of the area imaged. The LOM was taken following reaction for approximately 24 h at 600 mV. (From Lister et al., *J Electroanal Chem*, Vol. 579, (2005): p. 291. With permission.)

10 sequential 3 mm × 3 mm SECM images of a NiCrMoGd alloy sample biased at 600 mV in 10 mM KI [98]. The probe potential was set at −200 mV to detect corrosion products and/or I_2 formed by oxidation of I^- at active sites on the surface. Each image required 39 min to acquire, and the total acquisition time was approximately 6.5 h. Examination of the images in Figure 3.16 shows a number of corrosion sites that form and decay over the course of the experiment. These currents are believed to be generated at secondary phase sites where dissolution

of gadolinide particles has broken the protective oxide film. Close examination of the light optical microscopy (LOM) image shown in the figure reveals extensive pitting, with preferential removal of the secondary phase particles. As these particles dissolve, the underlying metal apparently passivates by the formation of an oxide film to prevent further corrosion. Of particular note is the swath of inactivity that cuts across the left side of the SECM images. This region corresponds to an area with a lower density of gadolinide particles. No active spots are seen in this area, yet SEM images show the particle density in the two regions to differ only by a factor of 2. This is an indication, therefore, that dissolving particles influence the dissolution of nearby particles. The authors propose that H^+ emitted from a dissolving particle affects the dissolution and repassivation of neighboring sites.

3.5   AC-SECM FOR THE STUDY OF CORROSION

3.5.1   An Early Approach

One of the earliest reports on the use of an AC perturbation in combination with SECM for the study of corrosion was by Tanabe and coworkers [60]. These workers employed several SECM modes, including what they termed an induced AC current mode, an ion selective potentiometric mode, a tip oscillation mode, and a double potential step chronoamperometric mode, all for the determination of H^+ and Cl^- in microscopic regions on the surface of austenitic SSs. The goal was to identify precursor sites for pitting on three SSs (SUS304, SUS316L, and HR8C) having differing corrosion resistance. Their AC mode applied a small amplitude sinusoidal potential perturbation (0.3 mV p-p, 1 kHz) superimposed on a DC potential to the steel substrate. The substrate and tip potentials were independently controlled using two potentiostats operated under floating ground conditions. Maintaining the tip potential at 0 V (vs. Ag/AgCl) and scanning at ca. 20 μm above the substrate surface, an AC current was induced at the tip (an electrochemically etched Pt wire insulated by Parylene C), detected by a lock-in amplifier, and used as the imaging signal. The increase in induced AC current was shown to correlate with a decrease in pH (as determined by potentiometric SECM) and a pit image obtained by this technique was in good agreement with that obtained by SEM. The pH over the center of a target pit was found to be as low as 1.6, compared to a bulk pH of 5.4 (in 0.6 M NaCl).

3.5.2   More Recent Work

More recent reports on AC-SECM have involved application of the AC perturbation to the tip electrode (rather than to the substrate) and a recent review of this approach has appeared [8]. Initially, this approach was used to facilitate tip positioning for potentiometric and amperometric SECM, making use of the observation that the solution resistance between the tip and a remote counter electrode depended on the tip–substrate distance (at appropriate frequencies and for electrolytes of sufficiently low conductivity) [99]. Subsequently, it was found that topographic as well as local electrochemical information could be obtained from this approach [8], first demonstrated by Katemann and coworkers, who imaged an array of band microelectrodes in redox mediator-free solutions of low conductivity [100]. A potential advantage of AC-SECM for corrosion studies is that a mediator is not required, making AC-SECM applicable to systems that might be sensitive to the potential (or other complications) imposed by such dissolved redox species. Several recent applications of AC-SECM to corrosion science have been reported [101–104,56,105–108], though many of these studies are on model systems used to demonstrate the proof of concept. One of these reports describes the use of AC-SECM to visualize defects (considered to be precursors for localized corrosion) in lacquered tinplates and is considered in more detail in the Section 3.6 on coatings [102].

AC-SECM has been used to visualize inhomogenities in the electrochemical activity of oxide-covered NiTi shape memory alloys [101,103]. Such alloys are used to manufacture orthopedic clamps, orthodontic wires, vascular stents, or staples, and corrosion causes the release of hypoallergenic or biotoxic Ni^{2+} and Cu^{2+} ions into the body [103]. Thus, these metal alloys are usually coated with a thin passivating film to enhance corrosion resistance. Using a 10-μm Pt disk microelectrode and an applied voltage perturbation of 150 mV peak–peak at 1 kHz in a 1 mM NaCl electrolyte (with the NiTi alloy at the OCP), images that revealed corroding spots on the surface of the NiTi substrate were obtained, attributed to a decrease in solution resistance between the substrate and the scanning tip [103]. The cause for this resistance decrease was not clear but was probably related to an increase in ionic strength within the gap between substrate and tip due to the release of dissolved metal species and/or to the breakdown of the oxide film. These spots were found to bear a high reactivity toward the onset of pitting corrosion, confirmed by polarizing the substrate to the pitting potential.

Ruhlig and Schuhmann combined anodic underpotential stripping voltammetry with AC-SECM to probe Cu^{2+}-ion release from metallic copper deposits on metal surfaces as a model for such release from copper containing alloys [104]. Two types of model substrates were used, one consisting of an array of 36 individually addressable gold microdisk electrodes (each 50 μm in diameter), some of which were coated with metallic Cu by electrodeposition, and the others consisting of metallic copper coated with a layer of nail polish through which holes were pierced using a small syringe needle (100–200 μm diameter). The substrates were maintained at the OCP throughout the experiments. AC-SECM was used to first detect the active sites of enhanced electrochemical activity on the surface of the substrates, employing a 15-μm gold-coated Pt microelectrode as the SECM tip (at 10–30 kHz with a 200 mV rms perturbation in 1 mM NaCl). Visualization and confirmation of Cu^{2+} release from these active sites were then accomplished using spatially resolved anodic underpotential stripping voltammetry (or stripping-mode SECM, SM-SECM) performed at the Au tip during SECM scanning. The AC-SECM and SM-SECM modes were synchronized using a communication module [104]. A detection limit of 0.15 nM Cu^{2+} was estimated for a 45 s accumulation. Although model substrates were used in this work, this approach could be useful for probing dealloying and Cu redistribution in alloys such as the Cu-containing aerospace aluminum alloy AA 2024-T3 where Cu redistribution plays an important role in the corrosion of this alloy. A recent study by our group has also demonstrated the power of SECM in combination with spatially resolved anodic stripping voltammetry for the study of Cu release from IMPs in AA 2024-T3 [109].

Another report from Ruhlig, Schuhmann, and coworkers [110] combined AC-SECM with SM-SECM to investigate cross sections of laser-fabricated welds between NiTi shape memory alloy and SS microwires of approximately 100 μm diameter (NiTi-SS, a heterogeneous joint) [110]. For comparison, SS–SS and NiTi–NiTi welds (homogeneous joints) were also examined. AC-SECM was used to reveal any variations in local interfacial chemical activity across the passivated weld and the base metals, and SM-SECM (underpotential deposition of Ni followed by anodic square wave stripping voltammetry) was used to examine Ni^{2+} release from the surfaces. The goal of this work was to understand the impact of laser welding on the surface passivation of the weld and the base metals in the heat-affected zone. The AC-SECM images revealed lower electrochemical activity above the weld for NiTi–SS than above the individual NiTi or SS wires (i.e., away from the weld). For the NiTi–NiTi and SS–SS substrates, no such differences in activity were observed above the weld compared to the parent wires. SM-SECM revealed higher Ni^{2+} concentrations above the NiTi wire, consistent with the AC-SECM results which showed this region to be less passivated. Energy dispersive elemental analysis in a scanning electron microscope revealed the coexistence of Ti and Cr in the mass weld, and it was suggested that the joint action of these two metals led to a more protective oxide and improved passivation in the weld region compared to the individual action of these metals in passivating their respective wires. The authors conclude that the AC- and SM-modes of SECM imaging were sufficiently sensitive to detect and visualize the impact of the changed surface passivation upon laser welding [110].

Gabrielli and coworkers combined SECM and an electrochemical quartz crystal microbalance (EQCM) to study the local breakdown of the passive layer and the resulting pitting corrosion of iron [56]. A twin electrode SECM probe was developed consisting of a 60-μm diameter Pt electrode and a 160-μm diameter Ag/AgCl electrode. The Pt electrode was used for an AC-SECM mode approach to the iron substrate, which was either a bulk iron sample or a 2–3 μm thick Fe film deposited by sputtering onto the quartz crystal of the EQCM. The Ag/AgCl electrode was then used to generate a local Cl^- concentration to initiate passive layer breakdown and pitting corrosion, which was followed by monitoring the substrate current (for both the bulk Fe and the Fe film substrates) or by the EQCM (for the Fe film substrate). The influences of the amount of Cl^- generated, the iron potential, the pH, and the tip-to-substrate distance were investigated. By initiating local pitting with the SECM, the mass loss over the entire EQCM electrode (Fe film) could be related to a local event and could be directly linked to the breakdown of the passive layer and to the dissolution of iron to Fe(II) in solution [56].

Pits were found to propagate faster in less basic solution than in highly basic solutions, and pit propagation was much faster if the SECM tip was withdrawn after the release of the Cl$^-$ ions. One final observation from this work is worth noting. The approach of the SECM tip to the Fe film substrate could be monitored in two ways, by the AC-SECM approach curve (i.e., the increase in electrolyte resistance as the tip approaches the substrate) and by the change in ECQM frequency as the tip contacts the substrate. These two approaches agreed to within 300 nm.

Freire and coworkers examined the electrochemical behavior of a low nickel SS (AISI 204Cu) in alkaline media and in the presence of chloride [106]. This SS is used in reinforced concrete subject to harsh environmental conditions, and the exposure conditions simulate those encountered in a marine environment. Cyclic voltammetry, AC-SECM, and critical pitting temperature measurements were used in conjunction with SEM and energy dispersive x-ray analysis (EDX) to compare the behavior of this steel with that of AISI 304 and AISI 434 SSs. Here we focus on the AC-SECM results. Localized corrosion was induced in an AISI 204Cu sample by immersing it in 0.1 M NaOH and 0.5 M NaCl and polarizing at +0.35 V (vs. SCE) for various times up to 100 min. A 500 µm × 500 µm area was scanned using a 10-µm Pt tip with a 100-kHz perturbation (the voltage amplitude was not specified). In spite of the high ionic strength of the immersion solution, well-defined current peaks in AC-SECM image were observed, attributing to increases in electrolyte conductivity due to pitting corrosion. A decrease in activity with time and even passivation of some sites were observed. From SEM/ EDX results, the authors conclude that the pitting originates at MnS inclusions and that repassivation occurs due to nodular deposits of metallic copper and copper cementation at the active corrosion sites [106].

3.5.3 CONSTANT DISTANCE IMAGING

The AC-SECM signal is sensitive not only to the tip-to-substrate distance but also to the electrochemical nature of the underlying substrate [8], and this limits its applicability as a distance control on electrochemically heterogeneous samples such as active/passive metal substrates. To overcome this difficulty, Eckhard and coworkers implemented piezo–piezo shear force detection to maintain constant tip-to-substrate distance and obtain AC-SECM images of electrochemical activity unimpaired by topography convolution [105]. Using a 5-µm diameter Pt sealed-in-glass SECM tip controlled by a piezo positioning cube, one piezo element was used to excite the tip at its resonance frequency and a second piezo transduced the readout of the vibration amplitude, which changed as a function of tip-to-substrate distance due to hydrodynamic shear forces near the substrate surface. A feedback loop was used to maintain constant tip vibration amplitude (thus, constant distance), and the z-displacement required was recorded to provide a topographical image of the substrate surface. The substrate was a 1 cm^2 piece of 316 Ti SS embedded in PVC.

Figure 3.17 shows images obtained in 1 mM KCl before (left column) and after (right column) induced corrosion. The images in the left column of

Figure 3.17 show shear-force based topography (a) and false color representations of AC-SECM current (b), as well as phase shift (c), all appearing rather feature-less (aside from the tilt observed in the topography image). With the substrate as working electrode and the SECM tip as counter electrode, local corrosion was then induced by positioning the SECM tip near the center of the imaging area (ca. 1 µm above the surface) and slowly ramping the potential from the OCP (ca. −40 mV) to 1.1 V vs. Ag/AgCl (200 mV positive of the pitting potential), where it was

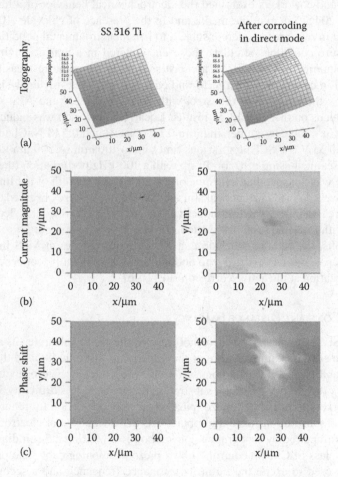

FIGURE 3.17 AC-SECM and simultaneous shear-force based topography measure-ments. (a) 3D representation of the topography, (b) 2D false colour representation of the local current magnitude, (c) 2D false color representation of the phase shift. (b) + (c) were mathematically corrected for a drift of the background. Left column: initial AC-SECM scan; right column: AC-SECM scan after initiation of localized corrosion in direct mode. Scans were carried out in constant distance with a piezo actuator frequency of 283.5 kHz. Electrochemical imaging was done with an excitation of f = 1 kHz, VPP = 250 mV in 1 mM KCl solution. (Reprinted from Eckhard et al., *Electrochem Commun*, Vol. 9, (2007): pp. 1793–1797, with permission from Elsevier.)

maintained for 15 min to ensure stable pit growth. The area was again scanned by AC-SECM, resulting in the images shown in the right column of Figure 3.17. Aside from some apparent roughening, the topography image is rather featureless at the resolution of the shear force measurement. However, the AC-SECM current image (Figure 3.17b, right) reveals a hot spot of 5–7 µm dimensions at the location where corrosion was induced. The AC-SECM phase shift image (Figure 3.17c, right) was even more pronounced and was somewhat larger in area, attributed to a stronger dependence of the phase shift on the ionic strength of the solution at the chosen frequency of 1 kHz [105]. It was concluded that the induced corroding pit continued to function as an ion point source (i.e., continued to actively corrode, increasing the ion concentration in its vicinity), even after the substrate was returned to the OCP for AC-SECM imaging. Improved spatial resolution should be possible by using electrodes of smaller dimension.

In a related approach to constant-distance imaging, Eckhard et al. described AC-SECM performed at an AFM tip with an integrated recessed ring microelectrode [111]. The capabilities of the approach were demonstrated by obtaining the topography and surface conductivity of Au/glass structures and of microelectrode arrays. Although no corrosion examples were given, this approach would be useful for corrosion studies where substrate surfaces often exhibit nonuniform topography.

3.5.4 OPTIMIZING CONTRAST

Most AC-SECM imaging has been performed at one fixed frequency of the AC voltage applied to the tip. However, the electrochemical contrast obtained with AC-SECM is highly dependent on the frequency chosen for the AC voltage [8]. Since the optimum frequency depends on such factors as the nature of the substrate, the ionic strength of the electrolyte, and the active diameter of the SECM tip, the frequency for optimum imaging contrast cannot be determined beforehand for unknown samples. To obtain data with optimum contrast and high resolution, Eckhard and coworkers describe an approach whereby the frequency is swept in a full spectrum (e.g., 50 frequencies between 100 Hz and 100 kHz) at each point in the scan plane, generating what they refer to as a 4D AC-SECM data set [107]. The potential applications to corrosion research were illustrated by imaging an actively corroding pit on 304 SS and by imaging local defects in a protective organic coating.

An example of the latter is shown in Figure 3.18 for low-carbon steel galvanized with tin and subsequently covered with a 9–15 µm layer of epoxyphenolic varnish (lacquered tin plate used in the food packaging industry). The polymer was deliberately damaged prior to the experiment, exposing the underlying metal and creating sites with increased electrochemical activity. The images in Figure 3.18a (left column) were obtained for a coating having a circular defect of about 220 µm in diameter (introduced by punching the polymer coating with a needle), whereas the images in Figure 3.18b (right column) were obtained for a coating having three parallel scratches introduced by a scalpel. The influence

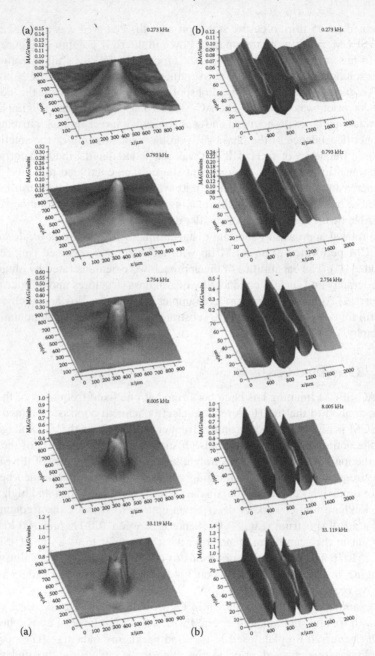

FIGURE 3.18 4D AC-SECM measurements on a lacquered tin plate; $d_{tip} = 25$ μm, $c = 1$ mM KClO4, $V_{pp} = 100$ mV, $f = 0.273$ kHz → 33.199 kHz in 27 logarithmic increments. From top to bottom: 0.273, 0.793, 2.754, 8.005, 33.199 kHz. (a) pinhole, scan area = 1000 μm × 1000 μm; (b) scan area = 2000 μm × 80 μm. (Reprinted from Eckhard et al., *Chem Eur J*, Vol. 14, (2008): pp. 3968–3976, Copyright Wiley-VCH Verlag GmbH & Co. KGaA. Reproduced with permission.)

of perturbation frequency on the images is readily apparent. Low frequencies lead to a significant broadening of the features, distorting the circular defect in Figure 3.18a and revealing only two of the three scratches in Figure 3.18b. This distortion was attributed to capacitive components that dominate the impedance at lower frequencies. Higher frequencies are clearly advantageous for this particular sample. Interestingly, the actively corroding pit on their 304 SS substrate was best imaged at lower frequencies, and the frequencies that led to the best electrochemical contrast in Figure 3.18 did not reveal the corrosion pit on the SS substrate at all. The ratio of tip capacitance to substrate capacitance (which in turn depends on the ratio of conducting to insulating areas of the substrate surface) apparently plays a key role in determining the optimum frequency. The AC perturbation frequency should be high in order to shortcut the capacitance of the tip, but at the same time it must not be so high as to shortcut the capacitance of large conducting substrate surfaces [107]. Thus, for unknown samples, the optimum frequency cannot be predicted, and the 4D AC-SECM approach provides a useful tool for obtaining optimal electrochemical contrast.

We conclude this section by mentioning a recent paper by Eckhard and coworkers describing another approach to optimizing contrast in AC-SECM [108]. Approach curves to two regions of the substrate having contrasting electrochemical activity (e.g., to a polymer coating and to exposed metal in a coating defect) are generated. Rather than plotting impedance magnitude |Z| as a function of tip-to-substrate distance as is conventionally done, the data are plotted in the impedance plane of im(Z) versus re(Z) at each approach distance (d). Such plots are generated for each of several frequencies (f). Three measures of imaging contrast are then computed at each d and at each f, defined as $\Delta|Z|$, $\Delta re(Z)$, or $\Delta im(Z)$, the differences Δ being taken between the two approach curves at corresponding d and f. The combination of d and f that leads to a maximum absolute value of one of the Δ's provides the optimum scanning conditions (d, f) and the optimum imaging variable (the corresponding |Z|, re(Z), or im(Z)). This approach was demonstrated by imaging a steel sheet covered with a zinc layer and an organic topcoat through which a scratch was made exposing the zinc layer. The best images were obtained using |Z| and re(Z), while the image using im(Z) was distinctly poorer. The authors attributed this observation to the fact that the scratch through the insulating polymer to the metal resulted in a drastic change in local surface conductivity. Thus, the imaging variables that are sensitive to resistive components of the impedance (namely, |Z| and re(Z)) displayed the highest contrast. A limitation of this method appears to be the need to know *a priori* two locations of contrasting electrochemical activity on the substrate surface.

3.6   SECM FOR PROBING COATING PROPERTIES

While SECM applications in corrosion have largely focused on reactions occurring at uncoated surfaces, several studies have investigated the properties of coatings used for corrosion control. Benzotriazole (BTAH), for example, is a well-known inhibitor for the corrosion of copper, yet several unanswered questions remain regarding the inhibition mechanism [112]. Mansikkamäki et al. have used SECM to investigate the formation of Cu(I)-BTA films on a copper substrate immersed in BTAH [112–114]. Using ferrocenemethanol as a mediator, probe approach curves were measured as a function of copper substrate potential and immersion time in BTAH. On the bare copper, approach curves indicate an ideally conducting substrate (as expected), but the surface becomes more insulating with immersion time as the BTAH film grows, thereby allowing the determination of the rate of film growth [112]. BTAH films were observed to grow much more rapidly at higher substrate potentials, leading to speculation that the presence of oxygen aids in film formation. This observation led to a follow-up investigation in which approach curves were acquired in controlled low-oxygen environments with the substrate at the OCP [114]. The rate of film formation decreased with decreasing oxygen content, and at the lowest oxygen concentration (<15 ppm) film growth was almost totally absent, indicating the important role oxygen plays in the mechanism of BTAH film formation.

A recent study by Walsh and coworkers used SECM to investigate the local electrochemical activity of thin coatings of Inconel 625 deposited on steel using high-velocity oxygen fuel (HVOF) thermal spraying [115]. Inconel 625 is a corrosion-resistant Ni–Cr–Mo alloy useful in extreme environments due to its strength and stability at high temperatures. HVOF is a deposition process that uses a supersonic jet to accelerate and heat small particles of feedstock, causing them to spread out and adhere to a surface upon impact. This process can significantly reduce the deposition times and increase the overall quality of the coating by increasing density and reducing porosity [115]. In this study, the electrochemical behaviors of HVOF thermal sprayed coatings of Iconel 625 were compared with the alloy in its wrought state and in a state prepared by sintering a powdered form of the alloy. SECM approach curves and images were obtained in the feedback mode using ferrocenemethanol as the mediator with the substrates at their OCP. The wrought state was found to behave as a pure insulator with no unique local features, while the sintered state behaved similarly except for large pores, which demonstrated high electrochemical activity. The thermal sprayed sample, on the other hand, displayed a high degree of heterogeneity in its behavior. All approach curves on this surface showed noninsulating activity, with the highest activity observed at small circular regions with diameters of 20–50 µm—dimensions similar to splat boundaries observed by SEM. It is believed that the overall increase in activity of the thermal sprayed sample results from the preferential oxidation of chromium that occurs during the spray process, causing the formation of Cr_2O_3-rich domains which are thinner (and thus more reactive) than the oxide coating on the wrought and sintered samples.

Polymeric coatings are widely used in the corrosion control of metals, therefore a growing amount of work has been performed utilizing the SECM to investigate both the integrity of these coatings in corrosive environments and the mechanism of corrosion that occurs at defects in these coatings. Souto et al. has used SECM in the feedback mode to examine the local integrity of paint coatings immersed in electrolyte solutions [116–118]. Galvanized steel samples coated with a chromate-containing primer and polyester topcoat were immersed in chloride and sulfate solutions containing ferrocenemethanol as the SECM mediator. The mediator was oxidized at the microelectrode tip as it scanned the surface, and any decrease in the magnitude of this oxidation current was interpreted as a decrease in the tip–substrate distance resulting from blister formation under the coating. Figure 3.19 follows the evolution of blister formation at various time intervals on a 100 μm × 100 μm area of a sample immersed in chloride solution [116]. It can be seen that while the surface was essentially featureless immediately after immersion, significant swelling due to the permeation of water and chloride ions through the coating was observed at both 21 min and 63 min. A similar procedure carried out in sulfate solution showed no blister formation even after 10 h of immersion time. Interestingly, a conventional EIS study of the chloride system by the same authors showed that more than 66 days were required to observe any decrease in impedance characteristics [119]; this result stands in contrast to the nearly immediate blistering detected by the SECM, thereby emphasizing the sensitivity of the SECM to local changes in coating performance.

More recently, a similar SECM investigation of this system by the same authors examined topographical changes in the polyester-coated steel samples when immersed in solutions of chloride, sulfate, and nitrate [118]. As in the previous report, only the chloride-immersed samples displayed signs of blistering, and these blisters were observed shortly after immersion. It was also reported, however, that the feedback current measured during imaging decreased steadily for samples in all three electrolytes with increasing immersion time. This was interpreted as a decrease in the tip–sample distance resulting from continuous swelling of the polymer caused by uptake of water into the coating. Finally, to address any questions regarding the role of the ferrocenemethanol in the observed blistering in chloride solution, similar negative-feedback imaging experiments were conducted using dissolved oxygen as the mediator. Surface roughening similar to that observed with ferrocenemethanol was detected, indicating that the mediator plays no role in blister formation.

The introduction of a defect in a polymeric coating often results in corrosion at the exposed underlying metal, and SECM has proven very useful in detecting localized products of corrosion reactions above such a defect. For example, as corrosion proceeds, there should be an observable decrease in the dissolved oxygen concentration above the defect site as oxygen is consumed in the cathodic half-reaction. Bastos et al. examined the effect of introducing a defect (scratch) into a polymeric coating on a steel sample immersed in aerated chloride solution [120]. The SECM tip was used in SG/TC mode to reduce O_2 and thereby monitor the relative oxygen concentration as a function of tip position. As the tip was scanned laterally 100 μm above the surface, a decrease in cathodic current was observed over the defect site, whereas no change in current was seen when the experiment was repeated 1,000 μm above the surface. This offered direct evidence of oxygen depletion in the solution directly in contact with the corrosion reaction. Similarly, the cathodic current increased when the tip was moved from 100 μm to 1,000 μm above the defect in the vertical

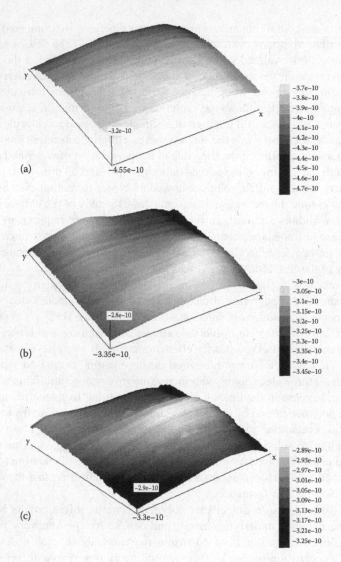

FIGURE 3.19 SECM images (100 μm × 100 μm) of painted steel sample immersed in 0.1 M KCl: (a) immediately after immersion, (b) 21 min after immersion, and (c) 63 min after immersion. (From Souto et al., *Corros Sci*, Vol. 46, (2004): p. 2621. With permission.)

direction, while no change in the current was seen when this procedure was repeated at a nondefective site on the coating.

In addition to the cathodic depletion of oxygen above a defect in a coating, Souto et al. also observed a corresponding enrichment of anodically generated Fe^{2+} above an exposed steel substrate [121]. In this work, a series of line scans across a 250-μm diameter defect in a polyurethane coating were performed, alternating between a potential to reduce (detect) O_2 and a potential to oxidize (detect)

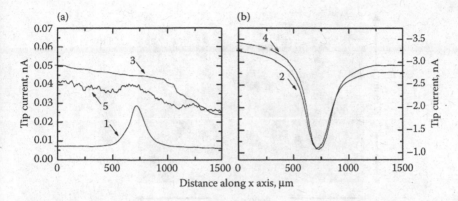

FIGURE 3.20 Line scans representing a single scan of the probe tip across the defect in the organic coating after immersion in 0.1 M KCl. The tip potentials were: (a) +0.60 V, and (b) −0.70 V vs. Ag/AgCl/KCl (saturated). Exposure times for each line scan: (1) 9, (2) 32, (3) 53, (4) 72, and (5) 128 min. (From Souto et al., *Corros Sci*, Vol. 47, (2005): p. 3312. With permission.)

Fe^{2+}. The results are shown in Figure 3.20 where scans 1, 3, and 5 show Fe^{2+} oxidation during line scans across the defect, and scans 2 and 4 show O_2 reduction. The enhanced Fe^{2+} oxidation signal in scan 3 is attributed to an overall increase in the Fe^{2+} concentration as corrosion proceeds, and the subsequent current decrease in scan 5 is believed to result from the precipitation of corrosion products, which hinders the release of metal ions into the solution. It is speculated that precipitation of iron-containing salts is caused by a local rise in pH due to OH^- production resulting from O_2 reduction.

A similar investigation by Völker and coworkers also detected anodically generated Fe^{2+} species at the SECM tip above the defects in steel coatings [122]. In this case, the samples consisted of tin/epoxy-coated steel (lacquered tinplate) and chromium-coated steel, both immersed in 10 mM H_2SO_4. Lacquered tinplate is used heavily in the food and beverage industry, so there is much interest in determining its corrosion characteristics. Katemann et al. applied AC-SECM (Sections 3.2.2 and 3.5) to the identification of defects in lacquered tinplates, which are presumed to be precursor sites for localized corrosion [102]. In AC-SECM the impedance modulus Z is dominated by the solution resistance R_{sol}, which in turn is strongly affected by the conductivity of the substrate. Therefore, as the tip is scanned in the x and y directions, a 2D map of the local conductivity is generated. Figure 3.21 shows AC-SECM images acquired from a lacquered tinplate sample into which several pinholes have been introduced as artificial defect sites. This image demonstrates the high spatial resolution of this technique, which has the added advantage of using no external redox mediator in solution.

While the anodic detection of Fe^{2+}-containing corrosion products demonstrated in several studies is relatively straightforward due to the ease of oxidation of these species at a noble metal electrode, the detection of other metal ions may be facilitated by the use of amalgam electrodes for cathodic detection. Janotta

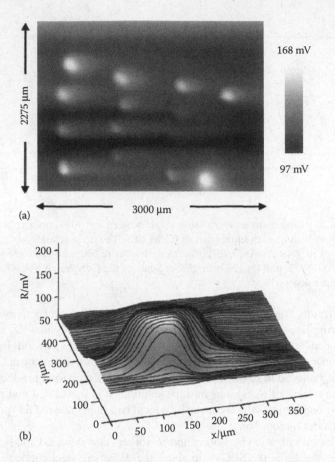

FIGURE 3.21 AC-SECM images of microscopically small pin holes in the coating of lacquered tinplates. (a) 2D image of an array of pin holes (overview). (b) 3D image of one pin hole at higher magnification. Images were obtained in 1 mM NaCl with AC-SECM in constant-height mode operating a 12.5-μm radius Pt microelectrode with a perturbation of 100 mV at 1 kHz frequency. (From Katemann et al., *Electrochim Acta*, Vol. 48, (2003): p. 1115. With permission.)

and coworkers employed anodic stripping voltammetry at a Au/Hg amalgam microelectrode for detection of Zn^{2+} in their examination of corrosion at ZnSe waveguides coated with diamond-like carbon (DLC) thin films [123]. This system is of interest due to the need for inert, IR-transparent coatings for corrosion protection of waveguides for ATR-based spectroscopic sensors in harsh chemical environments. Figure 3.22a shows an SECM image of a scratch in a DLC-coated ZnSe substrate imaged with a 25-μm Au/Hg amalgam microelectrode in feedback mode with O_2 as the SECM indicator in an air-saturated KCl solution. The increase in cathodic feedback current above the scratch is simply due to increased diffusion to the microelectrode resulting from a larger tip-sample separation distance. Figure 3.22b shows a single line scan across the scratch and the

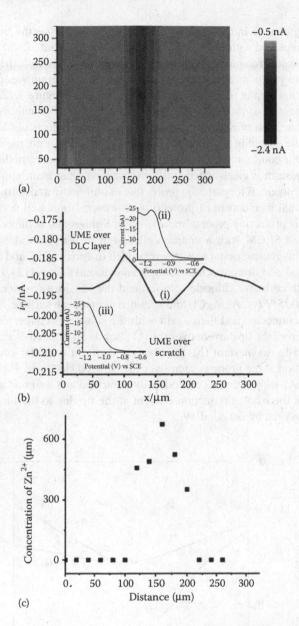

FIGURE 3.22 (a) SECM feedback mode image of a microscratch artificially created in the DLC coating. The image was recorded at an Au/Hg amalgam microelectrode with oxygen as the redox-active species in air-saturated KCl solution. (b,i) Change in the amperometric current across the DLC coating. (b,ii) Typical square-wave voltammogram at a position of $x = 160$ μm along the scan axis above the microscratch. (b,iii) Typical square-wave voltammogram along the scan axis recorded above the DLC coating. (c) Topography-corrected concentration profile of Zn^{2+} across the microscratch. (From Janotta et al., *Langmuir*, Vol. 20, (2004): p. 8634. With permission.)

corresponding change in the negative feedback current. After the line scan, the solution was replaced with H_2O_2 in an acetate buffer to initiate ZnSe dissolution, and a series of square-wave voltammograms were acquired with the microelectrode at various positions along the x axis. Each voltammogram was preceded by a 10-s deposition step at -1.4 V to reduce Zn^{2+} to $Zn(s)$. Figure 3.22c shows the resulting concentration profile of Zn^{2+} across the scratch. The authors conjecture that similar detection of Se(IV)-containing corrosion products could be carried out via cathodic stripping voltammetry at a copper amalgam microelectrode.

Just as zinc is commonly used as a sacrificial coating in the cathodic protection of iron, magnesium is electrochemically more active than aluminum and therefore offers the potential for cathodic protection of aluminum and aluminum alloys [124]. The actual mechanism of protection, however, is not well understood. To investigate this protection process more closely, Simões and collaborators in our laboratory used SECM with a unique cell design in which an aluminum alloy sample and a magnesium sample were embedded in epoxy resin, and then electrically connected and disconnected through an external switch [124,47]. The cell was filled with corrosive chloride solution and the SECM microelectrode probe was held at -0.75 V (vs. Ag/AgCl) and scanned across each surface, first with the substrates disconnected and then again with the substrates connected (coupled). Figure 3.23 shows the tip current over the Al alloy as an isolated electrode (a) and while coupled to magnesium (b). A relatively constant cathodic current due to oxygen reduction at the probe is seen over the isolated electrode. However, upon coupling, the Al alloy becomes cathodically polarized and a significant decrease is observed in the oxygen reduction current at the tip due to the increased consumption of oxygen by the Al alloy.

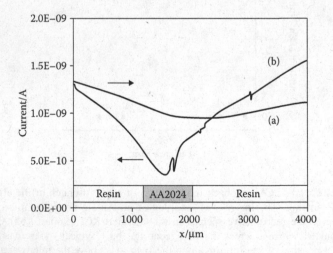

FIGURE 3.23 Amperometric scan curves obtained at -0.75 V vs. Ag/AgCl, over the Al alloy, scan direction as indicated: (a) isolated electrode and (b) electrode coupled to the Mg anode. (From Simões et al., *Prog Org Coat*, Vol. 63, (2008): p. 260. With permission.)

The SECM probe current above the magnesium electrode is seen in Figure 3.24. Interestingly, a local minimum in the cathodic current is seen over the metal whether or not it is coupled to the aluminum, indicating that the reduction of oxygen proceeds on magnesium even when it acts as an anode in the galvanic couple. This observation is attributed to the highly active nature of magnesium. While the corrosion potential of magnesium does shift to more noble potentials upon coupling to aluminum, O_2 reduction is still able to proceed under diffusion-limiting conditions at both potentials, and any decrease in the global reduction rate upon coupling results from a decrease in the rate of H_2O reduction [47].

Finally, SECM has also proven to be useful as an aid in determining the mechanism of corrosion protection for a polymeric coating. Seegmiller et al. created a blended coating for Al 2024 consisting of camphorsulfonate (CSA)-doped polyaniline (PANI) and polymethylmethacrylate (PMMA), and brushed this onto the alloy surface at a film thickness of 10 µm [125]. Open-circuit potential measurements in 10 mM H_2SO_4 found the OCP of the coated substrate to be −0.12 V (vs. Ag/AgCl), compared to −0.74 V for the bare alloy. A scratch was made in the coating to expose the underlying alloy, and the sample was immersed in 10 mM H_2SO_4 with 1.25 mM (dimethlyamino)methylferrocene (DMAFc$^+$) as the SECM mediator. SECM line scans were then performed across the scratch with the substrate held at the OCP and the probe held at +0.4 V (to oxidize DMAFc$^+$, as well as H_2 that may be generated at the surface). The experiment was then repeated using an alloy coated simply with PMMA (no PANI). The results of these experiments are presented in Figure 3.25. Curve A corresponds to a scan across the scratch in the fully blended coating (CSA-PANI-PMMA/Al 2024) at

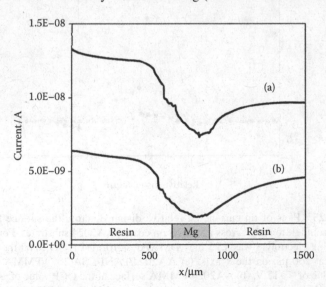

FIGURE 3.24 Amperometric scan curves obtained at −0.75 V vs. Ag/AgCl, over the pure magnesium electrode: (a) isolated and (b) coupled to the aluminum electrode. (From Simões et al., *Prog Org Coat*, Vol. 63, (2008): p. 260. With permission.)

the OCP of −0.12 V, while curve B is for the PMMA-only coated sample at its OCP of −0.74 V. Curve B shows a large increase in tip current attributed to the production of H$_2$ within the scratch, while curve A shows very little variation above baseline, indicating that the blend substantially inhibits cathodic activity within the scratch. Curve C is for the PMMA-only sample held potentiostatically at −0.12 V, and it is almost identical to the fully blended sample in curve A. The conditions of curve D were identical to those in curve B expect for the addition of CSA to the solution to determine the effect of solution-phase dopant ion on the evolution of hydrogen. It can be seen that curves B and D are very similar, with the difference in peak widths being simply due to different samples with different scratch dimensions.

Based on the SECM results of Figure 3.25, and along with the OCP measurements described above, the authors conclude that the fully blended coating most likely inhibits corrosion on Al 2024 by a combination of two mechanisms. First, interaction of the conducting polymer with the metal brings the potential of the metal into its passive region where a protective metal oxide film may form. Secondly, the more positive OCP of the coated substrate also acts to suppress both H$^+$ and O$_2$ reduction, as is clearly seen in a comparison of curves A and B.

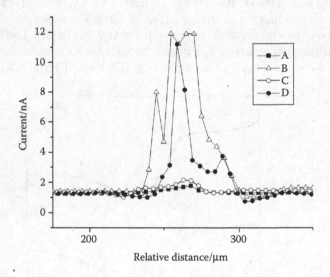

FIGURE 3.25 Plots of tip current vs. relative distance across the surface for SECM line scans from left to right across a scratch on coated AA2024 substrates. For all cases, the supporting electrolyte was 1.25 mM DMAFc$^+$ + 10 mM H$_2$SO$_4$, and the scratch is located at ca. 260 µm on the x axis. (a) AA2024/PANI-CSA(20%)/PMMA surface at the OCP value of −0.12 V, (b) AA2024/PMMA surface at the OCP value of −0.74 V, (c) AA2024/PMMA surface held potentiostatically at −0.12 V, (d) AA2024/PMMA at open circuit in a solution containing 0.1 M CSA + 10 mM H$_2$SO$_4$ with the pH adjusted to 2 with 0.1 M NaOH. (From Seegmiller et al., *J Electrochem Soc*, Vol. 152, (2005): p. B45. With permission.)

Finally, the observation that the addition of CSA to the solution has no effect on the evolution of H_2 (curve D) leads to the conclusion that the dopant ion does not combine with oxidized metal ions to form a protective precipitate that inhibits H_2 generation, as has been reported in investigations of other coating systems [125].

3.7 SECM FOR THE STUDY OF ELECTRON TRANSFER AT ACTIVE METALS

In this section, we will focus on SECM studies of electron-transfer reactions on passive metal oxide films under noncorrosion conditions. Electron transfer across the metal oxide/solution interface is of fundamental interest to corrosion studies. While passive oxide layers protect the underlying metal, it is the localized breakdown of these insulating films that is often recognized as the first step in corrosion. These breakdowns are believed to originate at heterogeneities in the passive layer that affect the local conductivity of the surface. Such heterogeneities may be due to structural defects in the film, differences in oxide stoichiometry, and/or differences in film thickness, and the localized nature of these sites makes them ideally suited for investigation with SECM.

3.7.1 TITANIUM

Following the initial work by Wei and Bard in which changes in SECM tip current were used to calculate the thickness of TiO_2 films grown at various anodization potentials [126], there has been considerable interest in using the SECM to better understand how and where redox processes occur on passive films of titanium and its alloys. Basame and White, for example, examined the electrochemical behavior of several redox-active species (bromide, iodide, ferricyanide, and ferrocyanide) on thin (~20 Å) TiO_2 films [91]. SECM images obtained in SG/TC mode showed that oxidation of bromide, iodide, and ferrocyanide occurred only at locally active sites. These sites were typically 10–50 μm in diameter, with a surface density of only 20 ± 5 sites/cm². Ferricyanide reduction, however, was observed to be spatially delocalized, occurring uniformly across the entire surface. This interesting behavior was explained in terms of the substrate behaving as an n-type TiO_2 semiconductor in which the oxide film is insulating when the electrode is biased at potentials positive of the conduction band edge (as in the oxidation reactions). The few sites where metal-like anodic activity was observed most likely correspond to regions where the oxide is sufficiently thin and direct electron tunneling can occur, or where the oxide possesses a different stoichiometry or impurity level that alters the local conductivity. However, when the electrode is biased at potentials negative of the conduction band edge (as in the reduction of ferricyanide), excess electrons accumulate at the surface, resulting in uniform conductivity over the entire electrode.

Fushimi et al. examined the reduction of $Fe(CN)_6^{3-}$ on thick anodic oxide films (10–50 nm) on polycrystalline Ti in pH 8.4 borate buffer [127]. SECM images and line scans were recorded as a function of increasing anodization potential (1–8 V vs. SHE) using TG/SC mode with ferrocyanide as the mediator such that $Fe(CN)_6^{2-}$ oxidized at the tip would be reduced at the substrate. While the overall cathodic activity of the oxide decreased with increasing anodization potentials, the decrease was much more pronounced on areas that were shown by optical microscopy to have particular grain patterns. It was concluded that the heterogeneity in cathodic response is a reflection of differences in oxide thickness that depend on the grain of the underlying substrate.

More recently, SECM has been applied toward investigations of passive films on industrially important titanium alloys. Zhu and coworkers used ferrocenemethanol as the mediator in TG/SC mode to identify cathodically active sites on the passive film of a grade-2 Ti alloy [128]. The spot density of these active sites (388 sites/mm²) was much larger than that reported by Basame and White on pure Ti/TiO_2 [91], covering approximately 10% of the surface. It was concluded that these active sites are the result of inherent Fe impurities in the alloy that are known to accumulate at grain boundaries. The most highly active spots were identified as triple points—sites where three grain boundaries meet—and when these points on the SECM image were connected, the resulting image resembled published metallographs that indicate grain boundaries and triple points. In a subsequent study, this same approach was applied to an investigation of a Pd-containing grade-7 Ti alloy [129]. Cathodically active sites were once again observed on the passivated surface, and in this case were identified as β-phase intermetallic compounds rich in Pd and Fe (impurity) that cosegregate to grain boundary locations. Quantitative kinetic measurements from probe approach curves revealed an apparent rate constant 100 times larger on the active locations (grain boundaries) than on the nonactive areas (grain surface).

Recent work by Wittstock et al. has focused on electron transfer on the biphasic Ti6Al4V alloy [130–131]. The redox behavior of this alloy is of interest due to its use as a biomaterial in load-bearing joint implants. SECM images and line scans identify significant differences in conductivity between native passive layers on the α and β phases [131]. By using probe approach curves to measure the kinetics of a number of different redox mediators, the authors were able to make a reasonable estimate of the flatband potential of the alpha phase [130].

3.7.2 TANTALUM

Basame and White investigated various electron-transfer reactions (I⁻ and $Fe(CN)_6^{4-}$ oxidation, $Ru(NH_3)_6^{3+}$ reduction) at Ta electrodes covered by a thin (~2.5 nm) native film of Ta_2O_5 [132]. Redox processes occurred at randomly distributed sites of 2–50 μm in radius, but an interesting selectivity of active sites toward the

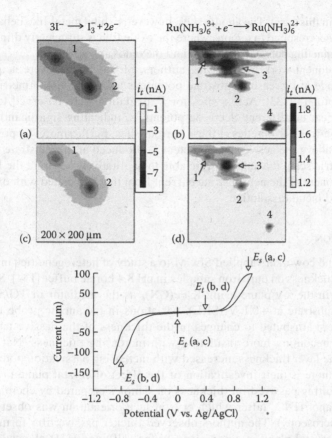

FIGURE 3.26 (Top) SECM images of a $200 \times 200~\mu m^2$ region of a Ta/Ta_2O_5 electrode in a solution containing both 2.5 mM $Ru(NH_3)_6^{3+}$ and 10 mM I^- (supporting electrolyte: 0.1 M K_2SO_4). Images were recorded sequentially in the order a, b, c, and d and are of the same area of the electrode surface. Potentials: $E_{substrate} = 1.0$ V and $E_{probe} = 0.0$ V in a and c; $E_{substrate} = -0.8$ V and $E_{probe} = 0.4$ V in b and c. (Bottom) Steady-state voltammetric response of Ta/Ta_2O_5 in the solution used for obtaining the SECM images. (From Basame and White, *Langmuir*, Vol. 15, (1999): p. 819. With permission.)

redox molecule was observed. The active site density for I^- oxidation, for instance, was determined to be 259 ± 51 sites per cm^2, while that for $Fe(CN)_6^{4-}$ oxidation was only 81 ± 9 sites per cm^2. Figure 3.26 illustrates this selective behavior for I^- oxidation and $Ru(NH_3)_6^{3+}$ reduction. The SECM images in the figure were taken from the same substrate area in sequential order (a, b, c, d) and show active sites for I^- oxidation (a and c) and $Ru(NH_3)_6^{3+}$ reduction (b and d). It can be clearly seen that sites 1 and 2 are active for both processes, while sites 3 and 4 are only active for $Ru(NH_3)_6^{3+}$ reduction, and that this behavior can reproducibly be switched on and off. The cathodic site behavior was determined to be consistent with an n-type Ta_2O_5 semiconductor biased at a potential negative of the conduction band

(~ −0.8 V in this case). The anodic sites, however, exhibit metal-like behavior and most likely correspond to regions where the oxide film is sufficiently thin to allow electron tunneling between the metal and the redox species.

In subsequent work by the same authors, SECM images were acquired in SG/TC mode at increasing substrate potentials along the voltammetric oxidation wave for I⁻ [133]. Active sites for I⁻ oxidation were observed to appear and disappear at different electrode potentials, indicating significantly different electron-transfer kinetics at these various sites. Furthermore, by positioning the probe above the active sites to detect I_3^- produced at the substrate during a voltammetric scan, the authors were able to qualitatively separate the I⁻ oxidation component of the net substrate current from that associated with oxide film growth and metal dissolution.

3.7.3 IRON

Fushimi and coworkers applied SECM to a study of heterogeneities in the passive film thickness on pure iron samples in pH 8.4 borate buffer [134]. Scanning the probe in the x-y plane using $K_4Fe(CN)_6$ as the mediator in TG/SC mode with the substrate at +0.1 V (SHE), variations in the anodic probe feedback current were attributed to changes in the thickness of the passive layer, with the oxide becoming more insulating with increasing thickness. Not surprisingly, oxide layer thickness increased with increasing passivation potential. Of particular note is their investigation of the effect of crystal plane orientation on the resulting passive film thickness on samples prepared by chemical etching in ethanol/HNO_3 (after which crystal face orientation was observable by optical microscopy). The authors observed thicker passive film formation on the (100) crystal plane in comparison to the (110) and (111) planes, and they speculate that this variation results from differences in the packing densities of the three crystal faces.

Mirkin and coworkers examined the effect of applied mechanical stress on the kinetics of electron-transfer reactions on SS [135]. This work was motivated by their interest in understanding the mechanism of stress corrosion cracking, which is induced by tensile stress in a corrosive environment. Rate constants for the oxidation of $Ru(NH_3)_6^{2+}$ on SS substrates were determined from probe approach curve data. In these experiments $Ru(NH_3)_6^{3+}$ was reduced at the probe tip, and current versus distance plots were measured at various substrate potentials. In the absence of mechanical stress, the standard heterogeneous rate constant for the reaction was found to be ~10^{-3} cm/s, a value approximately three orders of magnitude lower than for the same reaction on noble metal electrodes. As tensile stress was increased, the rate constant for the reaction decreased. This result could not be explained as a result of a simple disruption in the oxide layer, since exposure of nonpassivated steel would likely lead to an increase in the rate of electron transfer. Instead, the authors speculate that changes in the work function and the potential of zero charge affect the double layer potential profile and thereby influence the rate constant.

3.7.4 ALUMINUM

Lastly, several studies have focused on electron transfer across passive films of aluminum and aluminum alloys. Initial investigations in this area were performed by Serebrennikova and White [136–137]. Working in a nonaqueous (CH₃CN) environment to enhance the stability of the metal oxide, they utilized SECM in SG/TC mode to identify the areas of high conductivity in native Al_2O_3 films (2–3 nm thickness) on pure aluminum. Nitrobenzene (NB) was used as the mediator, and these high-conductivity areas were identified by their ability to reduce NB to NB⁻, which was then oxidized back to NB at the SECM probe. These areas were attributed to structural and/or electronic defects in the passive film. Defect sites ranging from 2.2 μm to 46 μm in radius were identified, and highly variable defect densities ranging from 3.3 to 1,000 sites per cm² were calculated.

Investigations in our own laboratory have identified both cathodic and anodic electron-transfer processes on pure aluminum in aqueous solution [109]. Our focus on anodic activity stems from our interest in understanding the mechanism for the mediated electrodeposition of conducting polymers on aluminum and aluminum alloys for corrosion control [138–141]. Sulfonated hydroxybenzenes have proven to be very effective electrodeposition mediators, and since the electrodeposition process begins with mediator oxidation, we have used these compounds as SECM mediators in our investigations of electron transfer on these substrates. Figure 3.27 shows SECM images identifying reduction sites (Figure 3.27a, TG/SC mode) and oxidation sites (Figure 3.27b, SG/TC mode) on the same 60 μm × 60 μm area of a pure aluminum sample using hydroquinone sulfonate as the mediator in an acidic sulfate solution. (Similar images were obtained in nonacidified solution as well.) Significant oxide growth has most likely occurred at the high substrate potential used to obtain the image in Figure 3.27b, but oxidative electron transfer still occurred in the same area as the reductive electron transfer identified in

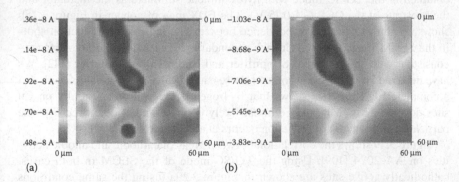

FIGURE 3.27 SECM images showing (a) reduction and (b) oxidation sites on pure Al using 0.01 M hydroquinone sulfonate in 0.1 M Na_2SO_4, 0.005 M H_2SO_4. (a) cathodic activity, TG/SC mode, $E_{probe} = 1.2$ V, $E_{substrate} = -0.7$ V. (b) anodic activity, SG/TC mode, $E_{probe} = 0$ V, $E_{substrate} = 2.0$ V. (From Jensen et al., *J Electrochem Soc*, Vol. 155, (2008): p. C324. With permission.)

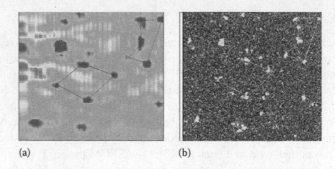

(a)   (b)

FIGURE 3.28   (a) SECM map of AA 2024 (200×200 μm) obtained in 0.01 M hydro-quinone sulfonate, 1.0 M Na_2SO_4, 0.005 M H_2SO_4. $E_{probe} = 1.2$ V, $E_{substrate} = OCP$ (TG/SC mode) and (b) elemental EDX map for copper. Lines are drawn to emphasize the patterns of correspondence. (From Jensen et al., *J Electrochem Soc*, Vol. 155, (2008): p. C324. With permission.)

Figure 3.27a (although a slight shift is noted in the region of highest activity). This is consistent with a model in which both oxidation and reduction take place at defect sites in the passive oxide layer.

As one would expect, much higher degrees of reactivity are observed when using SECM to investigate electron transfer at aluminum alloy surfaces. Seegmiller and Buttry examined the redox activity of AA 2024 in borate buffer (pH 8.5) and observed areas of high cathodic activity [142]. Analysis of these areas with SEM-EDX identified them as Cu-rich intermetallic inclusions in the Al matrix with an estimated density of 25,000 sites per cm^2.

Our own work with AA 2024, just as with the pure aluminum study described above, has focused on identifying the sites of both anodic and cathodic activities [109]. The SECM image in Figure 3.28a shows areas of high cathodic activity obtained in the TG/SC mode with hydroquinone sulfonate as the mediator and the substrate at the OCP (~ −0.75 V). The EDX map of copper in Figure 3.28b shows an almost exact correspondence between the most active reduction spots in the SECM image and the Cu-rich secondary particles on the alloy surface—consistent with the results of Seegmiller and Buttry in borate buffer [142]. We have detected H_2 evolution at some of these cathodically active areas as well, and a complete EDX analysis shows that hydrogen is evolved preferentially on Cu sites devoid of Mg, Mn, and Fe (most likely Θ-phase and/or dealloyed S-phase particles), while mediator reduction occurs at all Cu-rich sites [109].

Figure 3.29 shows the correlation between areas of cathodic and anodic activities on AA 2024 [109]. Using the TG/SC mode of the SECM in both cases, cathodically active sites are shown in Figure 3.29a (using the same conditions as in Figure 3.28a), and anodically active sites from the same 80×80 μm area are shown in Figure 3.29b. To identify the anodically active areas, yet minimize the amount of metal oxide formation, the oxidized form of a mediator with a low reduction potential was chosen (anthraquinone-2,6-disulfonate). Two areas of high anodic activity are apparent, and these correspond to two of the

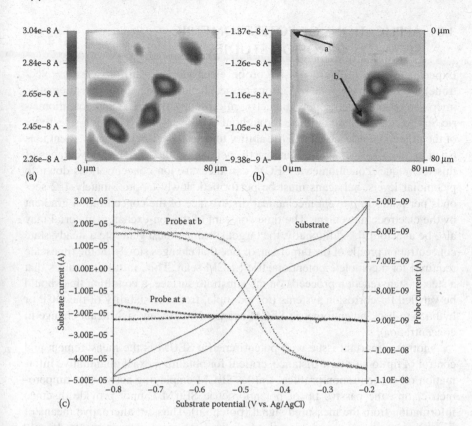

FIGURE 3.29 (a) SECM image identifying areas of cathodic activity on 80 × 80 μm area of AA 2024. Solution: 0.01 M hydroquinone sulfonate, 1.0 M Na$_2$SO$_4$, 0.005 M H$_2$SO$_4$. E_{probe} = +1.2 V, $E_{substrate}$ = OCP. (b) SECM image showing anodic activity of same area 80 × 80 μm area. Solution: 0.01 M anthraquinone disulfonate, 0.1 M Na$_2$SO$_4$. E_{probe} = −0.6 V, $E_{substrate}$ = −0.2 V. Arrows identify probe positions during voltammetry. (c) Cyclic voltammetry of solution in (b) with probe placed at two different positions above the surface. Scan rate: 100 mv/s, E_{probe} = −0.6 V, initial $E_{substrate}$ = −0.2 V. (From Jensen et al., *J Electrochem Soc*, Vol. 155, (2008): p. C324. With permission.)

cathodically active areas. The anodic activity of these spots is confirmed by the probe current (dotted) curves in Figure 3.29c. During cyclic voltammetry of the substrate, the probe was held at a reducing potential and placed ~5 μm above the surface at both an inactive location (a) and an anodically active location (b). The probe current at the active location mirrors the substrate current, while the current at the inactive location is featureless. The question arises as to why all cathodically active spots (Figure 3.29a) are not anodically active as well (Figure 3.29b). We speculate that these areas may have differing elemental compositions, or some passivation may occur during imaging of the anodic sites, but further investigation is needed.

3.8 APPLICATIONS OF POTENTIOMETRIC SECM TO CORROSION STUDIES

Experiments that utilize a passive probe electrode (i.e., an ion-selective electrode) for SECM have gone by various names, including scanning potentiometric microscopy, ion-selective potentiometric microscopy, the SIET, and potentiometric SECM, the term we will use throughout this discussion. The high selectivity of the potentiometric signal and the ability to detect electrochemically silent ions (such as the alkaline and alkaline earth metals) are the principal advantages of this technique. Potentiometric SECM can measure ion concentrations down to picomolar levels, but scans must be performed slowly (approximately 1–2 seconds per point) to prevent mechanical disturbance of the concentration gradient by the electrode movement. The time constant of the ion-exchange material may also be a factor [9]. Additionally, the target substrate must generate a steady-state concentration profile of the target ion, or one that changes slowly on the timescale required for a complete potentiometric SECM scan. This, in turn, requires that a steady-state reaction proceeds on the substrate surface, a condition that should be verified for corrosion systems (for example, from the stability of the OCP or by back-to-back potentiometric SECM scans to assess any significant change in concentration).

Another important issue with potentiometric SECM is the measurement and control of tip-to-substrate distance, critical for obtaining exact quantitative information on concentration profiles. Unlike SECM employing active (i.e., amperometric) tips, the passive tip of potentiometric SECM cannot provide distance information from the measured signal (potential). Thus, an alternative means of distance measurement and control is required for exact measurements. If only qualitative information is needed (e.g., an assessment of whether the pH or a metal ion concentration increases or decreases above some substrate surface feature), then distance control is not necessary, although some knowledge of the tip-to-substrate distance is certainly useful.

3.8.1 DISTANCE CONTROL IN POTENTIOMETRIC SECM

One of the earliest reports of an ion-selective tip or microelectrode for use in potentiometric SECM was by Horrocks and coworkers [99]. They described an antimony microdisk electrode of a few microns diameter that could be operated in both amperometric and potentiometric modes. The amperometric function of the tip was used to approach the surface and calibrate the distance or perform topographical imaging, based on the tip current from oxygen reduction in air-saturated buffer solution at −0.8 V versus Ag/AgCl, after which the potentiometric mode was used to perform pH imaging. They found that after distance measurement at negative tip potentials, the pH function deteriorated, but the original pH function could be completely restored by returning the tip electrode potential to 0 V for a few seconds before switching to the potentiometric mode. Using model systems, they demonstrated that pH changes occurring during electrochemical reactions could be mapped on the micrometer scale, and illustrated the approach by imaging pH profiles above a platinum microelectrode during water reduction, a corroding disk of silver iodide in aqueous potassium cyanide, a disk of immobilized urease that hydrolyzed urea, and a disk of immobilized yeast cells in glucose solution. The experimental pH profiles agreed reasonably well with theoretical profiles that were derived based on a constant flux approximation for generation of steady-state concentration profiles of the target ion at a disk-shaped substrate. Deviations at small tip–substrate separation were attributed to substrate screening and stirring of the diffusion layer by the tip.

Further developments in tip-to-substrate distance calibration in potentiometric SECM have been discussed by Tóth and coworkers [143]. In addition to the double function antimony microelectrode described above, they discuss two other approaches that use double-barrel micropipettes, with one of the barrels used for distance measurement and the other for selective potentiometric measurement. In one approach, the barrel for distance measurement was filled with low melting gallium and used in conjunction with the mediator $Ru(NH_3)_6^{3+}$ for distance control based on SECM feedback current–distance curves. The second barrel for potentiometric measurement was filled with a cocktail containing the ionophore for either Zn^{2+} or NH_4^+ with a Ag/AgCl wire inserted to complete the half-cell. In the second approach, the barrel for potentiometric measurement was prepared as above, but the barrel for distance measurement was simply fitted with an Ag/AgCl electrode to which was applied a potential of 50 mV. A conductivity-based distance measurement was performed with this open micropipette barrel, the resistance to current flow increasing as the tip approached an insulating or a conducting surface. The experimental approach curves exhibited good agreement with theory for both of these electrode configurations. However, the second approach may be of more use in corrosion research since a mediator is not required and the fabrication of double-barrel open-tip micropipettes is well established [143]. Measurements of ion concentration profiles (pH, Zn^{2+} and NH_4^+) were demonstrated for electrochemical model systems.

A constant-distance mode potentiometric SECM experiment for mapping local pH at microscopic defects in the passive oxide layer on the surface of NiTi shape memory alloys was described by Schulte et al. [144]. The distance control was achieved by vibrating the electrode and using shear forces between the tip and the substrate for feedback control. The anodic dissolution of Ni and/or Ti at sites of pitting corrosion results in proton release due to solvolysis of the metal ions, thus, altering the local pH at the active sites of pitting corrosion. To prepare the pH probes, carbon fiber disk electrodes were first prepared by sealing a 7.5-μm carbon fiber in a pulled glass micropipette and insulating the fiber extending beyond the glass sheath with an electrodeposited paint. The insulated fiber was then cut perpendicular to the longitudinal axis with a scalpel to expose the carbon disk microelectrode. Hydrous iridium oxide was electrochemically deposited onto the disk-shaped carbon fiber electrode, producing a pH probe with a diameter of ca. 10 μm. The stiffness of these probes was optimized by adjusting the micropipette length (by varying the pulling parameters) so that detection of the shear forces between the tip and the substrate was possible.

Constant-distance control was demonstrated by imaging the topography of a microstructured silicon wafer, and the pH mapping capability was demonstrated by imaging local pH changes at a 10-μm Pt-disk electrode at which OH⁻ was produced by electrolysis of water. Diffusional broadening of the pH images was noted, an inherent limitation of pH mapping due to the very high diffusion rates of H^+ and OH^- in aqueous solutions. The authors note that the effect of diffusional broadening can be minimized by choice of an appropriate buffer solution [144]. The probe was then used to visualize corrosion at a microscopic crack in the passive layer of the NiTi test sample immersed in 1 mM phosphate buffer (pH 6) containing 100 mM KCl, with the NiTi polarized at 1.5 V versus Ag/AgCl. The potential was shown to increase in close proximity to the crack, signifying a decrease in pH (not an increase as stated in the paper) due to dissolution of Ni and/or Ti. This group subsequently used AC-SECM to further probe the corrosion behavior of these alloys as discussed earlier in Section 3.5 [101,103].

3.8.2 ION MAPPING WITH POTENTIOMETRIC SECM

Several applications of potentiometric SECM to corrosion research have been reported. The majority of these reports have used the technique to map local pH variations across an active metal surface, since the redox reactions involved in corrosion usually engender either an increase (at local cathodes) or a decrease (at local anodes) in pH (see Section 3.3). One of the earliest such reports was by Tanabe and coworkers [60], who imaged the local distribution of H^+ and Cl^- near the surface of three austenitic SSs (SUS304, SUS316L, and HR8C) in an attempt to identify precursor sites for pitting. Their pH probe consisted of a 7-μm diameter carbon fiber mounted in a glass capillary, the probe being conditioned for at least 3 h in the test solution before measurement. Their chloride ion probe was prepared by electrodeposition of Ag onto a Pt microelectrode followed by anodic polarization in 1 M HCl. They estimated the selectivity coefficients of these probes to be unity for their respective ions in NaCl solutions, although (perhaps not surprisingly) there was considerable scatter in the potential–pH diagram for the carbon fiber based pH probe. The potentiometric measurements were complemented by measurements using additional modes of SECM, including an AC-SECM technique (discussed in Section 3.5), a tip oscillation mode, and a double potential step chronoamperometric mode with a negative potential step to detect H^+ reduction and a positive potential step to detect Cl^- oxidation (discussed in Section 3.4). The authors did not show a pH map obtained from their potentiometric SECM experiment, but they did show a Cl^- map for a 40 μm × 40 μm area that revealed an increase in Cl^- concentration near a target pit (coincident with a pH decrease detected by their AC-SECM technique). Higher resolution was obtained by the AC-SECM and double potential step chronoamperometric modes, but the potentiometric SECM experiment could provide more quantitative information using straightforward calibration techniques. Too many experimental details were missing to provide a critical evaluation of this work, but the authors claim that the precursor sites for pitting corrosion could be identified prior to passive oxide film breakdown, indicated by higher concentrations of both H^+ and Cl^- at those sites.

Park and collaborators used potentiometric SECM with ion-selective pH microelectrodes to measure hydroxide ion concentration profiles produced by the electrochemical reduction of oxygen on a model alloy of Al 6061 [145]. This alloy contains Al_3Fe intermetallic inclusions that are proposed to act as cathodes for oxygen reduction, causing a buildup of hydroxide ions around the inclusion which leads to pit initiation in the surrounding host metal. A model of the alloy was constructed by galvanically coupling a small piece of Al_3Fe (200 × 200 μm^2) with a larger piece of Al 6061 (1 cm diameter), both metals embedded in epoxy and immersed in 0.6 M NaCl (the OCP was −0.73 V vs. SCE). The coupling current was measured and the surface was scanned with a pH sensor consisting of a double barrel capillary, one barrel containing a H^+ sensitive LIX (Fluka 95293) and 0.6 M NaCl solution, the other barrel (a reference electrode) containing only 0.6 M NaCl. Ag/AgCl wires (250 μm diameter) were inserted into both barrels. The 7-μm tip of the micropipette was beveled and the probe was positioned 37° from the normal. No distance control was used, but a microscope was used to visually place the tip 33 ± 7 μm above the surface. A potentiometric SECM line scan across the coupled Al_3Fe and Al 6061 metals revealed a sharp increase in pH above the Al_3Fe, reaching pH 9.5 at its center, confirming that the Al_3Fe functioned as a cathode. The measured hydroxide concentration was typically 20%–30% of that calculated from the coupling current (40–50 nA). A slight increase in OH^- concentration was also observed above the Al 6061, somewhat surprising since a net oxidation of the alloy should lead to a pH decrease. Perhaps significant oxygen reduction also occurred at the Al 6061, sufficient to mask any pH change due to oxidation of the Al matrix. It was found experimentally that probe scan rates as high as 60 $\mu m/s$ could be used without perturbing the concentration profile.

Chloride ions are very influential to many types of localized corrosion such as pitting corrosion, crevice corrosion, and stress corrosion cracking. Even though it is accepted that Cl^- plays an important role in local breakdown of passivity and initiation of localized corrosion, there is little experimental evidence concerning the behavior of Cl^- in microenvironments close to the metal/solution interface. Lin and coworkers have addressed this issue by mapping in situ Cl^- ion distributions in several localized corrosion systems [13]. These workers recognized that the use of a single sensing electrode with a remote reference electrode could lead to error in measurement due to the contribution of the local corrosion potential, which may vary across the substrate surface (in fact, this is the basis for SRET). This error could be minimized by fabricating a combination electrode that placed the Cl^- probe and the tip of a reference electrode at approximately the same location, an approach taken by Park et al. in their work described above [145]. Using pulled double-barreled capillaries, they fabricated combination electrodes that contained either a solid Ag/AgCl or a liquid membrane Cl^- ion-selective microelectrode in one barrel and an Ag/AgCl reference electrode in the other. These microelectrodes were described as having micrometer-sized tips and were scanned ca. 5 μm above the substrate surface (determined with an optical microscope).

These Cl⁻ ion probes were then used for studies of localized corrosion of commercial SS (type 304 SS). A steel cylinder (6 mm diameter) was sheathed in epoxy, resulting in a crevice between the steel and epoxy. In a crevice corrosion experiment, the SS surface was exposed to a 0.1 M NaCl and polarized at 0.2 V (vs. Ag/AgCl) during the potentiometric SECM scan. The concentration of Cl⁻ ions along the junction between the SS surface and the epoxy sheath (i.e., at the crevice) was found to be up to three times higher than that in the bulk. The increase in Cl⁻ ion concentration was attributed to the migration of Cl⁻ to the corrosive (anodic) occluded zone to maintain charge balance. In another experiment, pitting corrosion was induced by exposing the SS surface to 0.5 M FeCl₃. Chloride ion mapping by potentiometric SECM revealed higher concentrations of Cl⁻ ion near sites where pitting corrosion occurred, with higher pitting corrosion activity resulting in higher Cl⁻ concentration. When pitting corrosion ceased, little variation in Cl⁻ concentration around the metal/solution interface was detected. Such accumulation of Cl⁻ ions in the anodic regions of crevice and pitting corrosion appears to play an important role in sustaining the propagation of localized corrosion [13].

In a series of papers, Ding and Hihara applied SVET for current density mapping and potentiometric SECM for pH mapping (along with other techniques such as SEM, EDX and AFM) to investigate the corrosion behavior of Al-based MMCs [146–149]. Their pH probes were fabricated from pulled, single-barrel, silanized glass micropipettes having a tip diameter of 5 μm and filled with a hydrogen ionophore (Fluka I-cocktail B) and with 100 mM KCl, into which a chlorinated silver wire was inserted [147]. The pH probe exhibited a linear response over the pH range 5–12. The SVET and potentiometric SECM instrumentation (from Applicable Electronics, Forestdale, MA) utilized no distance control, and the probes were positioned ca. 80 μm above the substrate. The sampling time for the potentiometric SECM data was ca. 8 s per point, requiring a total scan time of 65 min for a 20 × 20 grid scan.

Particulate 6092-T6 Al MMCs reinforced with 20 volume percent of B₄C, SiC, or Al₂O₃ were immersed in 0.5 M Na₂SO₄ solution exposed to air at room temperature. The goal of the study was to determine the effect of the electrical conductivity of the reinforcement particles on corrosion behavior [146–147]. Reinforcement particles of high conductivity are more likely to induce galvanic action with the Al matrix than those of low conductivity. The B₄C particles have rather low resistivity and, thus, could serve as local cathodes and induce corrosion of the surrounding Al matrix, whereas Al₂O₃ has a high resistivity and is not expected to galvanically couple with the Al matrix. Example SVET and potentiometric SECM maps of a B₄C reinforced composite after 40 h of immersion are shown in Figure 3.30. Indeed, a region of localized corrosion surrounded by a cathodic current trough is observed in the SVET map of Figure 3.30a (optical microscopy confirmed the formation of a circular region of corrosion products at the site of the anodic current). The pH map obtained by potentiometric SECM

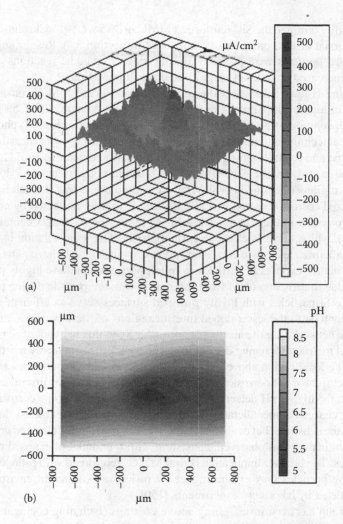

FIGURE 3.30 SVET corrosion current-density map (a) and potentiometric SECM pH map (b) of Al6092-T6/B4C/20p MMC after 40 h of immersion in 0.5 M Na₂SO₄ solution. (Reprinted from Ding et al., *J Electrochem Soc*, Vol. 152, (2005): p. B161. Reproduced with permission of The Electrochemical Society.)

verified that the pH was lowest (near pH 5) at the site of the anodic current peak, and was slightly alkaline (up to pH 8.5) in regions displaying cathodic current. The acidification in the anodic region was attributed to hydrolysis of Al^{3+} ions released from the matrix and alkalization in cathodic regions was attributed to the OH^- produced upon O_2 and/or H_2O reduction. The general conclusion of this work was that corrosion appeared to be greatest on the MMC reinforced with B_4C followed by that reinforced with SiC, both by galvanic interaction between the reinforcement particles and the matrix. The Al_2O_3 reinforced MMC exhibited the slowest corrosion and appeared to corrode by a different mechanism. A more recent study examined the SiC reinforced MMC in 3.15 wt. % NaCl solution [149] (see additional discussion of these composites in Section 3.4). Results suggested that the SiC particles were p-doped and the rate of cathodic reactions at these particles increased an order of magnitude upon illumination.

Ding and Hihara have also examined an Si reinforced MMC, consisting of 40 wt. % of high purity Si particles (98%–99%) in an Al matrix (99.25%), in 0.5 M Na_2SO_4 solution using SEM, EDX, SVET, potentiometric SECM, photoelectrochemical techniques, and scanning capacitance microscopy (for measuring the doping type and relative doping level of the Si reinforcements *in situ*) [148,150]. SVET and potentiometric SECM revealed that the net currents over localized corrosion regions (identified by SEM) were cathodic and that the solution near the localized corrosion region was alkalinized. This was explained by considering that corrosion initiates at a few Fe–Si–Al IMPs (the Al matrix contained ca. 0.25% Fe), which serve as effective cathodic sites. The local solution is thereby rendered alkaline, leading to corrosion of the surrounding Al matrix. Dissolution of the Al matrix exposes Si particles in relief, particles that have highly p-doped surfaces due to diffusion of Al from the matrix into the particle during processing. These Si particles with highly p-doped surfaces serve as efficient cathode sites, in contrast to the lesser-doped interior regions of these particles, and form a cathodic network. A galvanic couple forms between this network of Si particles and the Al matrix undergoing dissolution. The SVET measures only a net current flowing at a local region above the substrate, and in these experiments the local cathodic current at the corrosion site exceeded the local anodic current, in agreement with the higher pH detected by potentiometric SECM at the corrosion site. However, charge balance dictates that additional anodic current must flow elsewhere on the substrate, but evidence of this was not provided. Enhanced corrosion under illumination was observed, consistent with the model of exposed p-doped Si surfaces. The broader implication is that MMCs reinforced with p-doped semiconducting particles may exhibit higher corrosion rates in sunlit environments than predicted by laboratory experiments [150].

Our group has been investigating active coatings (including conjugated polymers and Mg-rich coatings) for corrosion control of Al alloys using a variety of scanning probe techniques, including SVET, potentiometric SECM, and SECM [47,151,45,48]. The use of SVET and potentiometric SECM to measure current density and pH, respectively, above a polypyrrole-coated AA 2024-T3 substrate was reported by He et al. [151]. A scribe (defect) extending through the polypyrrole film to the AA 2024-T3 surface was introduced so that any electrochemical interactions between the conducting polymer and Al alloy could be studied. Upon immersion of the coated substrate in dilute Harrison's solution (DHS, 0.35 wt. % $(NH_4)_2SO_4$ and 0.05 wt. % NaCl), SVET measurements revealed a rather intense oxidation current above the scribe (reaching a peak value of ca. 150 $\mu A/cm^2$) with the reduction current distributed more or less uniformly across the conducting polymer surface. This behavior is in contrast to that observed at a typical inert coating (such as epoxy primer) where both oxidation and reduction occurred within the scribe area [152]. The oxidation current was clearly due to the oxidation of the exposed Al alloy. However, the origin of the reduction current was not clear. It could arise from either an oxygen reduction reaction that occurred at the electrolyte/conducting polymer interface and/or from the reduction of the polypyrrole film itself. The reduction of O_2 (or of H^+ or H_2O) would lead to an increase in pH above the polymer surface. Reduction of the polypyrrole itself would lead to ejection of the dopant ion, which in this case was 4,5-dihydroxy-1,3-benzenedisulfonate anion (or Tiron), a weakly acidic species (pKa = 7.3), resulting in a slight decrease in pH above the polymer surface. Thus, local pH measurements were used to help elucidate the reduction process.

Ion-selective microelectrodes sensitive to protons were prepared by pulling single-barreled glass micropipettes (outer diameter 1.5 mm) to make electrodes with a tip diameter of ca. 3 μm. The micropipette tips were silanized, filled with a hydrogen ionophore (Fluka I-cocktail B) followed by a solution of 50 mM KCl and 50 mM KH_2PO_4, into which was inserted AgCl coated silver wire. The micro-pH electrode showed a linear response to pH changes over the range of 3–10 with a Nernst factor of 57.5 ± 1.0 mV. The external reference electrode for the potentiometric SECM measurements was an Ag/AgCl (3M NaCl) electrode. The micro-pH electrode was positioned approximately 100 μm above the polypyrrole-coated AA 2024-T3 surface. An example pH map is shown in Figure 3.31 (the bulk pH of the DHS was ca. 6). The decrease in pH just above the scribe, reaching a pH of 3.5 after 1.17 h immersion (Figure 3.31a), was attributed to hydrolysis of Al^{3+} ions generated by the oxidation process occurring in the scribe:

$$Al(H_2O)_6^{3+} + H_2O \rightarrow Al(OH)(H_2O)_5^{2+} + H_3O^+$$

$$2Al(H_2O)_6^{3+} \rightarrow Al_2O_3 \cdot 3H_2O + 6H_3O^+$$

The pH above the conducting polymer surface reached a value of ca. 8, higher than the bulk pH, suggesting that oxygen reduction occurred at the electrolyte/conducting polymer interface. Recent results from galvanic coupling experiments have confirmed this conclusion [153]. With immersion time, the pH above the conducting polymer surface decreased and that above the scribe area increased, eventually reaching a more or less uniform pH of ca. 6 across the entire substrate after 23 h immersion (Figure 3.31b). This pH change coincided with diminishing anodic and cathodic currents as measured by SVET, attributed to corrosion product build-up in the scribe area [151].

During the course of the work described above, the question arose as to how low the pH might be right at the bare alloy surface. The LIX micro-pH electrode used in the above work had a lower pH limit of approximately 3. Additionally, the closest approach of the probe was 100 μm. To estimate the surface pH, a solid state pH sensor was prepared (hydrated IrO_2 electrodeposited on a Pt microelectrode) that exhibited a linear response over the pH range 0.3–10. An approach of the electrode tip from bulk solution to the alloy surface (determined by tip contact) was made, recording pH as a function of distance. The experiment was performed

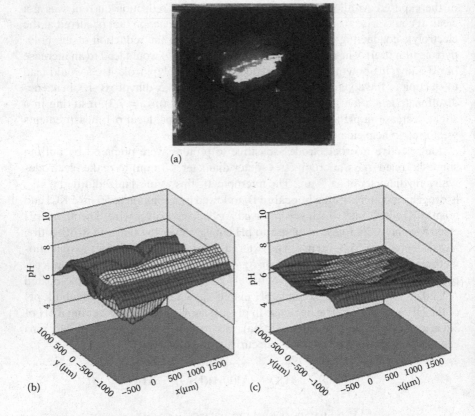

FIGURE 3.31 A typical potentiometric SECM result for polypyrrole-coated AA 2024-T3. (a) optical image of the sample with artificial defect. (b,c) the pH distribution above the sample after 1.17 and 23 h of immersion. (Reprinted with permission from He et al., in *New Developments in Coatings Technology*, 8 (2007), American Chemical Society. Copyright (2007) American Chemical Society.)

at a bare alloy held at a potential of −0.25 V, the OCP of the polypyrrole-coated alloy with defect. Note this potential is significantly positive of the OCP of the uncoated alloy, −0.65 V, due to ennobling by the conducting polymer. At the surface just before contact, the pH was measured to be 1.2. The same experiment conducted with the alloy at its OCP of −0.65 V indicated a constant pH (ca. 6) at all distances right up to the instant of contact (unpublished results). Thus, the solution near the surface within the scribe of the polypyrrole-coated AA 2024-T3 becomes extremely acidic, which explains the onset of hydrogen bubble formation often observed when polypyrrole is galvanically coupled with AA 2024-T3 [151].

Lamaka and coworkers describe the use of potentiometric SECM in combination with SVET for mapping the distribution of pH, Mg^{2+}, and ionic current density due to corrosion of a Mg-based alloy in aqueous chloride-containing solution [154]. This appears to be the first (and only, to date) example of using potentiometric SECM to map metal ion distribution in a corrosion system on the micron scale. The H^+ and Mg^{2+} electrodes were prepared from single-barreled glass capillaries pulled to a tip dimension of 2 μm. The LIX membranes were prepared in standard fashion using commercial ion exchangers (the magnesium ionophore II cocktail B from Fluka was used for the Mg^{2+} electrode) [154]. The local activities of H^+ and Mg^{2+} were mapped sequentially 10 μm above the substrate surface on a 25 × 25 point grid at a rate of 2 s per point (35 min was required for a complete scan of each ion). Linear response of the Mg^{2+} electrode was observed over the range 5×10^{-5} to 0.1 M $Mg(NO_3)_2$ with a slope of 28.5 ± 0.7 mV/decade, employing a background electrolyte of 0.05 M NaCl. The pH electrode had a linear range of 5–12 with a slope of 55.1 ± 1.4 mV/decade. The substrate was AZ31B magnesium-based alloy (containing 3% Al and 1% Zn) coated with an anticorrosive solgel film of 4 μm thickness, immersed in 0.05 M NaCl. Artificial defects through the coating were introduced with a sharp needle, producing ca. 120 μm diameter defects.

Figure 3.32a shows an optical micrograph of the coated substrate with the nearly circular defects clearly visible. Figures 3.32b through d show the SVET, pH, and pMg maps, respectively. The SVET scan was recorded after 4 h

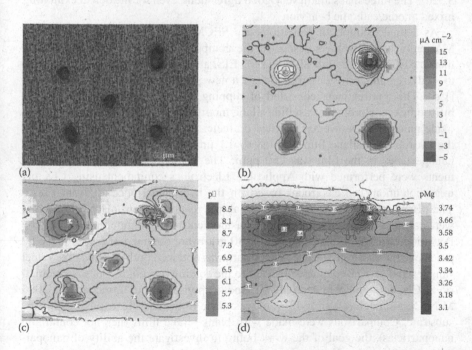

FIGURE 3.32 Optical microphoto (a), distribution of local ionic current density (b), activity of H^+ (c) and Mg^{2+} (d) ions, recorded during immersion of AZ31B magnesium-based alloy coated with thin solgel film in 0.05 M NaCl solution. (Reprinted from Lamaka et al., *Electrochem Comm*, Vol. 10, (2008): p. 259, Copyright (2008), with permission from Elsevier.)

immersion in 0.05 M NaCl, followed sequentially by the pH and pMg scans. A total of 90 min was required for all three scans, including switching from one instrument mode to the next and calibration of the electrodes. Clearly, a quasi-steady-state corrosion system is required for using this sequential combination of techniques if sensible correlation among the three maps is to be expected. Indeed, very nice correlation is observed in the example of Figure 3.32. Three of the five defects were indicated by SVET to exhibit a net cathodic current, one defect exhibited a net anodic current, and one defect exhibited spatially resolved anodic and cathodic current (Figure 3.32b). All reduction sites were associated with a local pH increase of up to 8.5, attributed to OH^- produced from H_2O reduction (Figure 3.32c). However, no mention is made of oxygen sparging in their experiments and rigorous control of oxygen in these scanning probe experiments is somewhat problematic. It is likely that oxygen reduction played a major role in the cathodic process. The anodic sites were associated with a pH decrease, attributed to Mg and Al dissolution and subsequent cation hydrolysis (Figure 3.32c). The pMg map reflects a depletion of Mg^{2+} in the cathodic regions due to reaction with OH^- to form $Mg(OH)_2$, while the highest Mg^{2+} concentrations were detected in the anodic regions (up to 0.8 mM) due to metal dissolution in these zones (Figure 3.32d). The three maps are in very good agreement, even for the defect exhibiting mixed anodic/cathodic behavior.

As a final example of potentiometric SECM applied to corrosion research, Montemor and coworkers described the combined use of SVET and potentiometric SECM (along with SEM, AFM, EIS, and potentiodynamic polarization) to study the anticorrosion properties of a new pretreatment for galvanized steel [155]. This pretreatment consisted of dipping the substrate into a solution of bis(triethoxysilylpropyl)tetrasulfide silane modified with $CeO_2 \cdot ZrO_2$ mixed metal oxide nanoparticles (average particle diameter of ca. 25 nm). The thickness of the hydrophobic silane film was 0.8 to 1.1 μm, and the final concentration of nanoparticles in the film was 250 ppm. The potentiometric SECM measurements were performed with Applicable Electronics equipment using LIX pH micropotentiometric electrodes based on the hydrogen I cocktail from Fluka. The pH electrodes had a linear potential–pH response from pH 5.5 to 12 and were calibrated before and after the measurements with commercial pH buffers. Potentiometric SECM maps on a 30 × 30 grid were obtained in a scan plane 50 μm above the substrate surface. SVET data was obtained both on galvanized steel samples and on zinc coupons subjected to the coating procedure, but potentiometric SECM data was obtained only on zinc coupons. Prior to immersion in 0.005 M NaCl, a scratch (1 mm × 0.1 mm) was made in the coating, exposing the zinc substrate. Comparisons were made with blank silane films (i.e., not containing nanoparticles), the goal of the work being to investigate the ability of nanoparticles to mitigate localized corrosion in coating defects.

SVET measurements revealed that, throughout the 48 h duration of the experiment, substrates coated with the silane film containing the nanoparticles exhibited an order of magnitude lower anodic current in the scratch than substrates coated with the blank silane film [155]. Potentiometric SECM revealed that the pH in cathodic (coated) areas reached as high as 10, attributed to reduction of oxygen, which apparently could pass through the coating to the metal surface, while the pH in the anodic area (scratch) remained nearly neutral (unlike Al or Mg ions, Zn ions do not undergo significant hydrolysis). It was suggested that the high pH in cathodic areas led to local decomposition of the silica network, releasing nanoparticles that precipitated on the zinc surface, perhaps forming complexes with charged Zn species that migrated from the anodic region. Additional potentiodynamic polarization experiments suggested that the inhibition effects were mainly due to the CeO_2 component of the nanoparticles. The authors seem to suggest that the nanoparticles function as a mixed inhibitor, slowing both the oxygen reduction reaction and the Zn oxidation reaction (by stabilizing the layer of protective Zn corrosion products). However, no elemental analysis of the metal surface exposed within the scratch was provided to support the notion that the nanoparticles were somehow transported to the scratch area.

In the majority of the applications discussed in this section, potentiometric SECM was used in combination with SVET to provide supporting evidence for interpreting the current density data and for postulating the corrosion mechanisms. Most of the reports involved pH mapping, with only one example of metal ion mapping (Mg^{2+}). Future corrosion studies could certainly benefit from metal ion measurements using potentiometric SECM, especially for understanding the corrosion behavior of complex metal alloys containing secondary phase particles.

3.9 ELECTROCHEMICAL MICROCELL AND SCANNING CAPILLARY MICROSCOPY TECHNIQUES

3.9.1 Stationary Microcapillary Techniques

Stationary microcapillary techniques permit the selection and interrogation of single metallurgical sites on the surface of heterogeneous alloys, often involving the measurement of currents in the pA to fA range. Typically, the electrolyte-filled capillary contains (or connects to) counter and reference electrodes with the end of the capillary sealed to the substrate (the working electrode). One of the earliest reports of this approach was by Böhni and coworkers who sealed a microcapillary to the substrate surface with silicone rubber, the microcapillary leading to a chamber containing a Pt counter electrode and a calomel reference electrode [156–157]. Using potentiodynamic polarization, potentiostatic current transients, and current noise analysis, they found that pit nuclei on SSs initiated due to oxidation and dissolution of (mainly) MnS inclusions, even in chloride-free environments. They were able to probe the electrochemical behavior of single inclusions on the micron scale.

More recent studies employing the stationary microcapillary approach (often in combination with other techniques) have appeared, including additional studies of the role of metallurgical heterogeneities in the pitting corrosion of SSs [158–161], the influence of wear (under sliding friction conditions) on the electrochemical behavior of heterogeneous phases of duplex SS [162–163], and the influence of applied strain on the electrochemical behavior and pitting susceptibility of sites containing secondary phase particles in AlCu4Mg1 cast aluminum alloy [17]. Recent reviews of the stationary microcapillary (or electrochemical microcell) approach have appeared [164–165].

3.9.2 Scanning Microcapillary Techniques (Scanning Droplet Cell)

More relevant to the discussions of this chapter are methods that scan the capillary over the substrate surface. The first report of this approach appears to be from 1940 by Dix [166] who scanned an electrolyte droplet ca. 1 mm in diameter at the end of a glass capillary over the surface of an alloy to measure the open-circuit potential of isolated anodic and cathodic regions. Lohrengel and coworkers extended the technique, employing a three-electrode arrangement for voltammetric measurements and improving the resolution to a few microns using small electrolyte droplets positioned on the substrate surface [167,19,168]. The capillary is positioned above the substrate (usually ca. 10 μm) and the droplet between the capillary tip and the substrate is held either by surface tension (free droplet, for non-wetting surfaces) or by a gasket ring of silicone rubber (enclosed droplet, for wetting or rough surfaces) [168]. In both cases, the wetted area defines the working electrode, with reference and counter electrodes connected by the capillary. The capillary assembly is mounted to a xyz positioning stage for scanning or for positioning the capillary above a substrate surface feature of interest. All common potentiostatic and galvanostatic techniques can be used (including impedance spectroscopy, cyclic voltammetry, or potential step chronoamperometry) to perform spatially resolved surface analysis or modification.

The resolution of this technique is limited by the detection limit of the current measurement circuit. For measurement of current densities down to 1 μA/cm², a minimum capillary diameter of 10 μm is recommended [168]. Capillaries of a few μm diameter position droplets of very small volume (<1 nL) on the substrate surface. As a result, a high rate of electrochemical reaction at the substrate surface can lead to severe contamination of the droplet or even to blockage of the capillary tip by insoluble reaction products. Furthermore, the stability of such a small droplet depends on the evaporation rate of the solvent, usually water. Electrolyte droplets of hygroscopic solutions (e.g., sulfuric or acetic acid) remain stable for several minutes, while other droplets can vanish within a few seconds [167]. Blowing a slow stream of solvent-saturated nitrogen across the droplet extends the stability to several thousand seconds. To overcome these difficulties of changing droplet volume and/or composition, flowing-electrolyte microcells have been developed using two-channel capillaries, one channel for electrolyte inlet, the other for electrolyte outlet, permitting circulation of electrolyte through the droplet [22–23]. Both theta-capillary [22] and coaxial capillary [23] geometries have been used. The coaxial double capillary, with the inner capillary employed as the electrolyte inlet and the outer capillary as the outlet, enabled use of a free standing electrolyte droplet to scan the specimen (i.e., no silicone gasket ring required) [23]. Experimental and theoretical comparisons between measurements made with the flowing microcapillary droplet cell and the closed microcapillary system (no flow – the microcell technique) have been discussed [23,169–170]. Not surprisingly, the current measured with flowing electrolyte was much higher (by a factor of 4 or more) and was found to depend on a number of parameters, including volumetric flow rate, kinematic viscosity of the electrolyte solution, and geometry of the droplet cell (capillary radius and wall thickness, and distance between substrate and capillary tip). Often, capillary (and droplet) diameters of several tens to hundreds of microns are used [171], limiting the spatial resolution of these techniques.

Though not yet used for corrosion studies, we mention one report describing the combination of scanning capillary microscopy with mass spectrometry [172]. Unlike the other techniques described in this section, the capillary and substrate (working electrode) are immersed in a bulk electrolyte containing reference and counter electrodes. A liquid stream containing an electroactive test species is fed through the outer chamber of a coaxial capillary where it radially traverses the substrate and interacts with the target location beneath the capillary tip. A portion of the product stream then passes through the inner chamber of the capillary that leads to an electrospray mass spectrometer. A potential-dependent image of redox product distribution versus location on the substrate is obtained, permitting imaging of nonuniform electrochemical activity of conducting substrates. Additional information about trace products, byproducts, and/or reaction intermediates generated at or near specific locations on the substrate surface can also be obtained [172]. The spatial resolution needs to be improved for this technique to find application to metal alloys and corrosion.

An SDC investigation on single grains of polycrystalline Fe7.5Al7Cr, a lightweight ferritic steel that shows a strong anisotropic dissolution behavior, was described by Lill and coworkers [173]. The partial replacement of Cr with Al in this steel results in weight reduction, a consequence of the lower atomic weight of Al compared with Cr and also a lattice expansion due to the larger radius of the Al atoms. Crystallographic orientation of individual grains was determined by electron-backscatter diffraction before and after annealing for 20 h at 900°C. Annealing was used to increase the grain size up to 300 μm so that dissolution of individual grains could be quantified. The single-barreled capillary (no electrolyte flow) had an inner diameter of 70 μm and the tip contacted the substrate with an unspecified sealer. Cyclic voltammetry was performed in 0.5 M H_2SO_4 on grains having (001) and (111) orientations, both orientations producing peaks at the same potential (E_{pp}), reflecting passivation of the grain. However, the peak current (i.e., the critical passivation current density, i_{crit}) and corresponding charge consumption were significantly higher for the (001) grains during the active-to-passive transition than for the (111) grains, meaning that the (001) grains undergo more extensive dissolution prior to passivation. The difference in reactivity was attributed to the less dense packing of atoms in the (001) orientation that enabled the atom layer beneath the surface atom layer to undergo reaction [173].

Major microstructural changes and changes in local residual stress can occur during welding, and Akid and collaborators evaluated the localized corrosion behavior of welded joints using the SDC technique in combination with SVET [174]. The SDC measurements of corrosion potential were performed by scanning a continuously refreshed droplet (300–500 μm diameter) in both the x and y directions over the surface of the weld, making potential measurements (vs. Ag/AgCl) every 250 μm and generating a 2D map of corrosion potential. For three types of welds (carbon steel–carbon steel, carbon steel laser weld, and copper–SS–aluminum bronze), the SDC and SVET were in agreement and consistent with the corrosion response observed with optical micrographs.

Fushimi et al. used the flowing electrolyte scanning-droplet-cell (f-SDC) to obtain the corrosion potential profile of the cross section of aluminum-alloy brazing sheets used in automotive heat exchangers [171]. Using a coaxial double-barreled capillary to continuously deliver fresh electrolyte to the surface, contamination of the electrolyte by species dissolved from the substrate surface was avoided. The substrate (brazing sheets) consisted of an Al–Zn alloy s.a. layer (to provide corrosion resistance), an Al–Mn–Cu core layer (to provide mechanical strength), and an Al–Si brazing filler (cladding) that permits attachment of various pieces by brazing (a process whereby the filler alloy is heated to melting, flows across the joint, and cools to form a solid bond). The heat generated during brazing causes the diffusion of Zn from the s.a. to the core layer, diminishing the performance of the s.a. Profiles of corrosion potential across the cross section would permit evaluation of the corrosion protection performance of the s.a. The cross section of the specimen was tilted (at an angle of 3.7°) and embedded into epoxy resin in order to expand the cross-sectional length (the ratio of the length of the tilted cross section d_t to the thickness of the sheet d_f was 15.5).

A reference electrode (Ag/AgCl) was positioned in the inner barrel of the coaxial capillary into which electrolyte solution (3 wt. % NaCl, pH adjusted to 3.0 with acetic acid, no de-aeration) was pumped at a flow rate of 4.2 mm^3/s (and pulled up into the outer barrel of the capillary with a vacuum of 5 kPa), forming a droplet 200 μm in diameter. Two aluminum-alloy brazing sheets of differing composition and differing layer thicknesses were examined, designated A and B [171]. Figure 3.33 shows the corrosion potential profiles obtained for the two specimens. Only small variations in potential were observed across the brazing filler (b.f.) and core layers, while significant variations were observed at the interface between the core layer and the s.a. layer (Figure 3.33). Sheet B with its thicker s.a. layer exhibited the sharpest drop in potential from the core to the s.a. layer, even after heating for 180 s, indicating that sheet B has a more effective s.a. layer for corrosion protection of the core layer, compared with sheet A. The changes in potential at the core/s.a. interface were attributed to the difference in the alloying elements, Zn making the s.a. potential more active (negative) and Cu making the core potential more noble (positive). Heat treatment reduced the potential gradient at this interface for both sheets A and B (Figure 3.33), an effect attributed to the diffusion of Zn from the s.a. layer to the core layer and, to a lesser

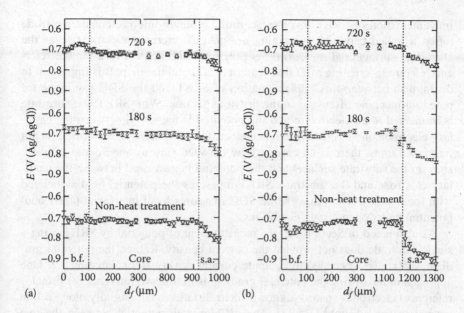

FIGURE 3.33 Corrosion-potential profile at the cross section of the sheet (a) A or (b) B when the droplet of 3 wt. % NaCl solution with a diameter of 200 μm was scanned in steps of 200 μm and an intermission of 30 s. The distance coordinate was transformed from the cross section (d_t) to the specimen thickness (d_f) by dividing by the ratio (d_t/d_f). The specimens were heated for 180 or 720 s at 868 K or nonheated. (Reprinted from Fushimi et al., *Electrochim Acta*, Vol. 53, (2008): p. 2529, Copyright (2008), with permission from Elsevier.)

extent, the diffusion of Cu from the core to the s.a. layer. This diffusion of alloying elements was confirmed by element profiling using rf-glow discharge optical emission spectrometry [171].

To summarize, capillary techniques may involve stationary or scanning capillaries, may utilize nonflowing or flowing droplets, and may be performed in potentiometric or voltammetric modes. In general, they suffer somewhat from lack of resolution compared with other scanning probe techniques, with droplet sizes often being in the range of tens to hundreds of microns. An advantage of the capillary techniques is the ability to perform localized electrochemical measurement (e.g., on individual grains of a polycrystalline metal or on heterogeneities of a metal alloy) without immersion of the entire substrate.

3.10 APPLICATIONS OF SRET AND SVET TO CORROSION STUDIES

3.10.1 SRET

As discussed in Section 3.2, ionic currents flow in the electrolyte between local anodic and cathodic sites on the surface of an immersed corroding metal substrate, creating a spatially varying potential field in the electrolyte. One implementation of SRET involves scanning a single microreference electrode (often a pulled capillary containing a Ag/AgCl reference electrode) near the corroding surface and measuring its potential with respect to a remote reference electrode, creating a 1D line scan or a 2D potential map. It is important to distinguish between this implementation of SRET and the SDC technique for potential mapping discussed in the previous section. With SRET, the substrate is immersed in an electrolyte and the potential is measured between the capillary electrode and a remote reference electrode. With SDC, the substrate is not immersed (thus, there is no current flow between various anodic and cathodic sites on the substrate surface) and the potential is measured between the capillary electrode and the substrate. SRET measures the potential field associated with local ionic current flow, while SDC measures local interfacial (corrosion) potential under open-circuit conditions.

As also noted in Section 3.2, the measurement of potential by SRET using a single electrode does not provide the current density. Rather, the potential gradient $\nabla \bar{V}(x, y, z)$ is needed to compute current density from Equation 3.2. One approach to measuring the potential gradient is to use two reference (or pseudoreference) electrodes spaced a known fixed distance apart, usually along a line normal to the substrate [175,27]. The difference in potential between the two electrodes divided by their separation distance provides the local potential field gradient (more specifically, the normal component of the potential gradient), from which the component of current density normal to the substrate surface can be determined. There are two limitations with this dual-electrode approach. First, the dual-electrode assembly is usually larger in size than the single (capillary) electrode, resulting in a loss of spatial resolution. The larger assembly might also partially shield the electrode surface or otherwise interfere with the current distribution. The second and perhaps more important limitation is the small potential difference that is generated between the two electrodes in a conductive solution, typically microvolts for current densities of several $\mu A/cm^2$. The noise and stability of the electrodes are also on the order of microvolts, so the poor signal-to-noise ratio greatly reduces the sensitivity of the technique.

Rosenfeld and Danilov appear to have been the first to measure current density above a corroding substrate using a dual reference electrode technique [175]. Their apparatus consisted of two independently positionable capillaries, each of 30–50 μm diameter and each leading to a nonpolarizable saturated calomel electrode. The substrate was mounted in a tank filled with electrolyte, which in turn was mounted to a device that permitted translation of the tank and substrate beneath the capillary electrodes as the potential difference between the electrodes was recorded. The two capillary tips could be positioned vertically or horizontally, permitting measurement of electric field strength and current density normal to or parallel to the substrate surface. Electric field strengths of up to ca. 2 V/cm and current densities of tens to hundreds of mA/cm^2 were recorded with a spatial resolution that appeared to be better than 100 μm. Using this approach, they characterized the origin and development of pitting corrosion of SSs in solutions containing $FeNH_4(SO_4)_2$ and NH_4Cl [175].

One of the earliest applications of single-electrode SRET in a corrosion study was by Isaacs and Kissel who monitored pitting sites on Type 304 SS exposed to 0.4 M FeCl$_3$ and 0.03 M HCl after various surface pretreatments [26]. The pretreatments included combinations of abrasion (with 600 grit SiC), electropolishing, cathodic polarization following electropolishing, and air oxidation at various temperatures following electropolishing or abrasion. These pretreatments influence both the composition and the thickness of the passive oxide layer. The number of pits as a function of time was determined by counting the positive potential peaks in the 2D maps, from which pit half-life (time taken for half of a given number of active pits to passivate) was determined. From this data they concluded that the abraded surfaces and the cathodically polarized electropolished surfaces were the more pit-resistant surface preparations, each yielding a half-life of less than 1 min. The electropolishing followed by oxidation at higher temperatures resulted in a much longer half-life, up to 480 min for oxidation at 250°C for 2 h, demonstrating that the probability of pits deactivating can be varied by over two orders of magnitude by changes in the thickness of the oxide. The half-lives of active pits increased with increased thickness of the oxide when grown thermally below 300°C, but at higher temperatures, significant dissolution of the oxide occurred, reflecting changes in the chemical properties of the oxide and reducing the half-life of active pits.

At the time of this writing, a literature search revealed approximately 38 journal publications citing the use of SRET for corrosion studies. Clearly an exhaustive review of these published applications of SRET is beyond the scope of this chapter. The majority of these applications use planar SRET as developed by Isaacs and coworkers, where the reference electrode is scanned in the x-y plane just above and parallel to a planar substrate. However, the use of a rotating cylinder of the substrate with a reference electrode that moves just above the surface along a line parallel to the axis of rotation has also been reported [176,29,177]. It is also important to note that while SRET is normally defined as employing one (or two) scanning but nonvibrating electrode(s) [5], techniques that use a scanning vibrating electrode have also been referred to as SRET [178].

A couple of interesting applications of SRET have appeared recently. Lin and coworkers integrated SRET with STM for *in situ* studies of localized corrosion of metals [179]. The hybrid technique employed an electrochemically-etched Pt–Ir tip coated with Apiezon wax for imaging both surface topography and local potential during the pitting corrosion of 18/8 SS (which has nominally 18% Cr, 8% Ni). Although Pt–Ir wire (electrochemically etched or mechanically cut) has been widely used as an STM tip, it was not clear that this tip would also be suitable for SRET measurements. However, the tip potential was found to be stable to within 3 mV during a 3 h immersion in 5% FeCl$_3$, the immersion solution used for initiating localized corrosion of the steel samples. In a SRET mode scan of the immersed steel sample, several pitting sites were identified in the 3 mm × 3 mm scan area, some of them disappearing with time as passivation occurred. To correlate an SRET image with an STM image, the tip was moved toward the surface until a tunneling current was sensed. The tip was then lifted ca. 200 nm from the surface and the specimen was scanned in SRET mode (3 mm × 3 mm scan area). When an active pit was identified from the SRET image, the tip was moved to its location and the topography was imaged in the STM mode around the pit site (30 μm × 30 μm scan area). In the example STM image shown by the authors, a distinct dissolution occurred at the pitting site detected by SRET and some corrosion product formed around the pit location. At only 200 nm above the substrate surface and without tip–substrate distance control, it was not clear how the authors avoided collision of the tip with the surface (due to sample tilt) or with corrosion product buildup on the surface during the SRET scan.

Tan and coworkers described an interesting combination of an array electrode with SRET to map current distribution from both the metal side of the interface (using individually addressable electrodes in the array) and from the electrolyte side using SRET [180–181]. Their array electrode consisted of 100 wires in a 10 × 10 array embedded in an epoxy matrix, a structure they refer to as the wire beam electrode (WBE). The wires were Al AA1100 (1.2 mm diameter) in one WBE [181] and mild steel (1.8 mm diameter) in another WBE [180]. Galvanic current flowing between any one selected wire and the remainder of the wires (shorted together) was measured along with the corrosion potential (OCP vs. SCE) of the selected wire, and this was repeated for all wires using an automatic switching device. Alternatively, current at individual wires under anodic polarization could be measured, generating a map of current distribution across the WBE. SRET measurements were carried out alternately with the WBE measurements to map current distribution from the electrolyte phase. During periods of no measurement, all wires were connected together to simulate a single metal substrate. They used a SP100 SRET system, which employed a two-electrode probe (1 mm separation between electrodes) positioned ca. 100 μm above the WBE. As noted above, the spatial resolution and current sensitivity of the twin-probe SRET are rather poor, but appeared adequate in these applications where the individual wires of the WBE were of 1–2 mm dimension and current densities approached 100s of μA/cm^2.

An example result is shown in Figure 3.34 for the steel WBE after immersion for 10 h in Evans solution (0.017 M NaCl, 0.008 M Na$_2$CO$_3$, pH 10.63) [180]. Six major anodes were detected by both the WBE measurements and by SRET, and their location did not change during the entire period of the experiment. The location of the anodes as mapped by these techniques also agreed well with the visual evidence of corrosion in the optical micrograph of Figure 3.34d. The cathodic sites as mapped by the two techniques were not in good agreement, however, with the cathodic sites detected by the SRET appearing to the right of each anode (for the left-to-right scan direction employed), whereas the cathodic sites detected by the WBE measurements were often distributed more uniformly across the WBE and/or localized in different areas (Figure 3.34). Furthermore, the locations of the cathodes detected by SRET were dependent on the SRET scan direction. The authors concluded that the movement of the (rather bulky) two-electrode probe of the SRET stirred the solution and enhanced mass transport over the WBE surface, thereby interfering with the steady-state cathodic

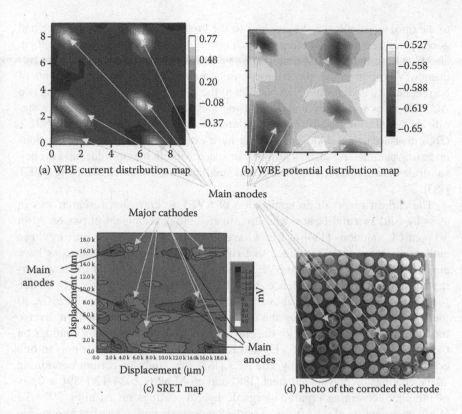

FIGURE 3.34 The WBE galvanic current (mA/cm^2) and potential (V vs. SCE) distribution maps (a and b), SRET map (c) measured from a mild steel WBE surface after about 10 h immersion in the Evans solution, together with the photo showing WBE surface after 4 days immersion (d). (Reprinted from Tan et al., *Corrosion Science*, Vol. 48, (2006): pp. 53–66, Copyright (2006), with permission from Elsevier.)

reactions [180]. As the pH was lowered in steps to 1.98 by addition of HCl, the number of anodes increased and the intensity of the anodic current decreased as the corrosion transitioned to a more uniform type of corrosion. During this transition, the correlation between the WBE measurements and SRET was lost, probably due to rapid and random interchange of anodic and cathodic sites. A similar type of study that used a WBE in combination with SVET for the study of Al alloy 2024-T3 was reported by Battocchi et al. [182].

3.10.2 SVET

To a great extent, the SVET overcomes the limitations described above for the two-electrode SRET. A single microelectrode, typically with a tip diameter of a few microns, is vibrated at a few hundred hertz either in one direction (the z direction, normal to the substrate) or in two directions (the x and z directions, each at different vibration frequencies) [28–29]. The vibration amplitude is comparable to the tip size, usually 10–20 μm, and so the probe is relatively nonintrusive and good spatial resolution is obtained. Each electrode vibration frequency results in an AC voltage induced at the tip that has a frequency equal to the vibration frequency. Vibration in two directions permits measurement of the components of current density both normal and parallel to the substrate surface. Each of the two AC voltages is separately measured using lock-in amplifiers, resulting in a significant signal-to-noise advantage compared to the single or dual electrode SRET (DC) measurements. Akid and Garma have evaluated the influence of various operating parameters, such as probe scan speed, vibration amplitude, and probe–substrate distance, on the resolution and smallest measurable signal using SVET [183].

The earliest report of an application of SVET to corrosion research was in 1984 by Ishikawa and Isaacs, who investigated the development of pits on Al in 30 ppm Cl⁻ solutions [184] (a Brookhaven National Laboratory report by these authors appeared in 1983). Since that time, over 198 journal publications have appeared on the application of SVET to corrosion, with ca. 148 of them appearing since the year 2000, evidence that SVET has become the method of choice for mapping current density above a corroding substrate. Thus far in 2009, 10 journal publications describing the use of SVET have appeared on such diverse topics as inhibitor-loaded Halloysite clay nanotubes for anticorrosion coatings on Cu and Al alloy [185–186], a SVET and Raman study of a processible polyaniline composite coating for Al alloy 2024-T3 [187], the study of cerium-containing silane coatings on galvanized steel [188] and on Al alloy 2024-T3 [189], a study of inhibitor-containing organic-inorganic hybrid coatings on Al alloy 2024-T3 [190], investigation of the effects of phosphate species on the localized corrosion of steel [191], a study of corrosion inhibition at galvanized steel cut edges by phosphate pigments [192], the use of surface-modified mesoporous SiO_2 nanoparticles as inhibitor containers for corrosion protection of Al alloy 2024-T3 [193], and characterization of the corrosion of welded pipeline steel [194]. In the remainder of this section, we describe a few example applications that illustrate the power of SVET for corrosion studies of various metals.

Prepainted galvanized steel sheets are often cut before use, creating a cut edge where the steel substrate and the zinc coating are exposed. The poor anode-(Zn)-to-cathode (steel) surface area ratio at the cut edge (1/100) could potentially lead to sacrificial protection of only short duration, resulting in blistering of the coating near the cut edge and corrosion of the steel. Thébault and coworkers studied this problem using SVET along with numerical simulations (using Comsol® software) to compare experimental and theoretical current distributions [195]. Because of the symmetry of the cut edge, only line scans of current density were required. As shown in Figure 3.35, the SVET measurements revealed three distinct kinetic regions over the cut edge surface: an anodic region localized over the zinc surface, a cathodic region localized on the steel but some distance from the Zn coating (area I), and a zone of partial inhibition of oxygen reduction on the steel adjacent to the zinc coating (area II, between the other two zones). The inhibition zone was attributed to the formation of an insulating ZnO/

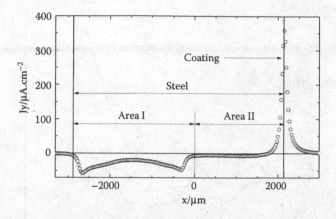

FIGURE 3.35 Experimental distribution profile of normal current densities at 50 μm above a sample of hot-dipped galvanized steel after 40 min of immersion in 0.03 M NaCl. (Reprinted from Thébault et al., *Electrochimica Acta,* Vol. 53, (2008): pp. 5226–5234, Copyright (2008), with permission from Elsevier.)

ZnO_2 layer on the steel surface from the interaction of cathodically generated hydroxide and anodically generated Zn^{2+} during the early stages of immersion when the anodic and cathodic current densities are a maximum. Thus, although both organic and metallic coatings are breached on the cut edge, the inhibiting properties of corrosion products that precipitate in this area provide a secondary anticorrosion mechanism. The results from numerical simulation demonstrated that the model of local inhibition explained both the normal and tangential current densities observed by SVET [195].

The corrosion control of aluminum alloy 2024-T3 by active coatings such as Mg-rich coatings [47] or conjugated polymers [45] has been investigated by SVET. Simões and coworkers used SVET in combination with EIS and SECM to investigate the mechanism of protection of this alloy at defect areas of a Mg-rich primer coating [47]. To simulate a coating defect, a cell was constructed in which a 1 mm × 2 mm Mg-primer coated piece of alloy was mounted (in a cylinder of epoxy resin) coplanar with and 2 mm away from a similarly sized piece of bare alloy, the two being electrically connected through a switch at the back of the sample holder. The cell was filled with DHS and SVET scans across the bare alloy before and after electrical connections with the coated alloy were performed. Before electrical connection, little activity was observed in the first hour of immersion, but after a few hours a pit typically nucleated and grew in intensity. Figure 3.36 shows line scans across such a pit before (curve a) and after (curve b) its development. Upon electrical connection with the coated alloy, the pit at the bare alloy rapidly decreased in intensity and cathodic current across the alloy increased (curve c). Three hours after connection, only cathodic current was observed at the bare alloy (curve d), being driven by the sacrificial action (oxidation) of the Mg particles in the Mg-rich primer. The indirect sensing of cathodic activity above both the bare and the coated substrates by SECM was discussed

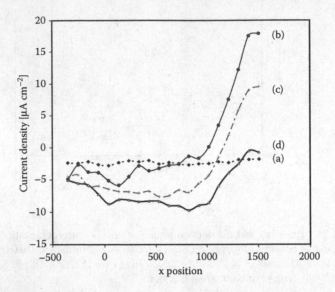

FIGURE 3.36 Ionic current density profiles over the Al alloy: (a) single electrode, 10 min after immersion; (b) with active pit, 21 h after immersion; (c) 5 min after coupling; and (d) 3 h after coupling. (Reprinted from Simões et al., *Progress in Organic Coatings,* Vol. 63, (2008) pp. 260–266, Copyright (2008), with permission from Elsevier.)

earlier in Section 3.6. The Mg-rich primer coating shows promise as a chromate replacement for corrosion protection of Al alloys.

Much of the research on conjugated polymers for corrosion control has been conducted with oxidized (p-doped) or occasionally neutral forms of the polymers on a variety of metals [45]. A recent report from our laboratory describes the first application of an n-doped conjugated polymer for the corrosion control of Al alloy 2024-T3 [48]. The polymer, poly(2,3-dihexylthieno[3,4-*b*]pyrazine) (or PC$_6$TP), was solvent cast on the alloy, forming a 2-μm thick coating, which was then electrochemically reduced (n-doped). SVET, galvanic coupling, and EIS measurements were described. For the SVET measurements, a defect was introduced through the coating to the alloy surface. Upon immersion in DHS, the SVET indicated an intense reduction current in the defect area as shown in Figure 3.37, attributed to oxygen reduction at the exposed alloy surface. Oxidation current was distributed more or less uniformly over the polymer surface, attributed to polymer oxidation from the n-doped state to the neutral state. The SVET result strongly suggests that the n-doped polymer cathodically protected the alloy exposed at the defect, a result that was corroborated by galvanic coupling measurements [48]. This work appears to be the first example of sacrificial protection of a metal by a fully organic coating and illustrates the power of SVET for probing interactions between active metals and active coatings.

Magnesium and its alloys are used in a range of aerospace and automotive applications, but its use is limited by its poor corrosion resistance. Williams and

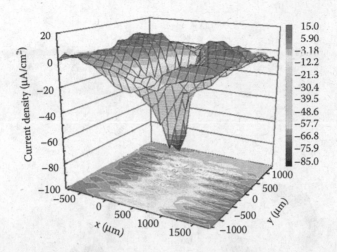

FIGURE 3.37 SVET current density map showing cathodic protection of AA 2024-T3 by an n-doped PC6TP film containing an artificial defect (scratch) after 5 min immersion in DHS (reduction current is negative). (Reprinted from Yan et al., *J Electrochem Soc*, Vol. 156, (2009): p. C360 (color online). Reproduced with permission of The Electrochemical Society.)

McMurray used SVET to study the localized corrosion of commercial-purity (CP) magnesium substrates at open circuit in aerated 5% NaCl electrolyte of pH 6.5 [196]. The SVET used a 125 μm Pt disk electrode (sealed in a glass sheath; total probe diameter of 250 μm) vibrated normal to the substrate surface with a vibration amplitude of ca. 30 μm at a height of 100 μm above the surface. The localized corrosion of Mg is commonly described as a disk-like morphology rather than the more usual pit morphology of iron and aluminum and their alloys. Upon initial immersion, an imperfect film of $Mg(OH)_2$ formed on the exposed Mg surface, a result of anodic dissolution to form Mg^{2+}, which in turn reacted with OH^- generated from cathodic hydrogen evolution. Localized currents detected by SVET during this period appeared to play a major role in the buildup of this semiprotective film and were used to estimate a film growth rate of 0.25 nm s^{-1} [196]. Typically, between 1 and 6 min following immersion, local breakdown of the surface film occurred, leading to SVET images such as those shown in Figure 3.38. These images revealed an expanding corrosion disk with intense anodic current at the circumference (approaching 40,000 μA/cm^2) and strongly cathodic current in the center, while current on the intact (uncorroded) film-covered Mg surface was negligible by comparison. The anodic ring was ca. 0.5 mm wide and the anodic current of the ring increased linearly with the radius of the corrosion disk, whereas the current density in the circular cathodic region remained largely unchanged. The uncorroded film-covered Mg surface appeared largely electrochemically inert and apparently played no part in the localized corrosion process. The localized corrosion mechanism involved anodic Mg dissolution constrained to a narrow ring at the (expanding) boundary between the

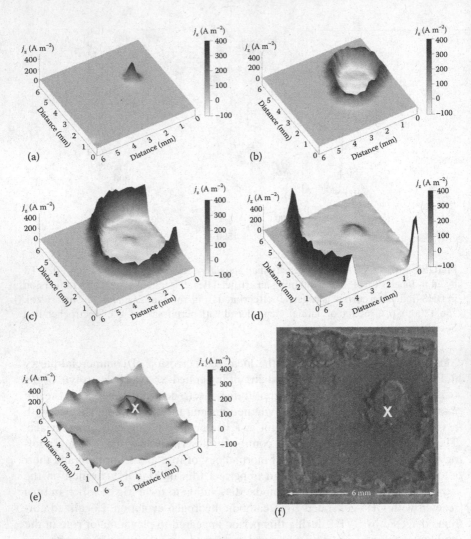

FIGURE 3.38 Surface plots showing the distribution of normal current density j_z above a CP magnesium sample freely corroding in aerated 5% w/v aqueous NaCl at 20°C. Data were obtained from SVET scans carried out (a) 6, (b) 24, (c) 36, (d) 54, and (e) 80 min after sample immersion. (f) The visual appearance of the same sample after 80 min immersion. A white cross on both e and f indicates the original point of breakdown. (Reprinted from Williams et al., *J Electrochem Soc*, Vol. 155, (2008): p. C340. Reproduced with permission of The Electrochemical Society.)

corroded surface and the intact (film-covered) surface, with the cathodic H_2 evolution reaction expanding to areas of previous anodic attack. This expansion of the cathodic reaction into areas that were previously anodic was attributed to a combination of factors, including surface area effects and ennoblement by iron impurity enrichment [196].

As a final example, we mention the work of Little and coworkers, who used SVET to study the correlation between localized anodic areas and biofilms of the marine bacterium Oceanospirillum on copper [197]. Copper and copper alloys have long been used in marine applications requiring corrosion resistance, mechanical strength and workability, thermal and electrical conductivity, and resistance to biofouling. In oxygenated seawater, a Cu_2O layer forms on all copper surfaces. Copper ions and electrons can pass through this layer, resulting in the precipitation of an outer layer of $Cu_2(OH)_3Cl$ which retards further anodic dissolution and slows the rate of oxygen reduction. It is the breakdown of this layer of $Cu_2(OH)_3Cl$ (by mechanical, chemical, or biological means) that leads to corrosion failure of copper and its alloys in seawater [197]. The bacterium Oceanospirillum colonizes copper-containing surfaces and secretes exopolymers that bind copper from the substrata, leading to a breakdown of the $Cu_2(OH)_3Cl$ layer and the formation of copper concentration cells, resulting in an accelerated rate of corrosion. The SVET experiments, using a 20-μm platinized SS probe, demonstrated a spatial correlation between Oceanospirillum biofilms (detected by fluorescent staining) and anodic currents. Since this early work, only two other papers have appeared describing the application of SVET to biocorrosion [198–199]. Clearly SVET can make significant contributions to our understanding of microbially influenced corrosion, and this is a fertile area for future research.

3.11 APPLICATIONS OF LEIS/LEIM TO CORROSION STUDIES

Although SVET and SRET have been used rather extensively in studies of localized corrosion, they are limited to systems that are electrochemically active, i.e., DC currents must flow in the electrolyte. Though not as widely used as SVET, local impedance techniques may be used for systems exhibiting little or no DC current flow, such as coated metal substrates or passivated metals. LEIS involves recording an impedance spectrum over a range of frequencies at a single location above the substrate, whereas LEIM provides spatially resolved changes in impedance (or its inverse, the admittance) across the substrate at a single frequency. These techniques have been applied to investigations of the impedance properties of pits, grain boundaries, welds, and coating defects.

3.11.1 First Techniques

The earliest attempts to develop scanning local impedance systems employed electrochemical microcells (of the type discussed in Section 3.9, but with the substrate immersed in electrolyte) to form a thin layer cell which locally polarized the area of interest and confined the current generated by the local polarization [200–202]. This approach produced qualitative results, but the thin layer geometry could significantly influence the natural corrosion process (for example, by altering mass transport) and its effectiveness in confining the current to the local area depended greatly on the impedance properties of the substrate [31]. A nonscanning approach described by Dehri and coworkers used an array of small cells to study the cut edge corrosion of polyester-coated galvanized steel [203]. Each cell was formed by gluing a 7-mm diameter PVC tube to the substrate, with 2-mm spacing between cells, and each cell contained a Pt counter/reference electrode for two-electrode EIS. Despite the rather poor spatial resolution of their approach, they were able to identify a greater number of defective coating areas near the cut edge than farther away.

3.11.2   TWO-ELECTRODE PROBE

An improved approach was described by Lillard, Moran, and Isaacs, who applied a sinusoidal AC perturbation to the substrate as in conventional EIS (employing a reference electrode and remote counter electrode) but obtained the local current density near the substrate surface from potential difference measurements using a two-electrode probe [31]. Their probe consisted of two parallel capillaries, each containing a platinized Pt pseudoreference electrode and terminating with an inside diameter of 100 µm. The capillary tips were displaced vertically from one another by ca. 900 µm and the probe was scanned 400 µm above the substrate surface. The local AC current density was derived from the measured AC potential difference between the microelectrodes, their known separation distance, and the solution conductivity. Local impedance was then obtained from the ratio of the (global) applied AC voltage and the local AC current density. The accuracy of the approach was demonstrated by the good agreement between the local and global (conventional) impedance spectra obtained for a pure (homogeneous) Al substrate. Although the spatial resolution of the technique was limited by the size of the probe, the ability to locate heterogeneities on substrate surfaces was demonstrated from impedance maps of a model electrode consisting of a pure Al substrate into which was embedded a 0.127-cm diameter molybdenum wire. Lillard and coworkers have also applied their technique to examine the effect of microstructure on passive film formation and breakdown of aluminum–tantalum alloys [204] and to locate and examine the electrochemical properties of artificial and natural defects in a heat-cured, urea-formaldehyde modified epoxy coating [205].

An order of magnitude improvement in spatial resolution (to 30–40 µm) was achieved by Zou et al. using two Parylene-coated Pt–Ir microelectrodes with sharpened tips of ca. 2 µm [206]. The tips were platinized to a final tip size of ca. 10 µm and the electrodes were mounted at an angle to the substrate with their tips adjusted one above the other so that a straight line passed through the two tips perpendicular to the substrate surface. Experimental measurements on model systems were made using Pt disks as current sources of defined geometry in an insulating plane. The experimental AC current distributions compared favorably with theoretical calculations. The influences of distance between the probe tips and the probe height above the substrate were investigated. For small tip separation, the measured AC potential difference between tips increased in proportion to tip separation, indicative of a linear potential gradient between the tips and a desirable condition for making measurements [206]. Not surprisingly, the measured AC potential difference increased as probe height decreased, but practical considerations (sample tilt, corrosion product buildup, etc.) usually limit the smallest probe–substrate distance to 50 μm or greater. Additionally, the optimum tip separation and probe height that maximizes both sensitivity and resolution of the measurement depend on the dimension(s) of the current source(s) on the substrate surface being examined, and this is usually not known *a priori*. This technique represents state of the art in local impedance measurements and has been applied to the study of heterogeneities in organic coatings on steel surfaces due to underfilm deposits and microblisters [207] and to the kinetics of pit initiation and growth on SSs [208–209].

3.11.3 Single Vibrating Probe

Keddam, Takenouti, and coworkers used a single vibrating electrode for measuring local electrochemical impedance by modifying a commercial SVET instrument from Applicable Electronics [210–211]. With a galvanostat and the generator signal from a frequency response analyzer (FRA), they applied an AC current of frequency ω to the substrate while vibrating the probe at a frequency Ω such that $\omega < \Omega$. Thus, the vibrating probe generates an AC voltage having two frequency components. The probe voltage versus a remote reference electrode is fed through the commercial SVET circuitry and to the X-input of the FRA, producing a signal that contains the local current density. The probe voltage versus the substrate (working electrode) is sent to the Y-input of the FRA, producing a signal that contains both local current density and local potential. After a specified calibration procedure, the transfer function between the two signals yields the local impedance at frequency ω [210]. The authors demonstrated the technique on a model electrode simulating a nonuniform sample, fabricated by embedding an iron wire (0.25 mm diameter) in a 316L SS matrix (1.5 cm diameter). Analysis of the spectrum obtained from 0.1 to 100 Hz above the center of the iron wire gave a reasonable value for the double layer capacitance of corroding iron (80 $\mu F/cm^2$) and the data in the capacitive region satisfied the Kramers–Kronig transformation.

There are two advantages to the vibrating probe approach. First, a single vibrating probe is smaller than a two-electrode probe, which should lead to increased spatial resolution while minimizing screening by the probe of the current density generated at the substrate. Second, since local current as well as local potential are measured, the true local impedance is obtained, in contrast to the approach using a two-electrode probe where local current density but global potential are used in computing impedance. Numerical simulations suggest that use of the two-electrode probe may lead to grossly inaccurate local impedance, especially if regions of the substrate exhibit large variation in polarization resistance (factors larger than 1,000), not uncommon for localized corrosion on passive substrates [212]. The disadvantages of the approach include the increased complexity of the instrumentation and the restriction that the frequency of the current perturbation ω must be less than the frequency of probe vibration Ω, although most of the useful information in corrosion systems occurs at lower frequencies. A comparison of the vibrating single probe and the stationary two-electrode probe for measuring local electrochemical impedance has been provided by Takenouti and coworkers [33].

3.11.4 RECENT APPLICATIONS

There have been a number of applications of LEIS/LEIM to corrosion studies of bare metals and/or alloys, such as Fe [213–217], Al [218], and Mg [218–221], but the majority of applications have been for coated metals for which DC techniques such as SVET may not be applicable. Three recent reports are discussed here to illustrate the power of this technique.

Galicia and coworkers used the two-electrode probe technique for studying the local impedance of the Mg alloy AZ91 [220–221]. This alloy has an average composition (wt. %) of 90.05% Mg, 8.90% Al, 0.85% Zn, and 0.20% Mn. The surface of this alloy consists of two phases, the α-phase (containing 96.5% Mg and 3.05% Al) and the β-phase (containing 69.60% Mg and 27.45% Al) [221]. The substrate was a 1-cm diameter disk of the alloy (prepared from rod laterally insulated with cataphoretic paint and encased in epoxy) immersed in aerated 1 mM Na_2SO_4. A pure Mg (>99.9%) disk substrate was prepared in the same fashion for comparison measurements. The two-electrode probe consisted of either two platinized Pt wires (20 μm diameter) or two Ag/AgCl wires (40 μm diameter). The height of the probe above the substrate was not specified. The global and local impedance spectra of the pure (homogeneous) Mg sample were in agreement except for a high-frequency inductive component in the LEIS spectrum, a behavior ascribed to cell geometry and not to the electrochemical behavior of the Mg sample [214].

LEIS performed at selected locations above the AZ91 alloy surface revealed that the amplitude of the impedance over the β-phase was about three times higher than that over the α-phase, and the time constant for a high-frequency capacitive

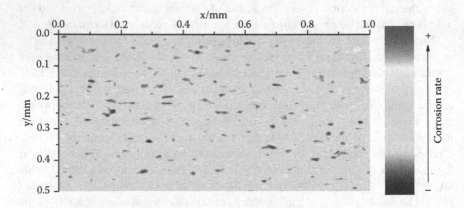

FIGURE 3.39 LEIS image of the AZ91 alloy after 24 h immersion in 1 mM Na_2SO_4 solution at the corrosion potential. The mapping was performed at a fixed frequency (100 Hz) by moving the probe over the alloy surface. (Reprinted from Galicia et al., *Corrosion Science*, Vol. 51, (2009): pp. 1789–1794, Copyright (2009), with permission from Elsevier.)

loop was 10 times smaller for the α-phase, indicating that the corrosion rate of the β-phase is lower than that of the α-phase. This difference was attributed to the increased content of Al in the β-phase [221]. A map produced by LEIM at 100 Hz is shown in Figure 3.39, revealing variations in corrosion rates across the surface that match the alloy microstructure as determined by SEM. This is a beautiful example of the ability of LEIM to reveal the influence of alloy microstructure on corrosion behavior. Simulated impedance spectra based on the dissolution model of pure Mg appeared to also describe the corrosion mechanism of the AZ91 alloy. It was concluded that the corrosion mechanism of the two phases of the AZ91 alloy was the same, at least at short immersion times, controlled by Mg dissolution, but the kinetics of dissolution of the β-phase was slower than that of the α-phase [221] due to its larger aluminum content that reduces its electrochemical activity. It should be noted that Baril and coworkers also examined AZ91 Mg alloy using LEIS [219], but the spatial resolution of their two-electrode probe was not sufficient to provide information on the influence of alloy structure.

Nguyen and collaborators used LEIS with a vibrating probe (together with SVET, *in situ* Raman spectroscopy, and galvanic coupling measurements) to study the corrosion protection of iron by the conducting polymers polypyrrole (PPy) and a composite film of poly(1,5 diamino-naphthalene) (PDAN) and polyaniline (PAni) [222]. Two types of substrate electrodes were used. One substrate electrode consisted of an insulated 0.25-mm diameter bare iron disk (simulating a defect) inside a 5-mm diameter polymer-coated iron ring with wires providing electrical coupling of the two. The other substrate electrode was a 5-mm diameter polymer-coated iron disk with a defect (pinhole) introduced through the coating. The goal was to determine the mechanism of protection within the defect area, the two prevailing theories being protection by formation of insoluble salts (active iron mechanism) or protection by formation of a passive layer of Fe_2O_3. Global impedance measurements were dominated by the response from the conducting polymer and revealed no information about the defect area. The shape of the LEIS spectra obtained over the defect corresponded well to that of an iron surface in the passive state, exhibiting a highly capacitive feature. Their conclusion that the iron in the defect area was passivated was corroborated by the results from the other techniques. They further suggested that when the defect area is large, there may be insufficient charge in the conducting polymer to passivate the defect, and/or the ohmic drop may obstruct such passivation, perhaps leading to the apparent contradictions regarding the mechanism of protection reported in the literature [222].

As a final example, we mention the efforts of Taylor and coworkers who have applied LEIS/LEIM to the study of the degradation of organically coated metals in aqueous environments [34,223–228], the goal being to understand the origins and growth mechanisms of underfilm corrosion in the early stages of exposure. Underfilm corrosion could initiate as a result of heterogeneities within the coating (e.g., pores or regions of low crosslink density that might facilitate ion transport) or as a result of heterogeneities within the substrate (e.g., intermetallic compounds that might lead to microgalvanic coupling). Perhaps a juxtaposition of these two types of heterogeneities is required. Their LEIS/LEIM system employed a two-electrode probe consisting of two Teflon-coated silver wires (125 μm diameter) with a vertical tip separation of 1.5 mm. The exposed tips of the silver wires were chloridized to form Ag/AgCl reference electrodes [34]. Using LEIM, they were able to detect subtle coating defects that led to corrosion before corrosion could be observed visually [34].

In one study, Mierisch and Taylor used LEIM to investigate the local underfilm corrosion on the Al alloy AA 2024-T3 coated with various organic films (neat epoxy, polyurethane, or vinyl resins of 5–20 μm thickness) immersed in NaCl solutions ranging in concentration from 0.1 to 0.6 M [228]. The LEIM experiments were performed with a 15 mV rms sinusoidal voltage perturbation (applied about the OCP) at a single frequency of 700 Hz applied to the alloy substrate. The probe height above the substrate was fixed at 25 μm and scanning was accomplished by moving the substrate (rather than scanning the probe as is usually done). Their results are plotted as admittance since defect sites and higher electrochemical activity then appear as peaks rather than valleys.

The admittance magnitude of the various coated substrates was monitored from initial immersion (when no underfilm corrosion was visible) to later stages of immersion when underfilm corrosion was clearly visible. Taylor and coworkers identified several types of underfilm corrosion after various exposure times, which they labeled according to the visible color of the resulting blisters (red, black, white, and clear). They had previously published the details on the local electrochemical and chemical characterization of these blisters using capillary electrophoresis, local pH, and local open-circuit potential measurements and reported that the red blisters were more electrochemically active than the others and caused aggressive pitting of the substrate [225]. The other blister types grew more slowly and caused little or no damage to the substrate. The red blisters appeared within 24 to 72 h after immersion and grew to 1–7 mm in diameter over 24 to 48 h after initial appearance.

One feature observed for growing red blisters was that they typically began as a discrete initiation site, forming a primary lobe. At later times, secondary lobes formed off the initial lobe, an example of which is shown in the optical image at the top of Figure 3.40 for AA 2024-T3 coated with polyester polyurethane and immersed in 0.6 M NaCl. An optical image of the region mapped by LEIM is also shown in the figure along with the admittance map. Two separate admittance peaks of differing magnitude were observed for the two lobes, with the peak at the primary lobe where the underfilm corrosion initially developed being significantly higher than the peak at the secondary lobe where the underfilm corrosion had spread (Figure 3.40), an observation consistently obtained for these multilobed red blisters. Thus, both LEIM and local OCP measurements indicate the initiation site to be more electrochemically active. SEM and energy dispersive spectroscopy of a blister site following coating removal revealed higher atomic ratios of Cu and lower atomic ratios of Al and Mg at the initiation site where the admittance in LEIM was the highest (again indicating a higher degree of corrosion or electrochemical activity), while the

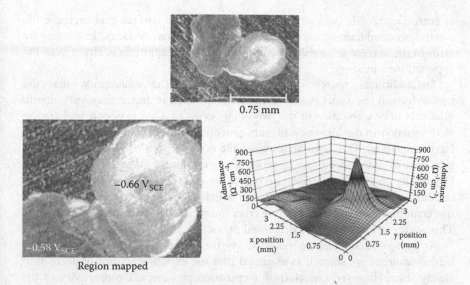

FIGURE 3.40 Red blister beneath polyester polyuethane coated AA2024-T3 in 0.6 M NaCl. The optical image (top) shows an example of multilobed blister containing an initiation site (lighter red lobe with open-circuit potential of -0.66 V$_{SCE}$) and a secondary lobe (darker red lobe with OCP of -0.58 V$_{SCE}$). The region mapped by LEIM is shown in the image on the bottom left. Local open-circuit potential values are also shown for each lobe. LEIM of the blister is shown in the plot on the bottom right. The initiation site appears to be more active (peak) than the secondary lobe. (Reprinted from Mierisch and Taylor, *J Electrochem Soc*, Vol. 150 (7), (2003): p. B303. Reproduced with permission of The Electrochemical Society.)

substrate at the secondary lobes showed no significant difference in composition from unaffected regions. These results demonstrate that admittance peaks can arise not only due to coating defects such as subsurface bubbles, under-film salt deposits, or pinholes [34], but also due to a substrate heterogeneity beneath a hydrated coating [228]. However, it remains unclear whether local variations in coating properties also played a role in the observed admittance measurements.

3.11.5 SOME PRECAUTIONS

We conclude this section by mentioning a few caveats in using the current mapping techniques of SRET/SVET and LEIS/LEIM. Calibration of these techniques usually involves injecting a known current density (typically from a point source) into the bulk electrolyte for the purpose of determining a collection of "constants" such as electrolyte conductivity, electrode spacing (for a two-electrode probe) or vibration amplitude(s) (for a vibrating probe), instrument gain, etc. It is then assumed that the electrolyte conductivity near the substrate surface is the same as that in the bulk electrolyte. However, surface reactions at corroding metals typically generate ions near the surface that increase the electrolyte conductivity, especially for highly reactive surfaces, leading to an error in the current measurement (the true current magnitude is larger than the apparent measured one).

An additional source of error arises from the assumption that the probe/electrolyte contact potential is constant, when in fact it may vary due to gradients in concentration of the two components of a redox couple and associated variation in the local equilibrium potential of the solution (as given by the Nernst equation). This error arises when the probe is positioned within the diffusion layer of redox species being produced or consumed at the substrate surface. Depending on the reactions occurring at the substrate surface, the gradient in equilibrium potential may be positive or negative, leading, respectively, to negative error (more cathodic) or positive error (more anodic) in the measured current. This source of error has been discussed previously [212].

As with global impedance spectroscopy, time is a critical factor in these localized measurements since it is assumed that the experiments are performed at steady state. However, especially for corrosion processes, the reactivity of the substrate surface may change during the course of the measurement, leading to results that are difficult to interpret. One simple way to verify the steady-state assumption is to perform consecutive (spatial or frequency) scans. Significantly differing results would suggest a breakdown in the steady-state assumption. Finally, it has been reported that frequency dispersion and imaginary contributions to ohmic impedance may result from cell geometry-induced current and potential distributions, and these factors as well as guidelines for the design of LEIS experiments have been discussed by Frateur et al. [214].

3.12   CONCLUSIONS

The scanning electrochemical probe techniques discussed in this chapter have contributed greatly to our understanding of localized corrosion mechanisms as well as the role of inhibitors and coatings in reducing corrosion. They have also led to an improved understanding of the nature of electron transfer at passive metal surfaces and at complex heterogeneous alloy surfaces. We have limited the discussion to those techniques for which the probe response involves an electrochemical phenomenon, thus requiring at least two electrodes and an electrolyte phase. The spatial resolution of the techniques is typically on the micron scale, although a few SECM measurements have been made at the submicron scale [7]. The temporal resolution for mapping is typically on the order of minutes, the time required to acquire a map of suitable spatial resolution over an area of the order of millimeters squared. These spatial and temporal resolutions are adequate for many applications, as has been demonstrated in this chapter. However, upon initial exposure of a bare or coated metal to a corrosive environment (e.g., by immersion), events occurring on the nm scale (e.g., at grain boundaries, at defects in an oxide layer, and/or at holidays in an organic or inorganic coating) and within the first few seconds of exposure are critical to the subsequent corrosion process. To study these early events, higher scan rates and smaller probe size will be required, and this will require new and innovative approaches to tip fabrication, tip insulation, tip positioning, and tip scanning [7]. Continued developments in combining SECM with other scanning probe techniques such as AFM and NSOM will almost certainly lead to fruitful results.

Complicating the application of SECM to corrosion studies is the stochastic nature of localized corrosion events. It is not yet possible to predict when or where on a (bare or coated) substrate surface a pit may develop and, thus, prepositioning of the probe in anticipation of such an event is not possible. However, amperometric SECM has the ability to modify local solution conditions (e.g., pH, Cl^- concentration, etc.) near the substrate surface, generating a locally aggressive chemical environment. This powerful approach permits initiation of a pit at a predetermined location so that its life cycle (propagation and perhaps passivation) can be studied, and several examples of this approach have been presented in Sections 3.4 and 3.5 of this chapter. While technically challenging, the use of multiple probes in a single scanning head (that permits initiation of local corrosion using one probe and interrogation of current, pH, metal ion concentration, local impedance, etc., using the additional probe(s)) is an attractive approach.

The traditional corrosion assessment techniques (e.g., weight loss measurements, EIS, and potentiodynamic polarization experiments) will continue to play an important role in corrosion research. However, these are global techniques that provide a surface-averaged response and do not provide sufficient spatial or temporal information on localized corrosion events. Within the past decade, there has been a rapid increase in application of the scanning electrochemical probe techniques discussed in this chapter to the study of corrosion. This trend is certain to continue.

ACKNOWLEDGMENTS

We thank our students, postdoctoral associates, and visiting faculty and collaborators, past and present, who have contributed greatly to our efforts to explore corrosion and corrosion control mechanisms through the use of scanning probe techniques. The support of our research by the Air Force Office of Scientific Research and the Concordia Chemistry Research Endowment is gratefully acknowledged.

REFERENCES

1. Jones, D. A., *Principles and Prevention of Corrosion*, 2nd Edition, Englewood Cliffs, New Jersey: Prentice Hall, 1996.
2. Department of Transportation, *Corrosion Costs and Preventative Strategies in the United States,* Tech Brief FHWA-RD-01-157, 2002, p. 16.
3. Miller, D., Corrosion control on aging aircraft: What is being done?, *Mater Performance,* Vol. 29, (1990): p. 10.

4. Bennet, J. A., and H. Mindlin, Metallurgical aspects of the failure of the point pleasant bridge, *J Test Eval,* Vol. 1, (March 1973): p. 152.
5. Kelly, R. G., J. R. Scully, D. W. Shoesmith, and R. G. Buchheit, *Electrochemical Techniques in Corrosion Science and Engineering,* New York: Marcel Dekker, Inc., 2003.
6. Niu, L., Y. Yin, W. Guo, M. Lu, R. Qin, and S. Chen, Application of scanning electrochemical microscope in the study of corrosion of metals, *J Mater Sci,* Vol. 44, (2009): p. 4511.
7. Bard, A. J., and M. V. Mirkin (eds.), *Scanning Electrochemical Microscopy,* New York: Marcel Dekker, Inc., 2001.
8. Eckhard, K., and W. Schuhmann, Alternating current techniques in scanning electrochemical microscopy (AC-SECM), *Analyst,* Vol. 133, (2008): p. 1486.
9. Denuault, G., G. Nagy, and K. Toth, Potentiometric probes, in: *Scanning Electrochemical Microscopy,* A. J. Bard, and M. V. Mirkin (eds.), p. 397, New York: Marcel Dekker, 2001.
10. Ammann, D., *Ion-Selective Microelectrodes: Principles, Design and Application,* Berlin, Germany: Springer-Verlag, 1986.
11. Thomas, R. C., *Ion-Sensitive Intracellular Microelectrodes. How to Make and Use Them,* London, England: Academic Press, 1978.
12. Maile, F. J., T. Schauer, and C. D. Eisenbach, Evaluation of corrosion and protection of coated metals with local ion concentration technique (LICT), *Prog Org Coat,* Vol. 38, (2000): p. 111.
13. Lin, C. J., R. G. Du, and T. Nguyen, In-situ imaging of chloride ions at the metal/solution interface by scanning combination microelectrodes, *Corrosion,* Vol. 56, (2000): p. 41.
14. Gyurcsanyi, R. E., A.-S. Nyback, A. Ivaska, K. Toth, and G. Nagy, Novel polypyrrole based all-solid-state potassium-selective microelectrodes, *Analyst,* Vol. 123, (1998): p. 1339.
15. Suter, T., and H. Böhni, The microcell technique, in: *Analytical Methods in Corrosion Science and Engineering,* P. Marcus, and F. Mansfeld (eds.), p. 649, CRC Press, 2006.
16. Suter, T., and H. Böhni, Microelectrodes for studies of localized corrosion processes, *Electrochim Acta,* Vol. 43, (1998): p. 2843.
17. Krawiec, H., V. Vignal, and Z. Szklarz, Local electrochemical studies of the microstructural corrosion of AlCu4Mg1 as-cast aluminium alloy and influence of applied strain, *J Solid State Electrochem,* Vol. 13, (2009): p. 1181.
18. Suter, T., and H. Böhni, A new microelectrochemical method to study pit initiation on stainless steels, *Electrochim Acta,* Vol. 42, (1997): p. 3275.
19. Lohrengel, M. M., A. Moehring, and M. Pilaski, Electrochemical surface analysis with the scanning droplet cell, *Fresenius J Anal Chem,* Vol. 367, (2000): p. 334.
20. Spaine, T. W., and J. E. Baur, A positionable microcell for electrochemistry and scanning electrochemical microscopy in subnanoliter volumes, *Anal Chem,* Vol. 73, (2001): p. 930.
21. Eng, L., E. Wirth, T. Suter, and H. Böhni, Non-contact feedback for scanning capillary microscopy, *Electrochim Acta,* Vol. 43, (1998): p. 3029.
22. Lohrengel, M. M., I. Kluppel, C. Rosenkranz, H. Bettermann, and J. W. Schultze, Microscopic investigations of electrochemical machining of Fe in $NaNO_3$, *Electrochim Acta,* Vol. 48, (2003): p. 3203.
23. Fushimi, K., S. Yamamoto, H. Konno, and H. Habazaki, Limiting current in a flowing-electrolyte-type droplet cell, *Chem Phys Chem,* Vol. 10, (2009): p. 420.
24. Lohrengel, M. M., Electrochemical capillary cells, *Corros Eng, Sci Technol,* Vol. 39, (2004): p. 53.

25. Evans, U. R., Report on corrosion research work at Cambridge University interrupted by the outbreak of war, *J Iron Steel Inst,* Vol. 141, (1940): p. 219.
26. Isaacs, H. S., and G. Kissel, Surface preparation and pit propagation in stainless steels, *J Electrochem Soc,* Vol. 119, (1972): p. 1628.
27. Isaacs, H. S., and B. Vyas, Scanning reference electrode techniques in localized corrosion, *ASTM Spec Tech Publ,* Vol. 727, (1981): p. 3.
28. Isaacs, H. S., Measurement of the galvanic corrosion of soldered copper using the scanning vibrating electrode technique, *Corros Sci,* Vol. 28, (1988): p. 547.
29. Sargeant, D. A., Electrochemical scanning probes: limitations, conflicts and observations. A direct comparison of two variants of the scanning reference electrode technique, *Corros Prev Control,* Vol. 44, (1997): p. 91.
30. Isaacs, H. S., and R. Jackson, Detection and performance of defects in aluminum coated steels, *Proc Electrochem Soc,* Vol. 84–83, (1984): p. 339.
31. Lillard, R. S., P. J. Moran, and H. S. Isaacs, A novel method for generating quantitative electrochemical impedance spectroscopy, *J Electrochem Soc,* Vol. 139, (1992): p. 1007.
32. Bayet, E., L. Garrigues, F. Huet, M. Keddam, K. Ogle, N. Stein, and H. Takenouti, Performances and limitations of current probes for LEIS measurements. A comparative study of vibrating (SVET) and double-electrode, *Proc Electrochem Soc,* Vol. 99–28, (2000): p. 200.
33. Takenouti, H., G. Galicia, M. Keddam, and V. Vivier, Vibrating single probe and stationary bi-probes to measure local electrochemical impedance spectroscopy, *Bulg Chem Commun,* Vol. 38, (2006): p. 165.
34. Wittmann, M. W., R. B. Leggat, and S. R. Taylor, The detection and mapping of defects in organic coatings using local electrochemical impedance methods, *J Electrochem Soc,* Vol. 146, (1999): p. 4071.
35. Philippe, L. V. S., G. W. Walter, and S. B. Lyon, Investigating localized degradation of organic coatings, *J Electrochem Soc,* Vol. 150, (2003): p. B111.
36. Stratmann, M., and G. S. Frankel (eds.), Corrosion and Oxide Films, in: *Encyclopedia of Electrochemistry,* A. J. Bard, and M. Stratmann (eds.), Vol. 4, Weinheim, Germany: Wiley-VCH, 2003.
37. Pourbaix, M., *Atlas of Electrochemical Equilibria in Aqueous Solutions,* 2nd Edition, National Association of Corrosion Engineers: Houston, Texas, 1974.
38. Wagner, C., and W. Traud, The interpretation of corrosion phenomena by superimposition of electrochemical partial reactions and the formation of potentials of mixed electrodes, *Zeitschrift fur Elektrochemie und Angewandte Physikalische Chemie,* Vol. 44, (1938): p. 391.
39. Bard, A. J., and L. R. Faulkner, *Electrochemical Methods: Fundamentals and Applications.* 2nd Edition, New York: Wiley, 2001.
40. Schultze, J. W., and A. W. Hassel, Passivity of metals, alloys, and semiconductors, in: *Corrosion and Oxide Films,* M. Stratmann, and G. S. Frankel (eds.), p. 216, Weinheim, Germany: Wiley-VCH, 2003.
41. Grundmeier, G., and A. Simoes, Corrosion protection of metals by organic coatings, in: *Corrosion and Oxide Films,* M. Stratmann, and G. S. Frankel (eds.), p. 500, Weinheim, Germany: Wiley-VCH, 2003.
42. Magnussen, O. M., Corrosion protection by inhibition, in: *Corrosion and Oxide Films,* M. Stratmann, and G. S. Frankel (eds.), p. 435, Weinheim, Germany: Wiley-VCH, 2003.
43. Flick, E. W., *Corrosion Inhibitors – An Industrial Guide,* 2nd Edition, Norwich, New York: William Andrew Publishing/Noyes, 1993.
44. Cohen, S. M., Replacements for chromium pretreatments on aluminum, *Corrosion,* Vol. 51, (1995): p. 71.

45. Tallman, D. E., and G. P. Bierwagen, Corrosion protection using conducting polymers, in: *Handbook of Conducting Polymers*, 3rd Edition, T. A. Skotheim, and J. R. Reynolds (eds.), p. 15/1, Boca Raton: CRC Press, 2007.

46. Hare, C., Corrosion control of steel by organic coatings, in: *Uhlig's Corrosion Handbook, 2nd Edition*, R. W. Revie (ed.), p. 1023, New York: John Wiley & Sons, 2000.

47. Simões, A., D. Battocchi, D. Tallman, and G. Bierwagen, Assessment of the corrosion protection of aluminium substrates by a Mg-rich primer: EIS, SVET and SECM study, *Prog Org Coat,* Vol. 63, (2008): p. 260.

48. Yan, M. C., D. E. Tallman, S. C. Rasmussen, and G. P. Bierwagen, Corrosion control coatings for aluminum alloys based on neutral and n-doped conjugated polymers, *J Electrochem Soc,* Vol. 156, (2009): p. C360.

49. Plieth, W., and A. Bund, Corrosion protection by metallic coatings, in: *Corrosion and Oxide Films*, M. Stratmann, and G. S. Frankel (eds.), p. 567, Weinheim, Germany: Wiley-VCH Verlag GmbH & Co. KGaA, 2003.

50. Mudali, U. K., H. S. Khatak, and B. Raj, Anodic and cathodic protection, in: *Corrosion and Oxide Films*, M. Stratmann, and G. S. Frankel (eds.), p. 393, Weinheim, Germany Wiley-VCH Verlag GmbH & Co. KGaA, 2003.

51. Wipf, D. O., Initiation and study of localized corrosion by scanning electrochemical microscopy, *Colloids Surf A,* Vol. 93, (1994): p. 251.

52. Still, J. W., and D. O. Wipf, Breakdown of the iron passive layer by use of the scanning electrochemical microscope, *J Electrochem Soc,* Vol. 144, (1997): p. 2657.

53. Fushimi, K., K. Azumi, and M. Seo, Use of a liquid-phase ion gun for local breakdown of the passive film on iron, *J Electrochem Soc,* Vol. 147, (2000): p. 552.

54. Fushimi, K., and M. Seo, Initiation of a local breakdown of passive film on iron due to chloride ions generated by a liquid-phase ion gun, *J Electrochem Soc,* Vol. 148, (2001): p. B450.

55. Gabrielli, C., S. Joiret, M. Keddam, H. Perrot, N. Portail, P. Rousseau, and V. Vivier, Development of a coupled SECM-EQCM technique for the study of pitting corrosion on iron, *J Electrochem Soc,* Vol. 153, (2006): p. B68.

56. Gabrielli, C., S. Joiret, M. Keddam, H. Perrot, N. Portail, P. Rousseau, and V. Vivier, A SECM assisted EQCM study of iron pitting, *Electrochim Acta,* Vol. 52, (2007): p. 7706.

57. Gabrielli, C., S. Joiret, M. Keddam, N. Portail, P. Rousseau, and V. Vivier, Single pit on iron generated by SECM: An electrochemical impedance spectroscopy investigation, *Electrochim Acta,* Vol. 53, (2008): p. 7539.

58. Luong, B. T., A. Nishikata, and T. Tsuru, Scanning electrochemical microscope study on pitting corrosion of iron, *Electrochemistry,* Vol. 71, (2003): p. 555.

59. Fushimi, K., and M. Seo, An SECM observation of dissolution distribution of ferrous or ferric ion from a polycrystalline iron electrode, *Electrochim Acta,* Vol. 47, (2001): p. 121.

60. Tanabe, H., Y. Yamamura, and T. Misawa, In situ ionic imaging for pitting corrosion sites on austenitic stainless steels with scanning electrochemical microscopy, *Mater Sci Forum,* Vol. 185–188, (1995): p. 991.

61. Misawa, T., and H. Tanabe, In-situ observation of dynamic reacting species at pit precursors of nitrogen-bearing austenitic stainless steels, *ISIJ Int,* Vol. 36, (1996): p. 787.

62. Zhu, Y., and D. E. Williams, Scanning electrochemical microscopic observation of a precursor state to pitting corrosion of stainless steel, *J Electrochem Soc,* Vol. 144, 1997, p. L43.

63. Williams, D. E., T. F. Mohiuddin, and Y. Y. Zhu, Elucidation of a trigger mechanism for pitting corrosion of stainless steels using submicron resolution scanning electrochemical and photoelectrochemical microscopy, *J Electrochem Soc,* Vol 145, (1998): p. 2664.

64. Williams, D. E., and Y. Y. Zhu, Explanation for initiation of pitting corrosion of stainless steels at sulfide inclusions, *J Electrochem Soc,* Vol. 147, (2000): p. 1763.

65. González-Garcia, Y., G. T. Burstein, S. González, and R. M. Souto, Imaging metastable pits on austenitic stainless steel in situ at the open-circuit corrosion potential, *Electrochem Commun,* Vol. 6, (2004): p. 637.

66. Bastos, A. C., A. M. Simões, S. González, Y. González-García, and R. M. Souto, Imaging concentration profiles of redox-active species in open-circuit corrosion processes with the scanning electrochemical microscope, *Electrochem Commun,* Vol. 6, (2004): p. 1212.

67. Paik, C. H., H. S. White, and R. C. Alkire, Scanning electrochemical microscopy detection of dissolved sulfur species from inclusions in stainless steel, *J Electrochem Soc,* Vol. 147, (2000): p. 4120.

68. Lister, T. E., and P. J. Pinhero, The effect of localized electric fields on the detection of dissolved sulfur species from Type 304 stainless steel using scanning electrochemical microscopy, *Electrochim Acta,* Vol. 48, (2003): p. 2371.

69. Lister, T. E., and P. J. Pinhero, Scanning electrochemical microscopy study of corrosion dynamics on type 304 stainless steel, *Electrochem Solid-State Lett,* Vol. 5, (2002): p. B33.

70. Lister, T. E., and P. J. Pinhero, Microelectrode array microscopy: Investigation of dynamic behavior of localized corrosion at type 304 stainless steel surfaces, *Anal Chem,* Vol. 77, (2005): p. 2601.

71. Fushimi, K., K. A. Lill, and H. Habazaki, Heterogeneous hydrogen evolution on corroding Fe-3 at.% Si surface observed by scanning electrochemical microscopy, *Electrochim Acta,* Vol. 52, (2007): p. 4246.

72. Malik, M. A., and P. J. Kulesza, Monitoring of conductivity changes in passive layers by scanning electrochemical microscopy in feedback mode: localization of pitting precursor sites on surfaces of multimetallic phase materials, *Anal Chem,* Vol. 79, (2007): p. 3996.

73. Malik, M. A., P. J. Kulesza, and G. Pawlowska, Surface analysis with scanning electrochemical microscopy in the feedback mode: Monitoring of reactivity and pitting precursor sites on the Nd-Fe-B-type magnet, *Electrochim Acta,* Vol. 54, (2009): p. 5537.

74. Simões, A. M., A. C. Bastos, M. G. Ferreira, Y. González-García, S. González, and R. M. Souto, Use of SVET and SECM to study the galvanic corrosion of an iron-zinc cell, *Corros Sci,* Vol. 49, (2007): p. 726.

75. Tada, E., S. Satoh, and H. Kaneko, The spatial distribution of Zn^{2+} during galvanic corrosion of a Zn/steel couple, *Electrochim Acta,* Vol. 49, (2004): p. 2279.

76. Büchler, M., J. Kerimo, F. Guillaume, and W. H. Smyrl, Fluorescence and near-field scanning optical microscopy for investigating initiation of localized corrosion of Al 2024, *J Electrochem Soc,* Vol. 147, (2000): p. 3691.

77. Díaz-Ballote, L., L. Veleva, M. A. Pech-Canul, M. I. Pech-Canul, and D. O. Wipf, Activity of SiC particles in Al-based metal matrix composites revealed by SECM, *J Electrochem Soc,* Vol. 151, (2004): p. B299.

78. Davoodi, A., J. Pan, C. Leygraf, and S. Norgren, In situ investigation of localized corrosion of aluminum alloys in chloride solution using integrated EC-AFM/SECM techniques, *Electrochem Solid-State Lett,* Vol. 8, (2005): p. B21.

79. Davoodi, A., J. Pan, C. Leygraf, and S. Norgren, Probing of local dissolution of Al-alloys in chloride solutions by AFM and SECM, *Appl Surf Sci,* Vol. 252, (2006): p. 5499.

80. Davoodi, A., J. Pan, C. Leygraf, and S. Norgren, Integrated AFM and SECM for in situ studies of localized corrosion of Al alloys, *Electrochim Acta,* Vol. 52, (2007): p. 7697.

81. Davoodi, A., J. Pan, C. Leygraf, and S. Norgren, The role of intermetallic particles in localized corrosion of an aluminum alloy studied by SKPFM and integrated AFM/SECM, *J Electrochem Soc,* Vol. 155, (2008): p. C211.

82. Davoodi, A., J. Pan, C. Leygraf, and S. Norgren, Multianalytical and in situ studies of localized corrosion of EN AW-3003 alloy–Influence of intermetallic particles, *J Electrochem Soc,* Vol. 155, (2008): p. C138.

83. Zhou, H. -R., X. -G. Li, C. -F. Dong, K. Xiao, and T. Li, Corrosion behavior of aluminum alloys in Na_2SO_4 solution using the scanning electrochemical microscopy technique, *Int J Miner Metall Mater,* Vol. 16, (2009): p. 84.

84. James, P. I., L. F. Garfias-Mesias, P. J. Moyer, and W. H. Smyrl, Scanning electrochemical microscopy with simultaneous independent topography, *J Electrochem Soc,* Vol. 145, (1998): p. L64.

85. Macpherson, J. V., and P. R. Unwin, Combined scanning electrochemical – Atomic force microscopy, *Anal Chem,* Vol. 72, (2000): p. 276.

86. Kranz, C., G. Friedbacher, B. Mizaikoff, A. Lugstein, J. Smoliner, and E. Bertagnolli, Integrating an ultramicroelectrode in an AFM cantilever: Combined technology for enhanced information, *Anal Chem,* Vol. 73, (2001): p. 2491.

87. Davoodi, A., A. Farzadi, J. Pan, C. Leygraf, and Y. Zhu, Developing an AFM-based SECM system; Instrumental setup, SECM simulation, characterization, and calibration, *J Electrochem Soc,* Vol. 155, (2008): p. C474.

88. Gilbert, J. L., S. M. Smith, and E. P. Lautenschlager, Scanning electrochemical microscopy of metallic biomaterials: Reaction rate and ion release imaging modes, *J Biomed Mater Res,* Vol. 27, (1993): p. 1357.

89. Casillas, N., S. J. Charlebois, W. H. Smyrl, and H. S. White, Scanning electrochemical microscopy of precursor sites for pitting corrosion on titanium, *J Electrochem Soc,* Vol. 140, (1993): p. L142.

90. Casillas, N., S. Charlebois, W. H. Smyrl, and H. S. White, Pitting corrosion of titanium, *J Electrochem Soc,* Vol. 141, (1994): p. 636.

91. Basame, S. B., and H. S. White, Scanning electrochemical microscopy of native titanium oxide films. Mapping the potential dependence of spatially-localized electrochemical reactions, *J Phys Chem,* Vol. 99, (1995): p. 16430.

92. Casillas, N., P. James, and W. H. Smyrl, A novel approach to combine scanning electrochemical microscopy and scanning photoelectrochemical microscopy, *J Electrochem Soc,* Vol. 142, (1995): p. L16.

93. Sukamto, J. P. H., W. H. Smyrl, N. Casillas, M. Al-Odan, P. James, W. Jin, and L. Douglas, Microvisualization of corrosion, *Mater Sci Eng A,* Vol. 198, (1995): p. 177.

94. James, P., N. Casillas, and W. H. Smyrl, Simultaneous scanning electrochemical and photoelectrochemical microscopy by use of a metallized optical fiber, *J Electrochem Soc,* Vol. 143, (1996): p. 3853.

95. Basame, S. B., and H. S. White, Scanning electrochemical microscopy: Measurement of the current density at microscopic redox-active sites on titanium, *J Phys Chem B,* Vol. 102, (1998): p. 9812.

96. Garfias-Mesias, L. F., M. Alodan, P. I. James, and W. H. Smyrl, Determination of precursor sites for pitting corrosion of polycrystalline titanium by using different techniques, *J Electrochem Soc,* Vol. 145, (1998): p. 2005.

97. Paik, C. H., and R. C. Alkire, Role of sulfide inclusions on localized corrosion of Ni200 in NaCl solutions, *J Electrochem Soc,* Vol. 148, (2001): p. B276.

98. Lister, T. E., P. J. Pinhero, T. L. Trowbridge, and R. E. Mizia, Localized attack of a two-phase metal, scanning electrochemical microscopy studies of NiCrMoGd alloys, *J Electroanal Chem,* Vol. 579, (2005): p. 291.

99. Horrocks, B. R., M. V. Mirkin, D. T. Pierce, A. J. Bard, G. Nagy, and K. Toth, Scanning electrochemical microscopy. 19. Ion-selective potentiometric microscopy, *Anal Chem,* Vol. 65, (1993): p. 1213.

100. Katemann, B. B., A. Schulte, E. J. Calvo, M. Koudelka-Hep, and W. Schuhmann, Localised electrochemical impedance spectroscopy with high lateral resolution by means of alternating current scanning electrochemical microscopy, *Electrochem Commun,* Vol. 4, (2002): p. 134.

101. Belger, S., A. Schulte, C. Hessing, M. Pohl, and W. Schuhmann, Alternating current scanning electrochemical microscopy (AC-SECM) studies on the surface of electro-chemically polished NiTi shape memory alloys, *Materialwiss Werkstofftech,* Vol. 35, (2004): p. 276.

102. Katemann, B. B., C. G. Inchauspe, P. A. Castro, A. Schulte, E. J. Calvo, and W. Schuhmann, Precursor sites for localised corrosion on lacquered tinplates visualised by means of alternating current scanning electrochemical microscopy, *Electrochim Acta,* Vol. 48, (2003): p. 1115.

103. Schulte, A., S. Belger, M. Etienne, and W. Schuhmann, Imaging localised corrosion of NiTi shape memory alloys by means of alternating current scanning electrochemi-cal microscopy (AC-SECM), *Mater Sci Eng A,* Vol. 378, (2004): p. 523.

104. Ruhlig, D. R., and W. Schuhmann, Spatial imaging of Cu^{2+}-ion release by combin-ing alternating current and underpotential stripping mode scanning electrochemical microscopy, *Electroanalysis,* Vol. 19, (2007): p. 191.

105. Eckhard, K., M. Etienne, A. Schulte, and W. Schuhmann, Constant-distance mode AC-SECM for the visualisation of corrosion pits, *Electrochem Commun,* Vol. 9, (2007): p. 1793.

106. Freire, L., X. R. Nóvoa, G. Pena, and V. Vivier, On the corrosion mechanism of AISI 204Cu stainless steel in chlorinated alkaline media, *Corros Sci,* Vol. 50, (2008): p. 3205.

107. Eckhard, K., T. Erichsen, M. Stratmann, and W. Schuhmann, Frequency-dependent alternating-current scanning electrochemical microscopy (4D AC-SECM) for local visualisation of corrosion sites, *Chem Eur J,* Vol. 14, (2008): p. 3968.

108. Eckhard, K., W. Schuhmann, and M. Maciejewska, Determination of optimum imag-ing conditions in AC-SECM using the mathematical distance between approach curves displayed in the impedance domain, *Electrochim Acta,* Vol. 54, (2009): p. 2125.

109. Jensen, M. B., A. Guerard, D. E. Tallman, and G. P. Bierwagen, Studies of elec-tron transfer at aluminum alloy surfaces by scanning electrochemical microscopy, *J Electrochem Soc,* Vol. 155, (2008): p. C324.

110. Ruhlig, D., H. Gugel, A. Schulte, W. Theisen, and W. Schuhmann, Visualization of local electrochemical activity and local nickel ion release on laser-welded NiTi/steel joints using combined alternating current mode and stripping mode SECM, *Analyst,* Vol. 133, (2008): p. 1700.

111. Eckhard, K., H. Shin, B. Mizaikoff, W. Schuhmann, and C. Kranz, Alternating cur-rent (AC) impedance imaging with combined atomic force scanning electrochemical microscopy (AFM-SECM), *Electrochem Commun,* Vol. 9, (2007): p. 1311.

112. Mansikkamäki, K., P. Ahonen, G. Fabricius, L. Murtomaki, and K. Kontturi, Inhibitive effect of benzotriazole on copper surfaces studied by SECM, *J Electrochem Soc,* Vol. 152, (2005): p. B12.

113. Mansikkamäki, K., U. Haapanen, C. Johans, K. Kontturi, and M. Valden, Adsorption of benzotriazole on the surface of copper alloys studied by SECM and XPS, *J Electrochem Soc,* Vol. 153, (2006): p. B311.

114. Mansikkamäki, K., C. Johans, and K. Kontturi, The effect of oxygen on the inhibition of copper corrosion with benzotriazole, *J Electrochem Soc,* Vol. 153, (2006): p. B22.

115. Walsh, D. A., L. E. Li, M. S. Bakare, and K. T. Voisey, Visualisation of the local electrochemical activity of thermal sprayed anti-corrosion coatings using scanning electrochemical microscopy, *Electrochim Acta,* Vol. 54, (2009): p. 4647.

116. Souto, R. M., Y. González-García, S. González, and G. T. Burstein, Damage to paint coatings caused by electrolyte immersion as observed in situ by scanning electrochemical microscopy, *Corros Sci,* Vol. 46, (2004): p. 2621.

117. Souto, R. M., Y. González-García, and S. González, Evaluation of the corrosion performance of coil-coated steel sheet as studied by scanning electrochemical microscopy, *Corros Sci,* Vol. 50, (2008): p. 1637.

118. Souto, R. M., Y. González-García, and S. González, Characterization of coating systems by scanning electrochemical microscopy: Surface topology and blistering, *Prog Org Coat,* Vol. 65, (2009): p. 435.

119. Souto, R. M., L. Fernández-Mérida, S. González, and D. J. Scantlebury, Comparative EIS study of different Zn-based intermediate metallic layers in coil-coated steels, *Corros Sci,* Vol. 48, (2006): p. 1182.

120. Bastos, A. C., A. M. Simões, S. González, Y. González-García, and R. M. Souto, Application of the scanning electrochemical microscope to the examination of organic coatings on metallic substrates, *Prog Org Coat,* Vol. 53, (2005): p. 177.

121. Souto, R. M., Y. González-Garcia, and S. González, In situ monitoring of electroactive species by using the scanning electrochemical microscope. Application to the investigation of degradation processes at defective coated metals, *Corros Sci,* Vol. 47, (2005): p. 3312.

122. Völker, E., C. G. Inchauspe, and E. J. Calvo, Scanning electrochemical microscopy measurement of ferrous ion fluxes during localized corrosion of steel, *Electrochem Commun,* Vol. 8, (2006): p. 179.

123. Janotta, M., D. Rudolph, A. Kueng, C. Kranz, H.-S. Voraberger, W. Waldhauser, and B. Mizaikoff, Analysis of corrosion processes at the surface of diamond-like carbon protected zinc selenide waveguides, *Langmuir,* Vol. 20, (2004): p. 8634.

124. Simões, A. M., D. Battocchi, D. E. Tallman, and G. P. Bierwagen, SVET and SECM imaging of cathodic protection of aluminium by a Mg-rich coating, *Corros Sci,* Vol. 49, (2007): p. 3838.

125. Seegmiller, J. C., J. E. Pereira da Silva, D. A. Buttry, S. I. Cordoba de Torresi, and R. M. Torresi, Mechanism of action of corrosion protection coating for AA2024-T3 based on poly(aniline)-poly(methylmethacrylate) blend, *J Electrochem Soc,* Vol. 152, (2005): p. B45.

126. Wei, C., and A. J. Bard, Scanning electrochemical microscopy, *J Electrochem Soc,* Vol. 142, (1995): p. 2523.

127. Fushimi, K., T. Okawa, K. Azumi, and M. Seo, Heterogeneous growth of anodic oxide film on a polycrystalline titanium electrode observed with a scanning electrochemical microscope, *J Electrochem Soc,* Vol. 147, (2000): p. 524.

128. Zhu, R., C. Nowierski, Z. Ding, J. J. Noel, and D. W. Shoesmith, Insights into grain structures and their reactivity on grade-2 Ti alloy surfaces by scanning electrochemical microscopy, *Chem Mater,* Vol. 19, (2007): p. 2533.

129. Zhu, R., Z. Qin, J. J. Noel, D. W. Shoesmith, and Z. Ding, Analyzing the influence of alloying elements and impurities on the localized reactivity of titanium grade-7 by scanning electrochemical microscopy, *Anal Chem,* Vol. 80, (2008): p. 1437.

130. Pust, S. E., D. Scharnweber, S. Baunack, and G. Wittstock, Electron transfer kinetics at oxide films on metallic biomaterials, *J Electrochem Soc,* Vol. 154, (2007): p. C508.

131. Pust, S. E., D. Scharnweber, C. Nunes Kirchner, and G. Wittstock, Heterogeneous distribution of reactivity on metallic biomaterials: Scanning probe microscopy studies of the biphasic Ti alloy Ti6Al4V, *Adv Mater,* Vol. 19, (2007): p. 878.

132. Basame, S. B., and H. S. White, Chemically-selective and spatially-localized redox activity at Ta/Ta$_2$O$_5$ electrodes, *Langmuir,* Vol. 15, (1999), p. 819.

133. Basame, S. B., and H. S. White, Scanning electrochemical microscopy of metal/ metal oxide electrodes. Analysis of spatially localized electron-transfer reactions during oxide growth, *Anal Chem,* Vol. 71, (1999): p. 3166.

134. Fushimi, K., K. Azumi, and M. Seo, Evaluation of heterogeneity in thickness of passive films on pure iron by scanning electrochemical microscopy, *ISIJ Int,* Vol. 39, (1999): p. 346.

135. Sun, P., Z. Liu, H. Yu, and M. V. Mirkin, Effect of mechanical stress on the kinetics of heterogeneous electron transfer, *Langmuir,* Vol. 24, (2008): p. 9941.

136. Serebrennikova, I., S. Lee, and S. White Henry, Visualization and characterization of electroactive defects in the native oxide film on aluminium, *Faraday Discuss,* Vol. 121, (2002): p. 199.

137. Serebrennikova, I., and H. S. White, Scanning electrochemical microscopy of electroactive defect sites in the native oxide film on aluminum, *Electrochem Solid-State Lett,* Vol. 4, (2001): p. B4.

138. Levine, K. L., D. E. Tallman, and G. P. Bierwagen, The mediated electrodeposition of polypyrrole on aluminium alloy, *Aust J Chem,* Vol. 58, (2005): p. 294.

139. Tallman, D. E., M. P. Dewald, C. K. Vang, G. G. Wallace, and G. P. Bierwagen, Electrodeposition of conducting polymers on active metals by electron transfer mediation, *Curr Appl Phys,* Vol. 4, (2004): p. 137.

140. Tallman, D. E., C. Vang, G. G. Wallace, and G. P. Bierwagen, Direct electrodeposition of polypyrrole on aluminum and aluminum alloy by electron transfer mediation, *J Electrochem Soc,* Vol. 149, (2002): p. C173.

141. Tallman, D. E., C. K. Vang, M. P. Dewald, G. G. Wallace, and G. P. Bierwagen, Electron transfer mediated deposition of conducting polymers on active metals, *Synth Met,* Vol. 135–136, (2003): p. 33.

142. Seegmiller, J. C., and D. A. Buttry, A SECM study of heterogeneous redox activity at AA2024 surfaces, *J Electrochem Soc,* Vol. 150, (2003): p. B413.

143. Tóth, K., G. Nagy, C. Wei, and A. J. Bard, Novel application of potentiometric microelectrodes: scanning potentiometric microscopy, *Electroanalysis,* Vol. 7, (1995): p. 801.

144. Schulte, A., S. Belger, and W. Schuhmann, Corrosion of NiTi shape-memory alloys: Visualization by means of potentiometric "constant-distance" scanning electrochemical microscopy, *Mater Sci Forum,* Vol. 394–395, (2002): p. 145.

145. Park, J. O., C.-H. Paik, and R. C. Alkire, Scanning microsensors for measurement of local pH distributions at the microscale, *J Electrochem Soc,* Vol. 143, (1996): p. L174.

146. Ding, H., and L. H. Hihara, Localized corrosion currents and pH profile over B4C, SiC and Al$_2$O$_3$ reinforced 6092 aluminum composites in artificial seawater, *Proc Electrochem Soc,* Vol. 2004–2014, (2005): p. 484.

147. Ding, H., and L. H. Hihara, Localized corrosion currents and pH profile over B4C, SiC, and Al$_2$O$_3$ reinforced 6092 aluminum composites. I. In 0.5 M Na$_2$SO$_4$ solution, *J Electrochem Soc,* Vol. 152, (2005): p. B161.

148. Ding, H., and L. H. Hihara, Localized corrosion of silicon-reinforced aluminum composites in Na$_2$SO$_4$ solution, *ECS Trans,* Vol. 2, (2007): p. 99.

149. Ding, H., and L. H. Hihara, Corrosion initiation and anodic-cathodic alternation of localized corrosion of SiC-reinforced aluminum-matrix composites in NaCl solution, *ECS Trans,* Vol. 3, (2007): p. 237.
150. Ding, H., and L. H. Hihara, Galvanic corrosion and localized degradation of aluminum-matrix composites reinforced with silicon particulates, *J Electrochem Soc,* Vol. 155, (2008): p. C226.
151. He, J., D. Battocchi, A. M. Simoes, D. E. Tallman, and G. P. Bierwagen, Scanning probe studies of active coatings for corrosion control of Al alloys, In *New Developments in Coatings Technology,* P. Zarras, T. Wood, B. Richey, and B. C. Benicewicz (eds.), American Chemical Society, 2007: p. 8.
152. He, J., V. J. Gelling, D. E. Tallman, G. P. Bierwagen, and G. G. Wallace, Conducting polymers and corrosion III. A scanning vibrating electrode study of poly(3-octyl pyrrole) on steel and aluminum, *J Electrochem Soc,* Vol. 147, (2000): p. 3667.
153. Yan, M., D. E. Tallman, and G. P. Bierwagen, Role of oxygen in the galvanic interaction between polypyrrole and aluminum alloy, *Electrochim Acta,* Vol. 54, (2008): p. 220.
154. Lamaka, S. V., O. V. Karavai, A. C. Bastos, M. L. Zheludkevich, and M. G. S. Ferreira, Monitoring local spatial distribution of Mg^{2+}, pH and ionic currents, *Electrochem Commun,* Vol. 10, (2008): p. 259.
155. Montemor, M. F., W. Trabelsi, S. V. Lamaka, K. A. Yasakau, M. L. Zheludkevich, A. C. Bastos, and M. G. S. Ferreira, The synergistic combination of bis-silane and $CeO_2 \cdot ZrO_2$ nanoparticles on the electrochemical behavior of galvanized steel in NaCl solutions, *Electrochim Acta,* Vol. 53, (2008): p. 5913.
156. Böhni, H., T. Suter, and A. Schreyer, Micro- and nanotechniques to study localized corrosion, *Electrochim Acta,* Vol. 40, (1995): p. 1361.
157. Böhni, H., T. Suter, and F. Assi, Micro-electrochemical techniques for studies of localized processes on metal surfaces in the nanometer range, *Surf Coat Technol,* Vol. 130, (2000): p. 80.
158. Webb, E. G., and R. C. Alkire, Pit initiation at single sulfide inclusions in stainless steel I. Electrochemical microcell measurements, *J Electrochem Soc,* Vol. 149, (2002): p. B272.
159. Vignal, V., H. Krawiec, O. Heintz, and R. Oltra, The use of local electrochemical probes and surface analysis methods to study the electrochemical behaviour and pitting corrosion of stainless steels, *Electrochim Acta,* Vol. 52, (2007): p. 4994.
160. Krawiec, H., V. Vignal, E. Finot, O. Heintz, R. Oltra, and J. M. Olive, Local electrochemical studies after heat treatment of stainless steel. Role of induced metallurgical and surface modifications on pitting triggering, *Metall Mater Trans A,* Vol. 35A, (2004): p. 3515.
161. Krawiec, H., V. Vignal, and R. Oltra, Use of the electrochemical microcell technique and the SVET for monitoring pitting corrosion at MnS inclusions, *Electrochem Commun,* Vol. 6, (2004): p. 655.
162. Krawiec, H., V. Vignal, P. Ponthiaux, and F. Wenger, Benefits of local electrochemical studies on worn surfaces, *ECS Trans,* Vol. 3, (2007): p. 355.
163. Krawiec, H., V. Vignal, O. Heintz, P. Ponthiaux, and F. Wenger, Local electrochemical studies and surface analysis on worn surfaces, *J Electrochem Soc,* Vol. 155, (2008): p. C127.
164. Suter, T., and H. Böhni, Microcell studies of localized corrosion on different systems, *Localized Corros, Proc Corros/2001 Res Top Symp,* (2001): p. 191.
165. Böhni, H., and T. Suter, Microelectrochemical techniques for high resolution corrosion studies, *Ochr Koroz,* Vol. 47, (2004): p. 245.

166. Dix Jr, E. H., Acceleration of the rate of corrosion by high constant stresses, *Trans AIME,* Vol. 137, (1940): p. 11.
167. Lohrengel, M. M., Spatially resolved electrochemical surface analysis by microcells, *Trans SAEST,* Vol. 33, (1998): p. 145.
168. Lohrengel, M. M., A. Moehring, and M. Pilaski, Capillary-based droplet cells: limits and new aspects, *Electrochim Acta,* Vol. 47, (2001): p. 137.
169. Krawiec, H., V. Vignal, and R. Akid, Numerical modelling of the electrochemical behaviour of 316 stainless steel based upon static and dynamic experimental microcapillary-based techniques: effect of electrolyte flow and capillary size, *Surf Interface Anal,* Vol. 40, (2008): p. 315.
170. Krawiec, H., V. Vignal, and R. Akid, Numerical modelling of the electrochemical behaviour of 316L stainless steel based upon static and dynamic experimental microcapillary-based techniques, *Electrochim Acta,* Vol. 53, (2008): p. 5252.
171. Fushimi, K., S. Yamamoto, R. Ozaki, and H. Habazaki, Cross-section corrosion-potential profiles of aluminum-alloy brazing sheets observed by the flowing electrolyte scanning-droplet-cell technique, *Electrochim Acta,* Vol. 53, (2008), p. 2529.
172. Modestov, A. D., S. Srebnik, O. Lev, and J. Gun, Scanning capillary microscopy/mass spectrometry for mapping spatial electrochemical activity of electrodes, *Anal Chem,* Vol. 73, (2001): p. 4229.
173. Lill, K. A., A. W. Hassel, G. Frommeyer, and M. Stratmann, Scanning droplet cell investigations on single grains of a FeAlCr light weight ferritic steel, *Electrochim Acta,* Vol. 51, (2005): p. 978.
174. Akid, R., P. Roffey, D. Greenfield, and D. Guillen, Application of scanning vibrating electrode technique (SVET) and scanning droplet cell (SDC) techniques to the study of weld corrosion, *Eur Fed Corros Publ,* Vol. 45, (2007): p. 23.
175. Rosenfeld, I. L., and I. S. Danilov, Electrochemical aspects of pitting corrosion, *Corros Sci,* Vol. 7, (1967): p. 129.
176. Hepworth, M. T., and K. J. Bhansali, Development of a corrodescope to test coatings and inhibitors, *J Paint Technol,* Vol. 46, (1974): p. 31.
177. Akid, R., and D. J. Mills, A comparison between conventional macroscopic and novel microscopic scanning electrochemical methods to evaluate galvanic corrosion, *Corros Sci,* Vol. 43, (2001): p. 1203.
178. Jin, S., E. Ghali, C. Blawert, and W. Dietzel, SRET evaluation of the corrosion behavior of thixocast AZ91 magnesium alloy in dilute NaCl solution at room temperature, *ECS Trans,* Vol. 3, (2007): p. 295.
179. Lin, C.-J., Y. Li, and X.-D. Zhuo, A hybrid scanning probe technique for in situ imaging surface topography and corrosion potential of localized corrosions, *ECS Trans,* Vol. 3, (2007): p. 313.
180. Tan, Y.-J., T. Liu, and N. N. Aung, Novel corrosion experiments using the wire beam electrode: (III) Measuring electrochemical corrosion parameters from both the metallic and electrolytic phases, *Corros Sci,* Vol. 48, (2006): p. 53.
181. Liu, T., Y.-J. Tan, B. Z. M. Lin, and N. N. Aung, Novel corrosion experiments using the wire beam electrode. (IV) Studying localized anodic dissolution of aluminum, *Corros Sci,* Vol. 48, (2006): p. 67.
182. Battocchi, D., J. He, G. P. Bierwagen, and D. E. Tallman, Emulation and study of the corrosion behavior of Al alloy 2024-T3 using a wire beam electrode (WBE) in conjunction with scanning vibrating electrode technique (SVET), *Corros Sci,* Vol. 47, (2005): p. 1165.
183. Akid, R., and M. Garma, Scanning vibrating reference electrode technique: a calibration study to evaluate the optimum operating parameters for maximum signal detection of point source activity, *Electrochim Acta,* Vol. 49, (2004): p. 2871.

184. Ishikawa, Y., and H. S. Isaacs, Electrochemical behavior of pits developed on aluminum in a dilute chloride solution, *Boshoku Gijutsu,* Vol. 33, (1984): p. 147.
185. Abdullayev, E., R. Price, D. Shchukin, and Y. Lvov, Halloysite tubes as nanocontainers for anticorrosion coating with benzotriazole, *ACS Appl Mater Interfaces,* Vol. 1, (2009): p. 1437.
186. Fix, D., D. V. Andreeva, Y. M. Lvov, D. G. Shchukin, and H. Moehwald, Application of inhibitor-loaded halloysite nanotubes in active anti-corrosive coatings, *Adv Funct Mater,* Vol. 19, (2009): p. 1720.
187. Gustavsson, J. M., P. C. Innis, J. He, G. G. Wallace, and D. E. Tallman, Processable polyaniline-HCSA/poly(vinyl acetate-co-butyl acrylate) corrosion protection coatings for aluminum alloy 2024-T3: A SVET and Raman study, *Electrochim Acta,* Vol. 54, (2009): p. 1483.
188. Montemor, M. F., R. Pinto, and M. G. S. Ferreira, Chemical composition and corrosion protection of silane films modified with CeO_2 nanoparticles, *Electrochim Acta,* Vol. 54, (2009): p. 5179.
189. Palomino, L. M., P. H. Suegama, I. V. Aoki, M. F. Montemor, and H. G. De Melo, Electrochemical study of modified cerium-silane bi-layer on Al alloy 2024-T3, *Corros Sci,* Vol. 51, (2009): p. 1238.
190. Raps, D., T. Hack, J. Wehr, M. L. Zheludkevich, A. C. Bastos, M. G. S. Ferreira, and O. Nuyken, Electrochemical study of inhibitor-containing organic-inorganic hybrid coatings on AA2024, *Corros Sci,* Vol. 51, (2009): p. 1012.
191. Reffass, M., R. Sabot, M. Jeannin, C. Berziou, and P. Refait, Effects of phosphate species on localised corrosion of steel in $NaHCO_3$ + NaCl electrolytes, *Electrochim Acta,* Vol. 54, (2009): p. 4389.
192. Simões, A. M., J. Torres, R. Picciochi, and J. C. S. Fernandes, Corrosion inhibition at galvanized steel cut edges by phosphate pigments, *Electrochim Acta,* Vol. 54, (2009): p. 3857.
193. Skorb, E. V., D. Fix, D. V. Andreeva, H. Moehwald, and D. G. Shchukin, Surface-modified mesoporous SiO_2 containers for corrosion protection, *Adv Funct Mater,* Vol. 19, (2009): p. 2373.
194. Zhang, G. A., and Y. F. Cheng, Micro-electrochemical characterization of corrosion of welded X70 pipeline steel in near-neutral pH solution, *Corros Sci,* Vol. 51, (2009): p. 1714.
195. Thébault, F., B. Vuillemin, R. Oltra, K. Ogle, and C. Allely, Investigation of self-healing mechanism on galvanized steels cut edges by coupling SVET and numerical modeling, *Electrochim Acta,* Vol. 53, (2008): p. 5226.
196. Williams, G., and H. N. McMurray, Localized corrosion of magnesium in chloride-containing electrolyte studied by a scanning vibrating electrode technique, *J Electrochem Soc,* Vol. 155, (2008): p. C340.
197. Little, B., P. Wagner, P. Angell, and D. White, Correlation between localized anodic areas and Oceanospirillum biofilms on copper, *Int Biodeterior Biodegrad,* Vol. 37, (1996): p. 159.
198. Basseguy, R., J. Idrac, C. Jacques, A. Bergel, M. L. Delia, and L. Etcheverry, Local analysis by SVET of the involvement of biological systems in aerobic biocorrosion, *Eur Fed Corros Publ,* Vol. 45, (2007): p. 52.
199. Iken, H., L. Etcheverry, A. Bergel, and R. Basseguy, Local analysis of oxygen reduction catalysis by scanning vibrating electrode technique: A new approach to the study of biocorrosion, *Electrochim Acta,* Vol. 54, (2008): p. 60.
200. Isaacs, H. S., and M. W. Kendig, Determination of surface inhomogeneities using a scanning probe impedance technique, *Corrosion,* Vol. 36, (1980): p. 269.

201. Hughes, M. C., and J. M. Parks, An a.c. impedance probe as an indicator of corrosion and defects in polymer/metal substrate systems, In *Corrosion Control by Organic Coatings*, H. Leidheiser Jr. (ed.), Houston. NACE, 1981: p. 45.
202. Standish, J. V., and H. Leidheiser Jr., The electrical properties of organic coatings on a local scale-relationship to corrosion, *Corrosion,* Vol. 36, (1980): p. 390.
203. Dehri, I., R. L. Howard, and S. B. Lyon, Local electrochemical impedance at the cut edge of coil-coated galvanized steel after corrosion testing, *Corros Sci,* Vol. 41, (1999): p. 141.
204. Kruger, J., R. S. Lillard, C. C. Streinz, and P. J. Moran, Effect of microstructure on passive film formation and breakdown of aluminum-tantalum alloys, *Faraday Discuss,* Vol. 94, (1992): p. 127.
205. Lillard, R. S., J. Kruger, W. S. Tait, and P. J. Moran, Using local electrochemical impedance spectroscopy to examine coating failure, *Corrosion,* Vol. 51, (1995): p. 251.
206. Zou, F., D. Thierry, and H. S. Isaacs, A high-resolution probe for localized electrochemical impedance spectroscopy measurements, *J Electrochem Soc,* Vol. 144, (1997): p. 1957.
207. Zou, F., and D. Thierry, Localized electrochemical impedance spectroscopy for studying the degradation of organic coatings, *Electrochim Acta,* Vol. 42, (1997): p. 3293.
208. Annergren, I., D. Thierry, and F. Zou, Localized electrochemical impedance spectroscopy for studying pitting corrosion on stainless steels, *J Electrochem Soc,* Vol. 144, (1997): p. 1208.
209. Annergren, I., F. Zou, and D. Thierry, Application of localized electrochemical techniques to study kinetics of initiation and propagation during pit growth, *Electrochim Acta,* Vol. 44, (1999): p. 4383.
210. Bayet, E., F. Huet, M. Keddam, K. Ogle, and H. Takenouti, A novel way of measuring local electrochemical impedance using a single vibrating probe, *J Electrochem Soc,* Vol. 144, (1997): p. L87.
211. Bayet, E., F. Huet, M. Keddam, K. Ogle, and H. Takenouti, Adaptation of the scanning vibrating electrode technique to a.c. mode. Local electrochemical impedances measurement, *Mater Sci Forum,* Vol. 289–292, (1998): p. 57.
212. Bayet, E., F. Huet, M. Keddam, K. Ogle, and H. Takenouti, Local electrochemical impedance measurement: Scanning vibrating electrode technique in ac mode, *Electrochim Acta,* Vol. 44, (1999): p. 4117.
213. de Lima-Neto, P., J. P. Farias, L. F. G. Herculano, H. C. de Miranda, W. S. Araujo, J.-B. Jorcin, and N. Pebere, Determination of the sensitized zone extension in welded AISI 304 stainless steel using non-destructive electrochemical techniques, *Corros Sci,* Vol. 50, (2008): p. 1149.
214. Frateur, I., V. M.-W. Huang, M. E. Orazem, N. Pebere, B. Tribollet, and V. Vivier, Local electrochemical impedance spectroscopy: Considerations about the cell geometry, *Electrochim Acta,* Vol. 53, (2008): p. 7386.
215. Li, M. C., and Y. F. Cheng, Corrosion of the stressed pipe steel in carbonate-bicarbonate solution studied by scanning localized electrochemical impedance spectroscopy, *Electrochim Acta,* Vol. 53, (2008): p. 2831.
216. Meng, G. Z., C. Zhang, and Y. F. Cheng, Effects of corrosion product deposit on the subsequent cathodic and anodic reactions of X-70 steel in near-neutral pH solution, *Corros Sci,* Vol. 50, (2008): p. 3116.
217. Zhang, G. A., and Y. F. Cheng, Corrosion of X65 steel in CO_2-saturated oilfield formation water in the absence and presence of acetic acid, *Corros Sci,* Vol. 51, (2009): p. 1589.

218. Jorcin, J.-B., M. E. Orazem, N. Pebere, and B. Tribollet, CPE analysis by local electrochemical impedance spectroscopy, *Electrochim Acta,* Vol. 51, (2006): p. 1473.

219. Baril, G., C. Blanc, M. Keddam, and N. Pebere, Local electrochemical impedance spectroscopy applied to the corrosion behavior of an AZ91 magnesium alloy, *J Electrochem Soc,* Vol. 150, (2003): p. B488.

220. Galicia, G., N. Pebere, B. Tribollet, and V. Vivier, Corrosion study of an AZ91 magnesium alloy by EIS and LEIS, *ECS Trans,* Vol. 1, (2006): p. 157.

221. Galicia, G., N. Pebere, B. Tribollet, and V. Vivier, Local and global electrochemical impedances applied to the corrosion behaviour of an AZ91 magnesium alloy, *Corros Sci,* Vol. 51, (2009): p. 1789.

222. Nguyen, T. D., T. A. Nguyen, M. C. Pham, B. Piro, B. Normand, and H. Takenouti, Mechanism for protection of iron corrosion by an intrinsically electronic conducting polymer, *J Electroanal Chem,* Vol. 572, (2004): p. 225.

223. Mierisch, A. M., and S. R. Taylor, Understanding coating and substrate heterogeneities using electrochemical impedance methods, *Mater Res Soc Symp Proc,* Vol. 500, (1998): p. 35.

224. Mierisch, A. M., and S. R. Taylor, Characterization of the electrochemical events at intrinsic breakdown sites on organically coated AA2024-T3, *J Corros Sci Eng,* Vol. 2, (1999): p. Paper 30.

225. Mierisch, A. M., J. Yuan, R. G. Kelly, and S. R. Taylor, Probing coating degradation on AA2024-T3 using local electrochemical and chemical techniques, *J Electrochem Soc,* Vol. 146, (1999): p. 4449.

226. Taylor, S. R., Incentives for using local electrochemical impedance methods in the investigation of organic coatings, *Prog Org Coat,* Vol. 43, (2001): p. 141.

227. Taylor, S. R., and A. M. Mierisch, Incentives for using LEIM in the investigation of corrosion initiation on organic coated alloys, *Mater Res Soc Symp Proc,* Vol. 699, (2002): p. 137.

228. Mierisch, A. M., and S. R. Taylor, Understanding the degradation of organic coatings using local electrochemical impedance methods, *J Electrochem Soc,* Vol. 150, (2003): p. B303.

4 Surface Interrogation Mode of Scanning Electrochemical Microscopy (SI-SECM): An Approach to the Study of Adsorption and (Electro)Catalysis at Electrodes

Joaquín Rodríguez-López

CONTENTS

4.1 INTRODUCTION: SI-SECM AS A COMPLEMENTARY TOOL FOR *IN-SITU* ELECTROANALYSIS

The electrochemical investigation of outer sphere electron transfer reactions coupled to complicated homogeneous reactions has become highly developed over the last decades because of the utilization of powerful techniques, like cyclic and ultramicroelectrode (UME) voltammetry and digital simulations [1–2]. However, understanding the inner sphere electrocatalytic reactions is more difficult because of their greater complexity and the difficulty in obtaining reliable quantitative data about intermediates and products on electrode surfaces. A variety of electrochemical and surface spectroscopic methods that can be employed with electrodes have been developed for this purpose [3]. For instance, mass sensitive devices such as the electrochemical quartz crystal microbalance allow the detection of small changes in the mass of an electrode during an electrochemical process. Infrared (IR) based techniques have been very successful at detecting carbonyl-based adsorbed species in electrocatalytic systems, second harmonic generation studies provide rich molecular information of chemisorbed species, and radiotracer methods have been used to distinguish mechanistic pathways in selected catalytic systems. However, it has been difficult to obtain the necessary versatility and sensitivity to address complex electrochemical processes in a direct quantitative manner. We review here an *in-situ* electrochemical technique based on the scanning electrochemical microscope (SECM) [4] for the detection and quantification of adsorbed species at electrodes that may be useful in this area [5]. This technique is based on the feedback mode of SECM, which is one of the first developed [6] and perhaps one of the most distinctive of this technique, because it can correlate an electrochemical signal at the SECM tip to topographical, chemical, and electrochemical features [7] in a variety of systems (e.g., imaging interdigitated arrays [8] or a leaf [9], detecting an immobilized enzyme [10] and measuring the heterogeneous kinetics on semiconductors [11]). The SECM tip can detect changes in the electrochemical signal produced by the concurrent production and collection of the species of a mediator pair (O/R, where only one of the species is present initially in solution); variations in the diffusive flux of these species caused by interaction with the substrate allow the current measured at the SECM tip to respond accordingly. One of the most useful features of SECM is the ability to calculate the tip current as a function of different solution and interfacial phenomena, i.e., electrochemical kinetics, by digital simulation [4]. We can thus consider the mediator to be an interrogation agent that reports to the SECM tip the state of the surface being examined.

In the feedback mode of SECM, a UME SECM tip is approached to a substrate; the type of substrate and the experimental conditions will impact the tip current, whose response can be used to describe the processes occurring *at the substrate* (in addition to those of the tip). The feedback theory considers two limiting cases for the tip current, namely negative (NF) and positive feedback (PF), as shown schematically in Figure 4.1a and b, respectively. In both cases, the SECM tip is approached perpendicularly to the substrate such that the flat

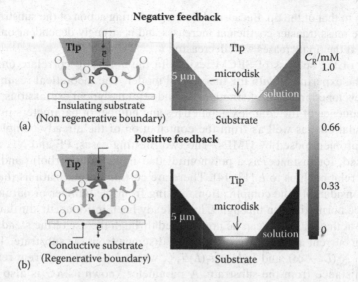

FIGURE 4.1 Depiction of NF and PF at a large substrate in 2D cross sections for the reduction of O to R; left: schematic, right: digital simulation of concentration profile of R. (a) In NF, the tip is approached to an insulating substrate where diffusion toward the microdisk is blocked and there is accumulation of the product, R, in the interelectrode gap. (b) In PF, the tip is approached to a substrate that performs the opposite reaction to that on the tip; the simulated concentration profile shows the buildup of a narrow diffusion layer for the product in the interelectrode gap.

end of the tip with the embedded microdisk is parallel to the substrate at distance d. This configuration maximizes the interaction of the diffusion layer at the UME with the substrate; a dimensionless distance can be expressed as $L = d/a$ where a is the radius of the embedded microdisk. A most important property of SECM is that at distances $d < a$ (or $L < 1$), the tip response exhibits a transition in its mass transfer coefficient, which is inversely proportional to d rather than to a [12]. Mass transfer to the tip, and therefore its electrochemical response, is a function of the interelectrode distance; this is a powerful feature of SECM, since the tip–substrate distance can be selected, and if made small, it can offer high quality data for electrochemical experiments such as kinetic measurements. It also relieves technical challenges in UME fabrication, i.e., it is easier to fabricate a 10 μm tip and approach it to 1 μm to a substrate than to fabricate a 1 μm UME.

In NF experiments, the tip is approached to an insulating substrate, or in more general terms, a substrate that cannot regenerate the consumed species in the tip–substrate gap. Since the diffusive flux of O is blocked by the presence of the substrate, the tip current decreases; the extent of this decrease will depend on the exact geometry of the substrate and on L. In PF experiments, the tip is approached to a conductive substrate, or more generally, a regenerating boundary. The substrate in this case is biased to a potential where it can carry out the reverse reaction to that of the tip. Because of this regenerating action of the substrate, the effective mass transfer coefficient increases and is strongly dependent on L; the current at the tip increases with decreasing L.

Part of the success of SECM resides in the ability to correlate quantitatively the experiment with the theoretical model. The theoretical results have been developed in terms of simulations and semianalytical expressions; this is a consequence of the complex geometries and number of parameters involved in calculations, as well as from the contribution of the already complex diffusion problem posed by UMEs. The two limiting cases, PF and NF, can be expressed, for instance, as a polynomial that includes hyperbolic and exponential relationships to L [13–14]. There are certain normalizations that alleviate considerably the complications arising from the number of parameters. Distance normalization through L has already been introduced; similarly, the current at the SECM tip can be normalized through the use of the steady-state limiting current at the UME at "infinite" distance from the substrate. That is, $i_{T,\infty} = i_{lim,ss}(L \rightarrow \infty)$, and $I_T(L) = i_T(L)/i_{T,\infty}$, where $i_T(L)$ is the current read at a given distance from the substrate. A parameter known as RG is also useful to describe the shape of the tip; here $RG = rg/a$, where rg is the radius of the flat end of the tip, including the radius a of the tip and the length of insulating material around the microdisk.

The more general case of finite heterogeneous kinetics at the substrate, that is, when the substrate does not operate at diffusion-limited conditions, has also been treated [15–16]. However, these cases have yielded complex semianalytical expressions. For instance, Lefrou has introduced solving procedures based on Bessel functions that yield, in certain cases, easy ways of estimating the response of the tip [17], but that upon introduction of parameters like the size of the substrate, inequalities in the diffusion coefficients, the effect of RG, or the consideration of accessibility of the redox mediator, become more complicated [16,18]. Since no exact analytical expression exists for SECM, the basis of comparison has been the results of numerical simulation (which has been thoroughly tested experimentally). Indeed, while the implementation of sophisticated numerical simulation may require a deal of mathematical training for the uninitiated, the availability of commercial finite element method (FEM) software has allowed a more straightforward exploration of SECM cases. The description of the SI-SECM mode was developed in such a way and is presented in Section 4.2. Figure 4.2a shows an example of the effects of heterogeneous kinetics at the substrate as obtained from simulation of a specific system. Normalization of the substrate rate constant, k_s, for regeneration of the species in the gap can be done through Equation 4.1, which also uses the radius of the microdisk and the diffusion coefficient of the mediator species, D [12]:

$$K = \frac{k_s a}{D} \qquad (4.1)$$

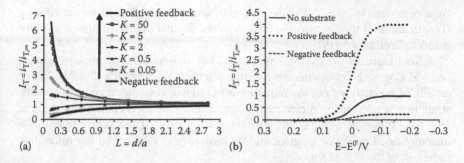

(a) $L = d/a$ (b) $E-E^{0'}/V$

FIGURE 4.2 Current–distance relationships in SECM and effects in voltammetry. (a) Family of normalized approach curves with finite substrate kinetics. Results for finite substrate kinetics obtained from simulation with $a = 12.5$ μm, $b = 25$ μm, $RG = 3$, $C_O^* = 0.5$ mM, $D_O = D_R = 6.2 \times 10^{-10}$ m²/s, and different values of $K = k_s a/D_O$; tip is operating at diffusion limited conditions. (b) Depiction of cyclic voltammetry at fixed distance ($L = 0.2$) for PF and NF at steady state and comparison to tip voltammetry at infinite distance from the substrate.

Note that in Figure 4.2a the response of the tip changes dramatically at $L < 1$ and that rate constants can be measured over several orders of magnitude. Moreover, at $L < 1$ there is an increasing contrast between PF and NF for smaller value of L. This property is used in the following sections for the detection of chemical reactivity in the SI-SECM mode.

For an SECM tip UME close to a substrate and assuming no kinetic complications at the tip for a given mediator electrochemical reaction, a steady-state cyclic voltammogram (CV) can be expressed as shown in Equations 4.2 and 4.3:

$$I_T(E,L) = \frac{I_T(L)}{\theta} \tag{4.2}$$

$$\theta = 1 + \frac{D_O}{D_R} e^{\left(\frac{nF}{RT\left(E-E^{0'}\right)}\right)} \tag{4.3}$$

$I_T(L)$ is the steady-state limiting current of the SECM tip at distance L, e.g., from Figure 4.2a. If the substrate operates completely under diffusion limiting conditions in the PF mode, the response will be that of a steady-state CV with an increased plateau current when compared to the voltammogram at infinite distance from the substrate. If the substrate is an insulator, it will exhibit a decreased plateau current. In both cases, however, the voltammogram should exhibit the characteristic Nernstian sigmoid shape of a UME voltammogram. Such contrast in CV is shown schematically in Figure 4.2b. The contrast obtained from the transition of the substrate from a "regenerative" (i.e., PF) to a "non-regenerative" (i.e., NF) behavior is of use in the SI-SECM mode. Kinetic complications at the tip can also be accounted for and this is discussed in more detail in Section 4.5 [19].

A final aspect of the feedback mode that has an impact in the implementation of SI-SECM is the effect known as open circuit (OC) feedback [20–23]. When an SECM tip carrying out the reduction of O to R approaches a conductive substrate that is isolated from electrical connection to the external circuitry, i.e., at OC, the tip current exhibits PF. This effect is clearer if the substrate exhibits uncomplicated mediator kinetics and its radius, b, is larger than the microdisk radius, a, of the SECM tip.

If the reaction is diffusion controlled at the substrate, the geometrical relation between a and b determines the extent of OC feedback. Figure 4.3 depicts the situation for OC feedback for a large substrate electrode $b \gg a$ and the same for a small substrate electrode $b < a$. The OC potential (OCP) of the unconnected conductive substrate is determined by the ratio of concentration of O to R at its surface according to the Nernst equation, Equation 4.4:

$$E = E^{0'} + \frac{RT}{nF} \ln\left(\frac{C_{O,s}}{C_{R,s}}\right) \tag{4.4}$$

where $C_{O,s}$ and $C_{R,s}$ refer to the concentrations of oxidized and reduced species, respectively, at the substrate. If only O is originally in the solution,

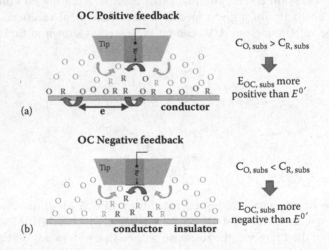

FIGURE 4.3 Depiction of the effect of substrate size on OC feedback for a mediator with Nernstian behavior at the substrate. (a) OC PF occurs when the conductive substrate is much larger than the microdisk of the tip; lateral charge transfer at the substrate can drive electrochemical reactions nonlocally. (b) OC NF occurs when the conductive substrate is of similar size or smaller than the microdisk on the tip; there is no option for nonlocal regeneration of the mediator.

with only traces of R, then the OCP of the substrate will be poised positive of $E^{0\prime}$. Depicted in Figure 4.3a, as the SECM tip, which is producing R and consuming O, is approached to the substrate, it will change the ratio of O to R locally at the surface of the substrate. If the substrate is much larger than the perturbation induced by the tip, then overall the OCP of the substrate will change only slightly and it will remain positive of $E^{0\prime}$. Although the substrate is isolated and no current can flow through an external circuit, there can be lateral charge transfer within its conductive phase. Since the OCP is positive, the system will tend to restore this condition and regeneration of R to O will occur locally below the tip; there cannot be a net flux of species at the substrate because there is no external circuit, so this reaction is compensated by the opposite process, reduction of O to R, at substrate positions away from the tip. If the substrate is substantially smaller, roughly of the size of the tip (e.g., $b < 2a$), the OCP of the substrate will now shift negative of $E^{0\prime}$ because of the larger excess of R with respect to O at its surface, as shown in Figure 4.3b. No current can flow through an external circuit, and no transformation of R to O can happen since there is no option to compensate with O to R nonlocally; thus, the response of the system is that of NF at the tip. Figure 4.4 from Xiong et al. shows the more general depiction of this effect as a function of tip to substrate radius [21]. The RG of the tip plays an important role in the extent of OC feedback due to differences in the accumulation of the electrogenerated species. For example, in a system with $b = 2a$ and $L \sim 0.15$, with an $RG = 10$, $I_T \sim 0.1$, while for $RG = 1.1$, $I_T \sim 0.8$.

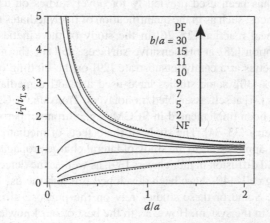

FIGURE 4.4 Effect of the substrate radius on the current approach curves with a disk UME probe with $RG = 10$. The solid lines are for $b/a = 30$, 15, 11, 10, 9, 7, 5, and 2 from the top. The upper and lower dotted lines represent theoretical approach curves with conductive and insulating substrates, respectively. (Adapted from Xiong et al., *Anal Chem*, Vol. 79, (2007): pp. 2735–2744.)

4.2 OPERATION MODE AND EXPERIMENTAL CONSIDERATIONS

The SI-SECM mode consists of a specific application of the feedback mode under transient conditions that allows for the quantification of finite amounts of reacting species on a substrate surface. In SI-SECM, the tip approaches a small substrate of the material that is to be studied. The tip must be concentrically aligned and approached to $L = d/a \leq 0.3$ (e.g., both electrodes would typically be 12.5 μm in radius, a, held at a short distance from each other, d ca. 1–3 μm). A small size of the tip is required for close positioning to the substrate, while a small substrate size is required to prevent OC feedback (Section 4.1). The sensing mechanism is based on bringing the substrate first to a potential to generate an intermediate or product on the electrode surface, then placing the substrate at OC, and allowing the tip-generated member of a redox pair O/R to react chemically with the adsorbed species at the substrate electrode. The tip current during this last stage carries the amount of adsorbate to the SECM tip through a feedback loop. The chemical aspects of this application, in a sense, are analogous to modulated beam relaxation spectroscopy (MBRS) used in vacuum for gas–solid studies of catalysis [24] in which a generated reactive species is allowed to interact with an adsorbate; the products of the reaction of the adsorbed species with the reactive molecular beam in MBRS are analyzed through different techniques, often mass sensitive ones. However, in the case of electrochemical setups, this is more challenging due to the presence of the electrolytic medium. Thus, SI-SECM is unique in that the reactive species generation and the detection scheme based on the feedback mode of SECM are integrated into the same electrochemical tip.

The SECM has been used previously for novel studies on a variety of surfaces and interfaces, such as in the quantification of intermediates released during an electrochemical reaction [25–26] in the study of the kinetics of adsorption [27] and desorption [28] at pH sensitive surfaces, and for the measurement of steady-state kinetics at a catalytic substrate [29] or the binding of metal ions at lipid monolayers [30]; some studies have used a reactive mediator to measure binding kinetics [31] at self-assembled monolayers. Transient SECM current measurements have been implemented in SECM, e.g., chronoamperometry [32] and cyclic voltammetry [33–34]. The study of the effects of mediator surface diffusion [28,35–36], applications to the study of lateral charge propagation in polymer films [37–38] and diffusion in monolayers [39], as well as the detection of reactive species in living cells [40], have been developed through the use of the transient feedback mode. Some of these studies rely on the purposeful integration of a reactive species in the system. However, to the best of our knowledge, no SECM approach to the detection and quantification of transient reaction intermediates at electrocatalytic and catalytic materials (i.e., potential or nonpotential dependent) had been reported before the introduction of SI-SECM. This surface interrogation (SI) method allows the detection and quantification of only reactive adsorbed intermediates formed upon operation of the substrate electrode, independent of their spectroscopic characteristics and their electrochemical reactivity at the same substrate.

Once two SECM probes, one being the substrate and the other the tip, are brought into close proximity, the feedback mode can be implemented. Total PF is obtained when the substrate regenerates the mediator at a rate only limited by diffusion in the tip–substrate gap. For example, for a reduction at the tip ($O + e \rightarrow R$), the reaction at the substrate electrode would be the opposite electrochemical reaction ($R-e \rightarrow O$). While most SECM studies are done at steady state, in the experiments presented here, a transient chemical reaction of a species on the substrate is used to generate the PF loop. Figure 4.5 (not to scale) illustrates the proposed mechanism for the SI of a reducible adsorbed species at the substrate. First, as depicted in Figure 4.5a, the substrate is pulsed or scanned to a potential where oxidation occurs and an adsorbed species, A, is formed; the

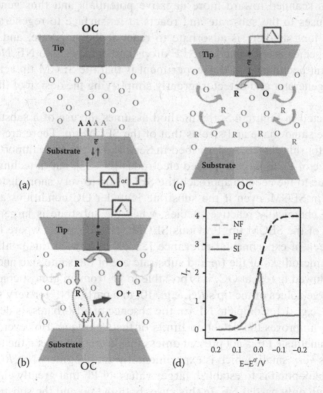

FIGURE 4.5 Schematics of the proposed mechanism for SI-SECM. (a) A reactive species is chemically or electrochemically adsorbed at the substrate upon a potential scan or step while the tip is at OC. (b) The substrate is taken to OC and the tip generates the titrant, which reacts at the surface of the substrate to support PF at the same tip. (c) After consumption of the adsorbate at the substrate, the tip experiences NF. (d) Expected current response at the tip following the events depicted in (a), (b), and (c) for SI at an arbitrary electrode setup. Schematic comparison is also made to PF and NF behavior. (Adapted from Rodríguez-López et al., *Am Chem Soc,* Vol. 130, (2008): pp. 16985–16995.)

tip at this time is kept at OC. The solution contains initially only the mediator in its oxidized state, O, which is stable under these conditions and does not participate in any reaction. In Figure 4.5b, the substrate electrode is taken to OC and the tip is scanned or pulsed to reduce O and generate R. After R diffuses across the tip–substrate gap, it reaches the adsorbed species A and reacts chemically with it. O is regenerated and A is consumed and transformed into a final product, P. Ideally, regeneration of O should only be possible through reaction with A present at the substrate and the system exhibits PF only for as long as this condition is met. As shown in Figure 4.5c, the final state of the system shows NF after all the A is depleted, and O reacts at the tip electrode at the rate governed by the hindered diffusion into the tip–substrate gap.

Figure 4.5d shows the expected voltammetric response for this system (exaggerated for clarity) and compares it to the case of PF and NF. The tip potential is scanned toward more negative potentials and thus generates R, which diffuses to the substrate and reacts at its surface to regenerate O. PF persists as long as there is adsorbate to react at the substrate, and once the adsorbed species is consumed, the PF drops and changes into NF. Notice that an important peculiarity in this experiment is that the SECM tip acts as both the titrant generator and detector, greatly simplifying the design of this *in-situ* technique.

As depicted in Figure 4.5, the method assumes the use of a substrate electrode of the same shape and size as that of the SECM tip. There are a number of reasons for this choice, as described in Section 4.1, the most important being the presence of PF at large unbiased electrodes [20–23], but in technical terms it is also due to the need to approach the SECM tip to very short distances. PF can occur in SECM, even if the substrate is under OC conditions and in the absence of chemically reactive species, when the substrate is larger than the UME disk of the SECM tip. Previous SECM surface studies where this situation has been of experimental relevance [22,29] have used this configuration. When the microdisks of the tip and substrate are of the same size and the substrate is allowed to rest at OC, it is possible to go from an almost complete NF for very large values of the tip's RG, e.g., 10, to a partial NF for very small values, of RG, e.g., 1.1. Complete NF (in the absence of adsorbates) is desirable as this greatly improves the detection limits of the technique. However, as stated earlier, an adequate PF is not present unless the distance between the tip and the substrate is very small, which is experimentally challenging when RG is large. A reasonable option is to establish larger values of PF that greatly enhance the contrast with only partial NF. In this case, both the tip and the substrate are of the same size and shape with an RG between 1 and 2. Typical SECM tips, e.g., with $a = 12.5$ μm, can be approached to distances within 1 to 3 μm in which PF with I_T between 4 and 8 (or higher at smaller L-values) is attainable. Finally, different techniques can be used to generate the titrant, R; CV is typically preferred but the use of chronoamperometry is also shown.

Implementation of the SI-SECM mode does not require special equipment other than that of conventional SECM operation. The relative positioning of the two electrodes is crucial to this experiment and is done by means of a combination of generation/collection (which is operative at long distances) and feedback (at short distances) modes. In a typical experiment, the tip is approached to the substrate and then scanned in the x and y directions in order to maximize either the feedback or generation/collection signal. The center of the substrate can typically be found within a 10% precision in the lateral resolution of the substrate as calculated from successive forward and backward scans in the x and y directions. The fine approach of the tip to the substrate is achieved in the feedback mode by a slow (e.g., 0.3 μm/s for $a = 12.5$ μm) approach curve. An approach is considered successful when a feedback of at least $I_T = 6$ and an estimated CE higher than 95% are achieved upon a CV scan under PF conditions. The initial orientation of the electrodes (tilt) and their quality as SECM tips play important roles in this experiment; as a rule of thumb, a good starting SECM tip for this type of experiment should be able to approach a tilt-corrected insulating substrate to at least $I_T = 0.15$ (where $I_T = i_T/i_{T,\infty}$; i_T is the tip current and $i_{T,\infty}$ is current at infinite distance).

Figure 4.6a shows a typical lateral scan for locating the center of the substrate electrode, in this case with the substrate producing $Ru(NH_3)_6^{2+}$ while the tip is oxidizing it back (collecting) to $Ru(NH_3)_6^{3+}$; it is desirable to obtain symmetrical features in the current profiles of both electrodes, and this indicates that the electrodes are well centered and devoid of tilt. The inset in Figure 4.6b shows a typical approach curve once the centers of both electrodes have been found and the electrodes aligned. A good indicator for a successful alignment of the electrodes is the collapse of the substrate and tip currents to the same curve, which indicates nearly 100% collection efficiency (CE). Upon a potential scan in the PF mode as shown in the main Figure 4.6b, the same high CE should hold. Furthermore, obtaining such a scan confirms that the electrodes are not physically touching (short circuited). Notice also that this voltammogram corresponds well to the Nernstian shape described in Section 4.1.

SI-SECM requires that the substrate be interrogated at OC so the experimenter must ensure that during the interrogation process the substrate is electrically isolated from the rest of the elements in the setup. CH Instruments SECM software allows automatic control of these features. Less convenient, manual switching can be used; i.e., once the tip and substrate are aligned, their leads can be connected to a switch that allows the working electrode in the SECM to be set either to the tip or to the substrate while the other remains disconnected from the setup. For a typical measurement, one would take the substrate electrode to the desired potential through a linear CV scan, then turn off the cell at that potential, and switch the working electrode lead to the tip and turn on the cell with the programming for the tip scan. For a reducing mediator, this scan can be started approximately 0.2 V above the mediator E^0 and run at the desired scan rate until -0.2 V versus E^0. The tip is typically run for two complete cycles within these potential limits to verify the consumption of the adsorbed species and to obtain a blank for subtraction of NF. The charge neutralized during the SI-SECM scan is easily calculated by

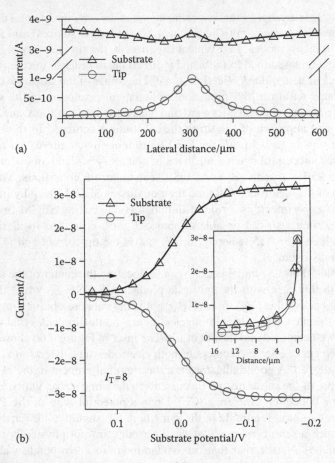

FIGURE 4.6 Experimental positioning and alignment of electrodes for SI-SECM. Solution is 1 mM in hexaaminneruthenium(III) ($E^0 = 0.05$ V vs. NHE). Substrate is generator, $E_S = -0.2$ V versus NHE and tip is collector, $E_T = 0.2$ V versus NHE. (a) Line scan of the tip over the substrate showing a desirable symmetry in the electrochemical response, the electrodes are centered at the correlated maxima of both curves. (b) Inset: approach curve at the centered position in (a) stopping at the highest current, the main curve shows a CV at this position in which the substrate is being scanned from 0.2 to −0.2 V versus NHE and the tip held at 0.2 V versus NHE. Since $a = b$, the electrodes are indistinguishable for positioning purposes. (Adapted from Rodríguez-López et al., *J Am Chem Soc*, Vol. 130, (2008): pp. 16985–16995.)

numerical integration of the background and baseline-subtracted interrogation voltammogram.

The nature of the system under study using SI-SECM will determine the type of mediator that is chosen. One has to decide if the adsorbed intermediate under study is prone to oxidation or to reduction; then one must also consider the availability of mediators that can accomplish the desired transformation (e.g., thermodynamically), their interaction with the electrolyte and with the substrate, as well as the possibility to electrogenerate them with an available tip material. Once these basic conditions are met, other factors, such as the reaction kinetics of the mediator with the adsorbed species or its selectivity toward it can also be considered. A good example of the selection process for an adequate mediator is that for the oxidation of adsorbed hydrogen on Pt in highly acidic media [41], for which, as shown in Section 4.6, a good interrogation mediator is N,N,N',N'-tetramethyl-p-phenylenediamine (TMPD) and its radical cation. The criterion used to determine the goodness of a mediator was largely based on the following characteristics: (1) high to moderate solubility in the medium, (2) stability of both forms of the redox pair, (3) a large difference in E^0 with respect to adsorbed hydrogen ($E^0 = 0$ V vs. NHE for the H^+/H_2 pair, but preferably $E^0 > 0.4$ V for the mediator to be outside the under potential deposition (UPD) region on Pt), (4) low electrochemical background when performing the SI technique (which reflects lack of interaction with the Pt substrate during the NF acquisition), and (5) quantitative reaction of the oxidized form with $H_{(ads)}$. Figure 4.7a shows the CV of Pt in 0.5 M H_2SO_4 as well as the position of the oxidation waves for three mediator candidates that comply well with criteria (1), (2), and (3): $Ru(bpy)_3^{2+}$, $Fe(phen)_3^{2+}$ (bpy = bipyridine, phen = phenanthroline), and TMPD. The oxidation products of these substances are good oxidizing agents that energetically overlap (to different degrees) in the region where the formation of oxides is observed in the CV of Pt; however, only TMPD complies well with criteria (4) and (5). Working in the SI mode with TMPD and upon dosage of the Pt substrate with $H_{(ads)}$, this mediator is able to interrogate it quantitatively (recovery of $H_{(ads)}$ is larger than 97% for 1 ML (monolayer) of $H_{(ads)}$) and its NF response shows a low background, roughly 0.3 $i_{T,\infty}$ (see Section 4.6).

While TMPD was an adequate mediator for the interrogation of $H_{(ads)}$ on Pt, the other mediators based on metal complexes displayed a less satisfactory behavior, mainly with respect to their interaction with the substrate. Figure 4.7b shows a comparison of the NF and PF scans for the oxidation of 0.1 mM $Ru(bpy)_3^{2+}$ by an Au tip in proximity to a Pt substrate; the difference between the NF and PF scans is so small that quantitative work would be impossible and even qualitative evaluation is severely affected. Moreover, upon interrogation of $H_{(ads)}$, the recovery was poor and often irreproducible. The large background observed in the NF scans can be rationalized in terms of the oxidized mediator reacting with the metal substrate to a small extent; in the case of platinum, this could proceed through the formation of platinum oxides (PtOH, PtO) along with the regeneration of the reduced form of the mediator, which can account for the recycling of the reduced species to give a PF-like response where otherwise a low background should be observed. In this case, the strong interaction between the oxidized form of the mediator and the reduced form of the surface limits the choice of the mediator. In fact, the use of oxidizing mediators in SECM studies has been applied extensively for the patterning and dissolution of substrates such as copper [22], silver [42] and silver nanoparticles [43], and semiconductors [44]. In this respect, TMPD is a special case that does not follow this trend, so it

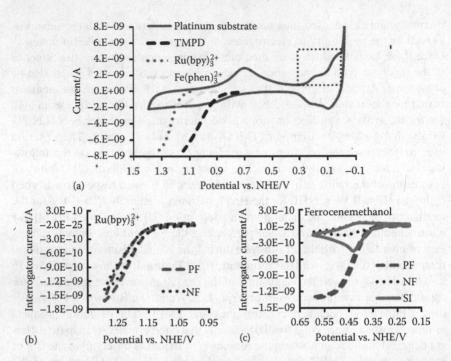

FIGURE 4.7 Experimental considerations for mediator selection. For all, interrogator tip electrode is Au ($a \approx 12.5$ μm) approached to $L \approx 0.2$ from a Pt substrate UME ($a = 12.5$ μm); All PF scans taken with E_{sub} at least 0.3 V more negative than $E_{1/2}$. (a) Energetics of mediators with respect to the behavior of Pt in acid media. CV of Pt substrate, $\nu = 50$ mV/s and comparison to PF voltammograms of 0.5 mM mediator (TMPD, Ru(bpy)$_3$(ClO4)$_2$ or Fe(phen)$_3$(ClO4)$_2$). Solution is 0.5 M H$_2$SO$_4$; $\nu = 50$ mV/s. (b) Comparison of PF and NF for 0.1 mM Ru(bpy)$_3$(ClO4)$_2$ in Ar-purged 0.5 M H$_2$SO$_4$; $\nu = 20$ mV/s. (c) SI of H$_{(ads)}$ at Pt with 1 mM FcMeOH in 0.1 M acetate buffer pH = 4. H$_{(ads)}$ was dosed by taking the Pt substrate through CV from 0.2 V versus NHE to −0.24 V versus NHE at $\nu = 200$ mV/s. Interrogator electrode scan $\nu = 20$ mV/s. Comparison to NF and PF for the same FcMeOH conditions, $\nu = 20$ mV/s. (Adapted from Rodríguez-López et al., *J Am Chem Soc*, Vol. 132, (2010): pp. 5121–5129.)

allows for the study of a wide potential range on the Pt substrate (≈ 0 to 0.8 V vs. NHE), which is optimal for the study of H$_{(ads)}$. The interrogation of Pt at higher pH, where more mediators are stable but formation of the surface oxide occurs at less positive potentials, was also studied and for most mediators with E^0 higher than 0.5 V at pH = 4, results similar to those shown for the Ru complex were obtained. Figure 4.7c shows the interrogation of H$_{(ads)}$ at pH = 4 in acetate buffer using ferrocenemethanol (FcMeOH) as mediator, which provides a milder oxidizing agent. While it was possible to establish a reasonably wide quantification window, recoveries with this mediator were about 90% and the reaction kinetics between FcMeOH$^+$ and H$_{(ads)}$ were sluggish, as shown by the broad and slow decay of the SI scan in Figure 4.7c. This sluggishness is also detrimental for quantitative work and it may be related to the smaller energetic gap between FcMeOH and $H_{(ads)}$. An energy–kinetic relationship is shown in Section 4.5 while the general evaluation of the kinetics involved in SI-SECM is discussed in Section 4.3. As a final word on choice of a suitable mediator for implementation of SI-SECM, steady-state background processes (in addition to transient ones as already discussed) may also play a role in designing an adequate quantification method. For example, the Pt surface is able to catalyze the reduction of proton into molecular hydrogen by a strong reducing mediator and to provide a steady-state feedback signal at the tip as a consequence [29]. This process is strongly dependent on pH and on the reducing power of the mediator and limits the choice of a suitable interrogating agent.

4.3   DIGITAL SIMULATION OF THE SI-SECM PROCESS

Digital simulations using implicit finite difference methods have been used since the first studies of SECM [15,16,45]; more recently, the advent of flexible commercial software has allowed an easier implementation of SECM problems. A popular example, software such as COMSOL Multiphysics [46] that uses the finite element method (FEM) has been useful for treating a variety of SECM problems; in some cases, numerical results obtained by such methods have been used as means of comparison to semianalytical expressions. The mathematical framework for FEMs is more complex than for finite difference methods; broadly speaking, while finite difference methods discretize the differential operators that describe a given phenomenon, FEMs preserve the original form of the differential operators and instead construct and test trial functions that satisfy the operators for the boundary conditions given [47–48]. This is done in a conveniently meshed simulation space (subdomain) that can be constructed using a computer assisted design (CAD) interface; changes in meshing and design can easily be accomplished, an important feature for SECM where geometrical factors such as the size of the tip, substrate, *RG*, and distances vary from one experiment to another. Programs such as COMSOL Multiphysics also allow the coupling of problems in different geometries and dimensions; for instance, a 1D problem (e.g., adsorption) can be treated simultaneously with a 2D problem with axial symmetry (e.g., diffusion toward a UME).

Simulation of SI-SECM requires the modeling of diffusion and reaction kinetics in the SECM geometry; provided the system has axial symmetry, the simulation space and dimensional transformation from a 3D problem to a 2D one can be done as depicted in Figure 4.8. The total simulation space consists of two relevant subdomains: the first corresponds to the SECM configuration with two UME electrodes that are concentrically aligned and the second one that represents the surface of the substrate. In the first subdomain, we consider only a diffusion problem in which the reduced species R is generated at the tip from the initially present oxidized species O; thus for transient conditions following Fick's second law as required for implementing linear voltammetry, we have Equations 4.5 and 4.6:

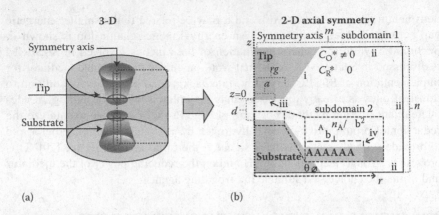

(a)   (b)

FIGURE 4.8 Description of the general SI-SECM simulation space and conditions. (a) The 3D geometry in which the tip and substrate are concentrically aligned can be transformed into (b) a 2D geometry with a symmetry axis and described by z and r. Boundary types: i, insulation; ii, bulk concentration (semi-infinite); iii, flux at the tip; iv, concentration of A at the substrate; $m = n = 20a$ and $\theta > 60°$. Other parameters include the interelectrode distance d, the size of the tip microdisk a, substrate microdisk b, and the flat end length rg.

$$\frac{\partial C_O}{\partial t} = D_O \nabla^2 C_O \qquad (4.5)$$

$$\frac{\partial C_R}{\partial t} = D_R \nabla^2 C_R \qquad (4.6)$$

where C_O and C_R are the concentrations of oxidized and reduced species, respectively, and are functions of r, z, and time, t. D_O and D_R are the diffusion coefficients of these species. The bulk concentration (C_O^* and C_R^*) of each species is set to the one used during the experiment and the adsorbed species A is not present in this subdomain. At the tip boundary, $z = 0$, $0 < r < a$, we write the inward flux condition following Butler–Volmer kinetics, Equations 4.7 and 4.8, and in the case of a linear potential sweep, we add Equation 4.9:

$$-D_O \nabla C_O = -k° e^{-\alpha f(E-E^{0'})} C_O(0,r) + k° e^{(1-\alpha)f(E-E^{0'})} C_R(0,r) \qquad (4.7)$$

$$-D_R \nabla C_R = k° e^{-\alpha f(E-E^{0'})} C_O(0,r) - k° e^{(1-\alpha)f(E-E^{0'})} C_R(0,r) \qquad (4.8)$$

$$E = E_{in} - \upsilon t \qquad (4.9)$$

where $k°$ is the heterogeneous rate constant, α is the transfer coefficient, f is equal to 38.94 V^{-1} at 298 K, and $E^{0'}$ is the formal potential of the reaction. E is the potential of the tip and in CV relates the scan to time by means of υ, the scan rate, and the initial potential E_{in}. A simple SI-SECM model can be considered by the reductive interrogation of adsorbed species A. The surface consists of a second subdomain in which A is contained and the reduction reaction shown in Equation 4.10 occurs [5,49]. The kinetic component to Equation 4.10 is assumed to be a second-order process associated with a kinetic constant k and Equations 4.11 through 4.13 operate in the substrate subdomain.

$$R + A \xrightarrow{k} O \tag{4.10}$$

$$\frac{\partial C_O}{\partial t} = kC_R C_A \tag{4.11}$$

$$\frac{\partial C_R}{\partial t} = -kC_R C_A \tag{4.12}$$

$$\frac{\partial C_A}{\partial t} = D_A \nabla^2 C_A - kC_R C_A \tag{4.13}$$

It is convenient to refer quantitatively to A in terms of the charge it carries; thus the charge Q is equal to the moles of A, n_A, times Faraday's constant, F, 96,485 C/mol. The surface coverage Γ_A in terms of charge can also be obtained by consideration of the area of the substrate. Figure 4.9 shows representative results for the simulations at different values of the kinetic constant k (for the reaction of titrant R and A) and for different charge amounts, Q, of adsorbed species at fixed k.

In Figure 4.9a, the kinetic constant was varied over three orders of magnitude and limiting behavior under these conditions is found when $k \geq 10^3$ m^{3}mol^{-1}s^{-1}

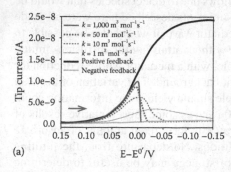
(a)

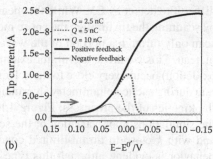
(b)

FIGURE 4.9   Representative LSV results of digital simulations for SI-SECM. $v = 50$ mV/s and $C_O^* = 1$ mol/m^3, O + e → R at the tip with no kinetic complications. Both tip and substrate $a = 12.5$ μm, $RG = 2$, and $L = 0.1$. (a) Variation of the reaction kinetic constant k at a fixed $Q = 8$ nC at the substrate. (b) Variation of the charge adsorbed at the substrate Q at fixed $k = 50$ m^{3}mol^{-1} s^{-1}. (Adapted from Rodríguez-López et al., *J Am Chem Soc*, Vol. 130, (2008): pp. 16985–16995.)

($mM^{-1}s^{-1}$) for this particular setup, as shown by the overlapping of the voltammograms with that of PF before it reaches a maximum and decays as the surface is depleted of A. This limiting behavior establishes that the chemical reaction cannot regenerate O at a rate faster than the diffusion of R to the substrate; this is denoted as a titrant-limited situation. When k drops below this value, there is a shift in the peak and the behavior is kinetically limited. Figure 4.9b shows the variation of the shape of the tip response with the simulated charge, i.e., total amount of A was entered in the simulation with kinetics close to titrant-limited conditions ($k = 50$ $m^3mol^{-1}s^{-1}$). The system responds proportionally to the amount of A following the PF curve until A is depleted, then it rapidly drops to NF behavior. Upon subtraction of the NF curve from these tip responses and integration of the current with time, the CE (defined as the ratio of the recovered charge by the tip to the input charge used in the initial charging of the substrate) was found to be larger than 98% for this simulation in which $a = b$ and $L = 0.1$. This is another feature that can only be achieved at very small tip-substrate distances when using a substrate of the same size as the tip, and provides the advantage of obtaining the titrated charge by direct integration of the tip current without the need of using a calibration factor. If the size of the substrate is larger, e.g., $b = 2a$, then the CE drops to approximately value of 80% at $L = 0.1$, and then this factor has to be taken into account. At larger interelectrode distances not only will the contrast between PF and NF decrease, but also the CE will be lower.

It is useful to have a titrant-limited system rather than a kinetically limited one. As shown in both parts in Figure 4.9, under titrant-limited conditions most of the collection takes place in a narrow potential region preceding the E^0 of the mediator and also exhibits very clear and distinguishable peaks (i.e., a larger contrast between PF and NF). An advantage of the SECM technique compared with conventional CV is that it decouples the generation of surface species from their detection and quantification. This allows one to detect species that would be difficult to detect by CV. While the measurements described here were all made by scanning the tip through the entire mediator wave, it would also be possible to scan only through a portion of the wave without attaining the diffusion-limited plateau. This would be useful, for example, with a mediator whose reduction (or oxidation) occurs very close to the solvent background. The variation of the scan rate during cyclic voltammetry in principle should yield some information about the kinetics of the processes; Figure 4.10a shows some representative results of the simulations upon variation of the scan rate for a fixed amount of adsorbate and with k close to titrant-limited conditions. No departure from the limiting behavior is evident, although this type of strategy may be useful to determine rate-limiting steps in sequential mechanisms of reaction of the adsorbed species. Of more interest is the observation that at higher scan rates the tip current is higher. This may be useful in improving the signal-to-background ratio for certain systems, but this may come at the expense of the quantification of the adsorbate, if not all can be titrated in a single scan.

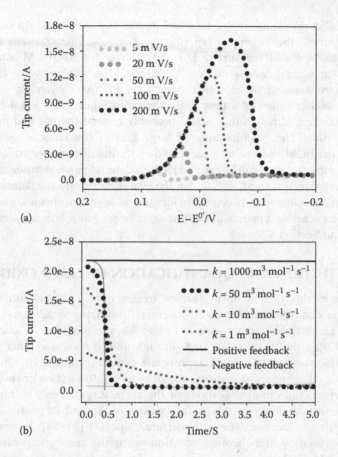

(a)

(b)

FIGURE 4.10 SI-SECM dependence of scan rate and in chronoamperometric mode. (a) Simulated tip scans at different scan rates for a constant charge of 8 nC and $k = 50$ m^3 mol^{-1} s^{-1}. (b) Simulated chronoamperograms at a constant charge of 8 nC and different values of k. $E_T = -0.2$ V, with $E^0 = 0$ V for the tip reaction. Same setup as used in Figure 4.9. (Adapted from Rodríguez-López et al., *J Am Chem Soc*, Vol. 130, (2008): pp. 16985–16995.)

Chronoamperometry, in which the potential of the tip is pulsed to generate a nearly constant flux of mediator, can also be simulated and is shown in Figure 4.10b. In this case, it is important to consider that most of the analytical signal is contained at the beginning of the step, where it may be convoluted with the (short) transient current decay of the tip. In the analysis of small amounts of adsorbate, this may result in a drawback since a small analytical signal is read over a large background. By contrast, the use of CV allows the detection of small amounts of adsorbate in a relatively quiet zone of the voltammogram (i.e., at the foot of the wave).

For SECM operating in the feedback mode, the closer the tip is located to the substrate, the larger the heterogeneous kinetic rate constant that can be distinguished and measured. This holds true for SI-SECM where the second-order heterogeneous process in SI-SECM is analogous to the first-order electrochemical process that occurs on PF. At shorter tip–substrate distances, higher values of k than those shown in Figures 4.9 and 4.10 might be distinguished, although this would probably require the use and modeling of smaller tips. The quantification of large kinetic constants is an ongoing challenge in SECM and has been addressed by the use of nanoelectrodes, where small tip–substrate gaps can be created [50]. To date, no nanoelectrode approach has been attempted with SI-SECM, but the first steps toward the determination of reaction mechanisms between the mediator and adsorbed species, quantification of their reaction kinetics as well as their impact in a broader context are discussed in Section 4.5.

4.4 DETECTION AND QUANTIFICATION OF GOLD OXIDES

Transition metals tend to form surface oxides upon contact with aqueous electrolytic conditions under electrochemical oxidizing conditions. For the specific case of noble metals such as Pt and Au, these oxides are chemically reversible, i.e., they can be formed through anodization and then reduced by switching the potential to an appropriate cathodic value [51]; while other changes may accompany this process such as metal dissolution or roughening of the surface, the initial conditions of the surface are largely restored [52]. The exact stoichiometry of the surface oxides of Au and Pt is still a matter of debate due to the coexistence of different species [53–54]; however, to a first approximation, the chemical reaction occurring can be expressed as in Equation 4.14 for Au:

$$2Au + 3H_2O \rightarrow Au_2O_3 + 6H^+ + 6e \tag{4.14}$$

SI-SECM was introduced with the quantification of gold and platinum oxides at neutral buffered pH [5]; this case represents a classical example of the stable chemisorption of oxygen and formation of surface oxides. This section will discuss SI-SECM of Au oxides as an introductory topic, while Section 4.5 will discuss in further detail the kinetic analysis of the reaction of Pt oxides with different mediators.

To demonstrate a tip reaction with the necessary PF for SI-SECM, experiments with the $Ru(NH_3)_6^{3+/2+}$ pair ($E^0 = 0.05$ V vs. NHE) [55] were carried out to evaluate the feasibility of the reduced form $Ru(NH_3)_6^{2+}$ reacting with electrogenerated oxide on Au. In its simplest scheme, we assume that the reaction (surface titration) proceeds as shown in Equation 4.15:

$$Au_2O_3 + 6Ru(NH_3)_6^{2+} + 6H^+ \rightarrow 2Au + 6Ru(NH_3)_6^{3+} + 3H_2O \tag{4.15}$$

where three electrons are transferred per each gold atom. Because the CV of gold in phosphate buffer at pH = 7 shows the onset of oxidation at approximately 0.9 V vs. NHE, then the reaction with $Ru(NH_3)_6^{2+}$ is allowed thermodynamically ($\Delta E^0 \geq 0.85$). This is further confirmed in Figure 4.11 where a gold tip was anodized at 1.2 V for 30 s; when the potential scan was reversed, a characteristic reduction wave is observed. However, if an anodized electrode was immersed in a solution containing $Ru(NH_3)_6^{2+}$ (chemically generated from a 1 mM solution of $Ru(NH_3)_6^{3+}$ in phosphate buffer at pH = 7 by passing through a Zn-packed syringe stored in an Ar-flushed vial) and returned to the electrochemical cell with a scan starting from the onset of the reduction wave, this oxide reduction process was not observed. If the anodized electrode was instead immersed in a control solution of $Ru(NH_3)_6^{3+}$, the reduction wave was still present with some losses that we attribute to instability of the oxide upon changing environments, reconnection to the potentiostat circuitry, or traces of reduced species in the solution.

Once the electrodes are aligned and positioned, the oxidation of the substrate and its subsequent interrogation can be performed. Figure 4.12 shows a typical result obtained from oxidation of a gold substrate at 1.4 V, followed by bringing it to OC and then by reduction with electrogenerated $Ru(NH_3)_6^{2+}$, which shows a PF response at the tip. The general features expected from digital simulation (Section 4.3) are present: the titration of the gold oxide by the reducing agent follows the curve of PF until the consumption of all of the surface oxide causes a sharp decrease in the feedback current. Most of the charge neutralization occurs well before reaching $E_{1/2} = 0.01$ V versus NHE for this mediator and after this, the current drops to the level of the expected NF. This confirmed early in the development of SI-SECM that regardless of the specific reaction mechanism encountered for a given system, the general idea of detecting the transient reaction of the mediator with the adsorbed species is possible.

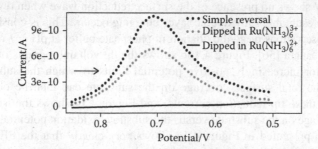

FIGURE 4.11 Test of the reactivity of electrogenerated Au oxides with $Ru(NH_3)_6^{2+}$. A gold tip is oxidized for 30 s at 1.2 V in pH = 7 phosphate buffer 0.3 M. The voltammograms correspond to the comparison of electroreduction just following oxidation against the case when they are dipped in solutions containing either $Ru(NH_3)_6^{2+}$ or $Ru(NH_3)_6^{3+}$, taken back to the cell and run the reductive scan. Initial potential 0.85 V versus NHE, $v = 50$ mV/s, dip time in mediator was 2 s. (Adapted from Rodríguez-López et al., *J Am Chem Soc*, Vol. 130, (2008): pp. 16985–16995.)

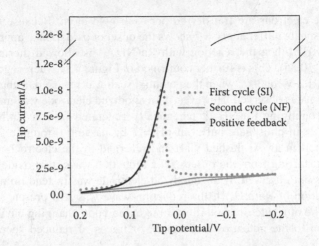

FIGURE 4.12 Representative SI-SECM result of mediated reduction of Au oxide with hexaamminneruthenium(II). Au substrate was oxidized for 30 s in TRIS buffer 0.3 M pH = 7 at 1.4 V versus NHE. The first cycle (SI scan) shows a clear transient feedback that follows the trace of an independent PF experiment under the same conditions. The NF is represented by a second cycle in the absence of substrate oxidation. Both tip and substrate are Au microelectrodes $a = b = 12.5$ μm. Scans ran at $v = 50$ mV/s, solution contained 1 mM Ru(NH$_3$)$_6^{3+}$, electrode gap $L \sim 0.1$. Tip potential versus NHE. (Adapted from Rodríguez-López et al., *J Am Chem Soc*, Vol. 130, (2008): pp. 16985–16995.)

A powerful feature of SI-SECM is that it allows the OC study of the adsorbed species formed at different potentials of operation of the substrate. Figure 4.13a shows the CVs of gold at selected reversal potentials; the amount of oxide is quantifiable above 1.0 V by integration of the surface reduction wave present between 0.8 and 0.5 V upon the cathodic scan. With CV, the oxidation of the substrate below 0.9 V shows no presence of the surface reduction wave when reversing the scan, and presumably only double layer charging occurs. This is consistent with previous observations of gold oxidation in phosphate buffer at pH = 7 reported by Oesch and Janata [56]. Figure 4.13b shows the tip voltammograms obtained by SI-SECM for increasingly positive potential limits to which the substrate was brought. The total surface coverage on the substrate can then be obtained by integrating these tip reduction currents, and Figure 4.14 shows the isotherm for oxide coverage on the substrate versus the substrate oxidation potential limit. The comparison presented in Figure 4.14 shows very clearly that the SECM-based technique senses a reactive species adsorbed at less positive substrate potentials where the CV of the substrate electrode does not, in both the results obtained here or those previously reported [56]. We suggest this as evidence for the formation of the so-called "incipient oxide" or a singly oxygenated species such as Au–OH or Au–O, which has recently been proposed to explain the catalytic activity toward CO oxidation present at noble metals at the liquid–solid [57–60] and gas–solid

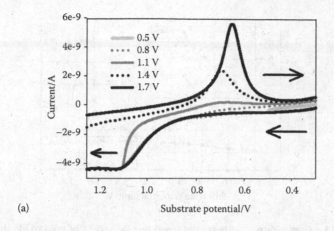

(a)

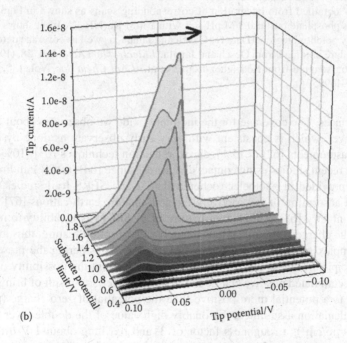

(b)

FIGURE 4.13 Reactivity of Au oxides by CV and by SI-SECM with $Ru(NH_3)_6^{2+}$. Comparison of substrate voltammetry to the interrogation of Au in phosphate buffer 0.3 M pH = 7 with 1 mM $Ru(NH_3)_6^{3+}$. (A) Voltammetry of the Au substrate, ν = 50 mV/s, showing curves at different reversal potentials. (B) Spectrum of SI-SECM tip scans where Au oxides were formed by taking the Au substrate to different substrate potential limits through a potential scan at ν = 50 mV/s. Both tip and substrate were Au microelectrodes a = b = 12.5 μm. Scans ran at ν = 50 mV/s, electrode gap $L \sim$ 0.1. Tip and substrate potentials versus NHE. (Adapted from Rodríguez-López et al., *J Am Chem Soc,* Vol. 130, (2008): pp.16985–16995.)

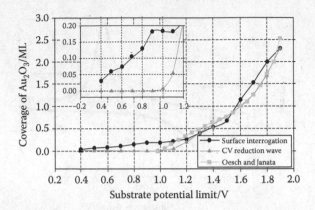

FIGURE 4.14 Comparative isotherm obtained by CV and SI of gold oxide in pH = 7. SI and CV obtained from integration of corresponding scans as shown in Figure 4.13. Solution is phosphate buffer 0.3 M pH = 7 with 1 mM $Ru(NH_3)_6^{3+}$. Inset shows blowup of the early potential region. (Data for Oesch and Janata, as well as coverage determination procedure obtained from Oesch and Janata, *J Electrochim Acta*, Vol. 28, (1983), pp. 1237–1246; Adapted from Rodríguez-López et al., *J Am Chem Soc*, Vol. 130, (2008); pp. 16985–16995.)

[61] interfaces. The coverage for the incipient oxide we observe is about 20% of a monolayer, which is consistent with very early observations done by capacitance measurements [62] ($\approx$15%), electroreflection techniques [63] (10%–20%), and more recent studies using contact electroresistance [64] ($\approx$15%); indirect evidence is provided in photoelectrochemical [65] and SERS [66] studies, by the enhanced adsorption of radioactively labeled alkaline earth cations [67] and by AC voltammetry [68]. An inherent feature of SI-SECM is its ability to measure directly the charge that corresponds to such an adsorption feature, thus avoiding more complex treatments to estimate coverage. In considering the presence of these "incipient oxides," we must also take into account the possibility of titrating the accumulated positive charge in the double layer as a result of bringing the substrate to a potential more positive than the potential of zero charge (pzc). A simple calculation assuming a reasonably high value of the double layer capacitance (30 μF/cm^2), a roughness factor of 3, and reaching about 1 V difference from the pzc yields a value of 0.4 nC for double layer charging, which represents roughly 4% of the charge for a monolayer of gold oxide. This is just slightly above our limit of detection as will be explained later, and while it may explain some effects observed at OC or at very low potentials it does not account for the amounts found of incipient oxide. An interesting feature we found when obtaining blanks (where the substrate was not taken purposely to any oxidation potential) and when running the second CV after the main interrogation (e.g., Figure 4.12), is that the appearance of a small peak shown in Figure 4.15 that, as most of our responses, follows well PF. This feature can even be present in very reducing environments, although in very small quantities, in our measurements, less than 2% of

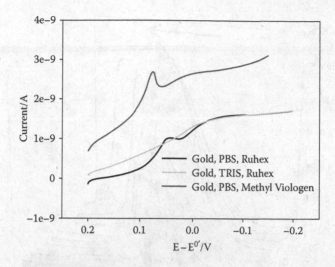

FIGURE 4.15 Blank curves for Au oxide SI experiments with different reductive mediators and buffers. The horizontal axis is set as the difference in potential versus the standard potential of the redox pair in order to compare the different mediators, Ruhex $(Ru(NH_3)_6^{3+/2+})$ and Methyl Viologen (MV^{2+}/MV^+); current response was different for all experiments based on differences in alignment, mediator, concentration of mediator and traces of oxygen. (Adapted from Rodríguez-López et al., *J Am Chem Soc*, Vol. 130, (2008): pp. 16985–16995.)

a monolayer, which is our limit of detection. Marichev [64] observed features at potentials as negative as −0.6 V and Lertanantawong et al. [68] also discuss this in view of the role of surface defects in providing high energy features that allow the formation of these oxides under such conditions. This peak was only present when the experimental setup described in the main text was used. This peak is also present for platinum and for other mediators used in this study, although not necessarily in the same amount. Another possible origin for this small peak could be the adsorption of very small amounts of mediator at the surface of the electrode; this cannot be discarded although it seems unlikely in light of the high solubility and hydrophilic behavior of the mediators used in this study, e.g., both $Ru(NH_3)_6^{3+}$, and methyl viologen (*vide infra*) are ionic species and can be dissolved at least in an amount an order of magnitude higher than the one used here.

Figure 4.13b also shows a transition between 1.4 and 1.5 V from a single-peaked tip response to a double-peaked one. From Figure 4.14, this transition corresponds to the coverage reaching more than half of a monolayer. The finding of a change in behavior upon reaching discrete coverage values has been treated in the literature as an indicator of the presence or transformation of different oxygen species (e.g., surface hydroxyl species vs. place exchanged oxides) [53]. However, in the present case, it is difficult to discern between an actual surface behavior and a kinetic effect in the reduction reaction. As explained later, these features change with conditions, such as the mediator chosen. From a kinetic viewpoint however, the second peak in the

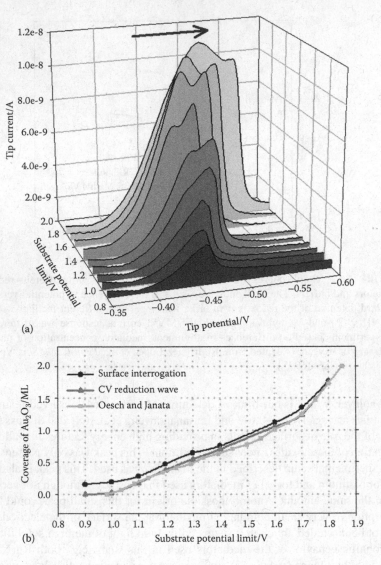

(a)

(b)

FIGURE 4.16 SI-SECM of Au oxides in phosphate buffer mediated by $MV^{2+/+}$. (a) Spectrum of SI-SECM scans in 0.3 M phosphate buffer at pH = 7, ν = 50 mV/s. Substrate taken to potential limit at ν = 50 mV/s. Mediator is 1 mM MV^{2+}. (b) Isotherm obtained from results in (a) including comparison to electrochemical reduction of the oxide and results from Oesch and Janata, *J Electrochim Acta*, Vol. 28, (1983), pp. 1237–1246. Both tip and substrate were Au microelectrodes $a = b = 12.5$ μm, electrode gap L ~ 0.1. Tip and substrate potentials versus NHE. (Adapted from Rodríguez-López et al., *J Am Chem Soc,* Vol. 130, (2008): pp. 16985–16995.)

tip SI-SECM scans that appears at $E_S > 1.5$ V in Figure 4.13b shows a faster rate for reaction with the reduced mediator than the first peak, indicating that the first process is conditioned for its appearance, e.g., a sequential instead of a parallel mechanism for the reduction of two different oxide species would have to be invoked. This could involve, for example, a rearrangement of the oxygen moieties [51], either spatially within the surface or chemically, e.g., the addition of a proton [54]. Section 4.5 deals with the reactivity of different oxide species.

Other mediators can be used to carry out the SI-SECM process; Figure 4.16 shows the case of the interrogation of Au oxides with methyl viologen. We selected the methyl viologen pair $MV^{2+/+}$ ($E^0 = -0.44$ V vs. NHE, experimental $E_{1/2} = -0.46$ V vs. NHE) [69] whose reduced form is a stronger reducing agent than $Ru(NH_3)_6^{2+}$. Figure 4.16a presents the tip scans for interrogation of gold oxide in phosphate buffer at pH $= 7$ and Figure 4.16b the corresponding isotherm, which shows good agreement with the literature reference results [56] and with the CV surface reduction wave. As with $Ru(NH_3)_6^{3+/2+}$, the presence of two processes in the tip scans is observed with $MV^{2+/+}$, although slightly different in shape and proportion. These results indicate that although the mechanism of the surface reduction may be different with different mediators, the quantification of Au_2O_3 is similar, indicating complete detection of submonolayer amounts of the oxide. The larger reducing power of MV^+ may also play a role in the kinetics of reduction, and thus, affect the shape of the SI-SECM scans.

The use of SI-SECM in the CV mode has yielded so far kinetic diagnostic features in the voltammograms, reproducible results, and good figures of merit; in the case of gold, the estimated limit of detection (taken as three times the average standard deviation of our measurements) is 0.24 nC, which, for the roughness factors found with our gold electrodes (typically between 2 and 3), represents 2.1% to 3.4% of a monolayer. This figure is also equivalent to 49 $\mu C/cm^2$ of geometric area (16 to 25 $\mu C/cm^2$ of real area considering the above mentioned roughness factors); however, it also depends on the system under study.

4.5 DETERMINATION OF THE REACTIVITY OF PLATINUM OXIDES WITH DIFFERENT MEDIATORS AND ITS IMPLICATIONS IN ELECTROCATALYSIS

The effect of electrode surface oxidation on the rate of apparently outer sphere electrode reactions (e.g., $R \rightarrow O + n\bar{e}$) has long been of interest. In such cases, at least two routes are possible: (1) reaction of the oxide directly with R and (2) tunneling of electrons from R through the oxide. The feedback mode of SECM, operating at steady state, can be used for measuring the kinetics of chemical and electrochemical processes while SI-SECM, using transient feedback, can be used for measuring the amounts and kinetics of adsorbed reactants on the electrodes. In this section, we describe both steady-state and transient SECM measurements to evaluate the reactivity of Pt oxides and draw on their correlation [49].

Figure 4.17a depicts the kinetic characterization of the reaction of chemisorbed oxygen at the Pt substrate with an electrogenerated mediator produced at an SECM tip, which can be written also as Equation 4.16:

$$2\mathbf{R} + \text{PtO} + 2\text{H}^+ \xrightarrow{\ k_{SI}\ } 2\mathbf{O} + \text{Pt} + \text{H}_2\text{O} \tag{4.16}$$

In this case, the kinetics and mechanism reaction (Equation 4.16), characterized by an overall reaction constant k_{SI}, will determine the shape of the response curve; fast reactions will give sharp CV curves that decay quickly (E is a function of time) compared to slower ones. The mechanism and rate constants can then be obtained by comparison with digital simulation of this transient response.

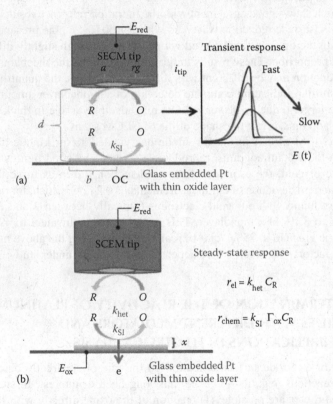

FIGURE 4.17 Schematic diagram (not to scale) of the system used to study oxide reactivity with mediators by transient and steady-state feedback. (a) SI-SECM used to evaluate the OC chemical reactivity of Pt oxide. The tip "titrates" the oxide through a potential scan; sharper feedback curves indicate higher kinetics of reaction for the mediator with the oxide. (b) SECM in the feedback mode can be used to evaluate the steady-state electron transfer across the oxidized Pt; two routes are possible: an electrochemical one with rate r_{el}, and a chemical one, with rate r_{chem}. (Adapted from Rodríguez-López et al., *J Phys Chem C*, Vol. 114, (2010): pp. 18645–18655.)

An advantage of the SECM over other electrochemical techniques is that it allows decoupling of the analytical measurement from the operation of the substrate; this case is depicted in Figure 4.17b. Here, the SECM tip generates the reduced species R at steady state while the substrate is biased to a potential at which it generates an oxide layer. At sufficiently positive potentials, the substrate will regenerate O by oxidizing R, thus providing a feedback current at the tip. Changes in the kinetics of this oxidation reaction, which may be brought about by the formation of oxide, can be followed through measurement of the tip current, which is a reflection of how fast the substrate is feeding back oxidized mediator, O [6,15,18]. This type of measurement on the oxidation rate of a mediator is done *in-situ* and at the desired potential of the substrate, in contrast to conventional single electrode measurements where the potential of the substrate would otherwise be changed to measure charge transfer kinetics of a redox mediator. By combining measurements done with the SI-SECM and feedback modes, under certain conditions the steady-state substrate response can be described by a combination of chemical and electrochemical routes (r_{chem} and r_{el} respectively in Figure 4.17b for the oxidation of R), rather than a purely electrochemical path. The measurement of chemical processes at steady state with the SECM has been exploited since the introduction of the technique [22,29,70–71]; integrated approaches exist for the determination of substrate kinetics for combined electrochemical and chemical systems such as self-assembled monolayers [72–73]. Our approach uses information obtained through transient measurements to decouple the possible contributions of the chemical route from those of the electrochemical route, as depicted in Figure 4.17b, in an electrocatalytic system.

A particularly interesting case in the oxidation of species on oxide-covered Pt electrodes is the observation that the electrochemical response for some processes, as evaluated from the heterogeneous rate constant for electron transfer, is decreased upon oxide formation [74]. Thick oxide passive layers on valve metals are known to act as insulating barriers for electron transfer [75–76]; the decrease in the rate constant due to such layers can often be modeled in general terms according to Equation 4.17:

$$k_{het,eff} = k_{het}e^{-\beta x} \tag{4.17}$$

where $k_{het,eff}$ is the effective rate of electron transfer across the insulating layer of thickness x; k_{het} is the rate constant for electron transfer in the absence of insulating layer, and β is the tunneling constant (typically taken as ~1.1 Å^{-1}, although smaller values have also been reported) [72]. Pt oxides decrease the rate of inner-sphere electrochemical reactions, such as the oxidation of hydrogen or alcohols or the reduction of protons or oxygen [77], where competitive adsorption plays a major role in the reaction mechanism [78–79]. In other processes, such as the oxygen evolution reaction (OER), the formation of these oxides is central to the mechanism of reaction, and determining the effect of adsorbed oxygen is clearly of interest [78–81]. A limited number of reports exist on the effect of Pt oxides on outer-sphere reactions, e.g., the oxidation of Fe(II) and As(III) [74] or Ce(III) and Mn(II) species [82], or $Fe(CN)_6^{4-}$ [83–84], where the participation of chemisorbed oxygen on the reduction mechanism was proposed, although no direct study of it was made. Our initial observations on the reactivity of Au oxides through the SI-SECM technique (Section 4.4) where reaction of the oxide with an *in-situ* electrogenerated redox mediator produces a transient feedback response motivated us to explain further the impact of the Pt oxide reactivity on the shape of the obtained tip electrochemical responses. This information can be used to explain the decrease in reactivity observed with certain mediators, like Fe(II)EDTA, at different stages of oxide formation on Pt. In a larger context, the introduction of this *in-situ* SECM methodology to long-standing topics in electrocatalysis such as the higher-valence oxide mediation of the OER in metal oxides (e.g., IrO_2, RuO_2 or Co_3O_4) [85–86] would be especially useful; such *in-situ* measurements are still being developed and combined with new and powerful methods of characterization [86–87]. In this section, we thus provide a first example of the capabilities of SECM in this research area with a simpler system.

There is a large body of literature on Pt oxides and adsorbed oxygen on Pt. Studies of the electrochemical reactivity of Pt oxides formed at different stages of anodization have been proposed based on the shape of the surface reduction response observed in cyclic voltammetry [51,88], where electrochemical irreversibility is observed after the initial formation of a reversible oxide. This irreversibility has been ascribed to the formation of a "place exchanged" oxygen species, i.e., oxygen buried into the surface layers of Pt [53,89]. The characterization of the structure of surface Pt oxides, however, is challenging and still not well understood, even at single crystal electrodes [78]. In fact, there is no general agreement on the type of species of which Pt oxide consists; early CV studies suggested that PtOH and place exchanged OHPt were the relevant species in the initial stages of oxide formation [88]. Other measurements done *ex-situ* such as XPS and Auger spectroscopy suggest the formation of anhydrous PtO in varying amounts [90–91]. For example, electrochemical quartz microbalance studies claim that 0.5 ML of PtO ($2\bar{e}$ per equivalent) is formed rather than 1 ML of differently coordinated PtOH ($1\bar{e}$ per equivalent) [92] for the first monolayer. Here we focus on the <2 ML PtO, i.e., equivalent to <880 $\mu C/cm^2$, or to roughly four times an adsorbed hydrogen ML, Q_H (210 $\mu C/cm^2$). Thus we assign $\theta_{OX} = 1$ (1 ML) at $Q_{OX} = 2Q_H$ (formally PtO) and 2 ML ($\theta_{OX} = 2$ at $Q_{OX} = 4Q_H$) (formally PtO_2) [89].

In this case, the approach described in Section 4.3 for the simulation of the interrogation reaction at the substrate in the SI-SECM mode has to be replaced by one that considers the formation of species that may have different reactivity toward the mediator and that is contained in limited amounts at the surface, i.e., with a finite coverage, as well as the possibility that these species react in a stepwise manner. A simple model that described the quantification and reactivity of Pt oxide well assumed that the limiting coverage (or equivalent charge) of a given species at the electrode was 0.5 ML of oxide as PtO ($\theta_{OX} = 0.5$) and that species with different reactivity could be formed at different potentials (or stages) of oxide formation. In terms of the simulated coverage $\Gamma_{ox,n}$, and for submonolayer amounts of PtO (e.g., two oxide species are modeled, where species n = 2 is formed after species n = 1 reaches limiting coverage), Equations 4.18 through 4.21 summarize the interrogation process:

$$\Gamma_{ox,2} + \Gamma_{ox,1} = \Gamma_{ox,max} \tag{4.18}$$

$$r_{ox,2} = -k_2 \Gamma_{ox,2} C_R \tag{4.19}$$

$$r_{ox,1} = -r_{ox,2} - k_1 \Gamma_{ox,1} C_R \tag{4.20}$$

$$r_{chem} = r_O = -r_R = k_2 \Gamma_{ox,2} C_R + k_1 \Gamma_{ox,1} C_R \tag{4.21}$$

Here $\Gamma_{ox,max}$ and $r_{ox,n}$ represent the limiting coverage and the reaction rate of the adsorbed species, respectively, r_{chem} represents the rate of chemical reaction of the mediator with the oxide, and r_R and r_O represent the rate of reaction of the reduced and oxidized forms of the mediator, respectively. Notice that in this specific example, each modeled species is associated with its own reaction rate constant to account for differences in the reactivity of the modeled species. Species 1 and 2 in this case are also consumed in a stepwise fashion, e.g., consumption of an equivalent of species 2 generates one equivalent of species 1, this being taken by assuming possible transformations such as that of PtO into PtOH, the latter being an intermediate state upon gaining an electron and a proton, or due to spatial reorganization of place-exchanged oxide species.

Traditional CV studies have suggested that changes in the kinetics of reduction occur when forming ~200 μC/cm² (θ_{OX} ~ 0.5) (formally PtOH); these observations are also complemented by changes in ellipsometric parameters at that coverage [88]. LEED studies performed *ex-situ* suggest that the place exchange process starts in the range 200–300 μC/cm² [93]. *In-situ* x-ray techniques also revealed morphologic changes at specific coverage, although identification of the involved species (i.e., adsorbed oxygen or hydroxyl) is not possible; place exchange in this case was said to occur between 0.6 ē/Pt and 1.2 ē/Pt, afterward coexisting with further formation of oxide on top of this structure until the surface disorders and roughens beyond 1.7 ē/Pt [94–95]. Alternative models for oxide formation (which make predictions about their limiting growth) have suggested that the unit cell for two monolayers of PtO has a limiting thickness of 0.534 nm [96], before formation of an outer phase of hydrated Pt oxide at more positive potentials [97].

The reduction of different amounts of Pt oxide at OC, obtained after bringing the Pt substrate to different potentials, was studied by performing SI-SCEM with different mediators: methyl viologen (+/2+) (MV), Ru(NH$_3$)$_6^{3+/2+}$ (Ruhex), iron(II/III) EDTA (FeEDTA), and Fe(CN)$_6^{3-/4-}$ and fitting the corresponding scans to the simulation model. The maximum distinguishable rate constant for reaction of mediator with Pt oxide was k_{SI} ~ 100 m³mol⁻¹s⁻¹, (different mediators were run at slightly different distances, d and have different diffusion coefficients). The lowest measurable rate constant for the typical maximum coverage was ~0.1 m³mol⁻¹s⁻¹. The accessible rate constants in this case cover the same three-order of magnitude range observed for SECM studies of heterogeneous electron transfer using steady-state feedback at L ~ 0.2.

SI-SECM scans of MV at pH = 11.3 are shown in Figure 4.18a where increasing amounts of Pt oxide (i.e., formed at increasing positive potentials) are titrated by electrogenerated MV^+ produced from the originally present MV^{2+}. MV^+ is a strong reducing agent with a standard potential $E^0 = -0.45$ V versus NHE. We chose to use this mediator in basic solution since MV^+ also can participate in the pH-dependent reduction of H^+ to H_2 at the Pt surface [29]. SI-SECM scans were not possible at pH 6.8 due to high steady-state backgrounds observed in blank measurements due to this process. At pH = 11.3 the background levels were reasonable and the SI-SECM scans reveal the fast rate of the reaction of the mediator with the surface oxide species as shown by the sharp curves that follow the PF scan and decay quickly after consuming most of the substrate adsorbate. Figure 4.18b confirms the fitting parameters by exploring the scan rate dependence of the SI-SECM signal; good general agreement is observed in both cases with different amounts of oxide and different interrogation conditions with two kinetic fitting parameters (i.e., k_1 and k_2 as described in Equations 4.19 and 4.20).

An important feature observed in Figure 4.18a and b and with most of the mediators in this study was the observation of a small distinguishable signal that appeared even in the background scans obtained when the substrate was interrogated after resting at OC for a long time in the presence of traces of air (which are difficult to avoid in the partially open SECM cell, even with an Ar blanket), i.e., when the substrate was not purposely electrochemically oxidized. We have previously discussed these signals in terms of "incipient oxides" formed by chemical routes [5]. Reactive forms of oxygen species derived from adsorbed water have been suggested on Pt. For example, sum frequency generation spectra revealed the presence of OH stretch signals at pH ~ 1 that start at 0.4 V versus NHE, peak at 0.6 V, and disappear beyond 0.9 V where the main Pt oxide starts to grow [98]. Activated oxygen species at potentials less positive than 0.9 V have been proposed in experimental and theoretical studies to be involved in processes such as the electro-oxidation of carbon monoxide [99–101]. Coverage by this incipient, or OC, oxide depended on the pH as found in our experiments: pH = 4, $\theta_{OX,OC} \approx 0.025$ (Ruhex); pH = 6.8, $\theta_{OX,OC} \approx 0.15$ (Ruhex); pH = 6.8, $\theta_{OX,OC} \approx 0.06$ (FeEDTA); pH = 11.3, $\theta_{OX,OC} \approx 0.23$ (MV). The amount of these incipient oxides was previously reported to depend on pH and increase with increasing pH [102], as observed here, although finding it by the SI-SECM technique also depends on the reducing power of the mediator used. Ferricyanide, the weakest reducing agent in this study, did not detect this type of oxide. Note also that the theoretical fits deviate slightly from the measured ones, especially during the final stages of the interrogation process, i.e., experimental transients decayed less sharply than the simulated ones. This probably reflects effects of the diverse population of the submonolayer species, e.g., PtOH with different configurations of OH, bridged to different numbers of Pt atoms depending on the coverage, where repulsive interactions between them could generate coverage-dependent adsorption energies

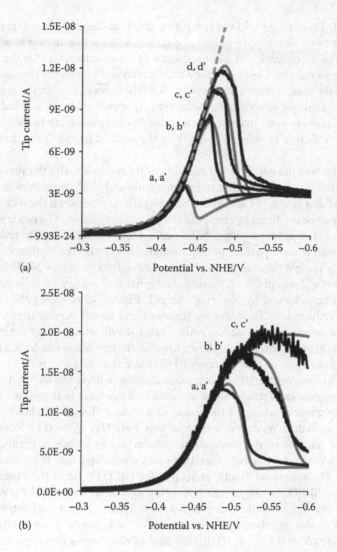

(a)

(b)

FIGURE 4.18 SI-SECM of Pt oxides using methyl viologen as mediator. (a) Potentials for oxidation of substrate: a = OC, b = 1.22 V, c = 1.42 V, d = 1.62 V; substrate held for 70 s at the indicated potential versus RHE and then taken to OC for the interrogation; Tip scan rate ν = 20 mV/s, tip potential in the NHE scale. Labels a', b', c' and d' (light curves) indicate the simulated curves for the corresponding curves a, b, c and d (dark curves); PF indicated by light dashed curve. (b) SECM of Pt oxide using MV at different scan rates: a = 20 mV/s, b = 50 mV/s and c = 100 mV/s (dark curves) as well fitted responses labeled as a', b' and c' respectively (light curves). Oxide generated by holding the substrate for 70 s at 1.87 V versus RHE to then open the circuit; For both (a) and (b) a Au tip (a = 12.5 μm) was used at d = 2 μm from a Pt substrate (b = 25 μm). Solution was 1.1 mM in MV(ClO$_4$)$_2$ in pH=11.3, 0.1 M phosphate buffer. Fitted rate constants: k_1 = 20 m^3 mol^{-1} s^{-1}, k_2 = 100 m^3 mol^{-1} s^{-1}. (Adapted from Rodríguez-López et al., *J Phys Chem C*, Vol. 114, (2010): pp. 18645–18655.)

[78,88,95]. These changes could result in a coverage-dependent kinetic constant, where our model simply represents a constant k_1 as a first approximation. The formation of these species as a consequence of water activation over the metallic Pt surface would also imply that they are formed and consumed transiently near the end of the scan, a process that would inevitably generate tailing in our electrochemical signal. We introduce k_1 values only as approximations; k_2 values as will be shown, provide more interesting and reliable information about the kinetics of the oxide reduction process, especially in the case of higher (place exchanged) oxide.

The effects of the energetics of reaction of the mediator with the surface oxide can be tested by choosing other reducing agents and conditions. Provided the E^0 of the mediator is not pH sensitive, the energetic gap between the mediator and the oxide can be modified by changing the pH of the solution. The oxide reduction process is pH dependent; in the NHE scale, electrochemical oxide reduction is accomplished at more positive potentials and at lower pH. The mediator reduction is typically not pH dependent, for instance $E^0 = 0.05$ V versus NHE for Ruhex between pH $= 2$ and pH $= 12$, therefore a higher free energy for reduction of the oxide can be achieved by lowering the pH. Figure 4.19a shows the SI-SECM scans for reduction of Pt oxides by Ruhex; these reveal very sharp curves that denote fast interrogation kinetics, while Figure 4.19b, which shows the interrogation results at pH $= 6.8$, reveals slower kinetics. In fact, upon reaching a substrate potential limit close to 1.32 V versus RHE, there is a clear deviation of the scans from PF. At this potential $\theta_{OX} > 0.5$, the reduction of higher oxides is dominating and the decrease in kinetics can be ascribed to a decrease in the value of k_2.

The observed decrease in the value of k_2 was followed by choice of a less reducing mediator. When the mediator was FeEDTA ($E^0 = 0.12$ V vs. NHE), the rate of reaction of the oxide decreased further as shown in Figure 4.20 for high oxide coverage values, where there is a clear departure from the ideal PF behavior. The measured kinetic constants for FeEDTA lie in the middle of the measurable SI-SECM range, which is advantageous in terms of the goodness of the fit; in this range subtle changes in k_1 and k_2 have a profound impact on the simulated shapes and thus, a better estimate of these can be given. In the case of ferricyanide ($E^0 = 0.4$ V vs. NHE), the rate of oxide reduction is so slow, compared with the full extent of PF, that one might consider the system as nonreacting; nonetheless, the corresponding kinetic constants can be reasonably extracted from the model fits.

The free energy of reaction between the mediator and the oxide can be represented by the parameter ΔE, which equals the potential separation between the E^0 of the mediator and the peak potential of the electrochemical oxide reduction wave at $\theta_{OX} \approx 0.5$ at a given pH. A plot of ln k_2 versus ΔE (Figure 4.21) shows an approximately linear relationship. Homogeneous outer sphere reactions are often modeled by parabolic relationships with respect to the free energy of activation of the process in the context of Marcus and related electron transfer theories [103–106], where the free energy of activation, ΔG^*, is expressed in terms of the driving force of the process ΔG^0 (in our case proportional to ΔE) and the

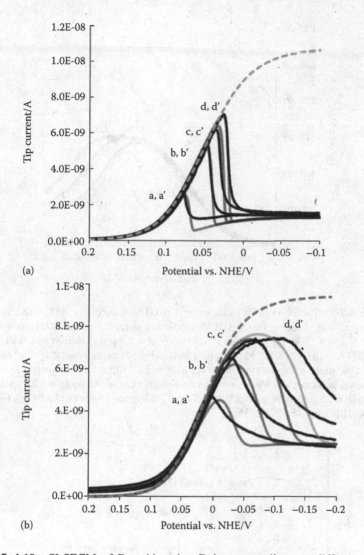

(a)

(b)

FIGURE 4.19 SI-SECM of Pt oxide using Ruhex as mediator at different pH. (a) pH = 4, 0.6 mM in $Ru(NH_3)_6(ClO_4)_3$ in 0.15 M acetate buffer. Substrate oxidation done by holding for 70 s at the indicated potential versus RHE: a = 1.12 V, b = 1.32 V, c = 1.52 V, d = 1.62 V. Tip scan rate ν = 10 mV/s. d = 2.3 μm; Fitted rate constants: k_1 = 100 m^3 mol^{-1} s^{-1}, k_2 = 100 m^3 mol^{-1} s^{-1} and shown in curves a', b', c', d' (light) for their respective experimental scans (dark curves). (b) pH = 6.8, 0.82 mM in $Ru(NH_3)_6(ClO_4)_3$ in 0.1 M phosphate buffer. Substrate oxidation done by holding for 70 s at the indicated potential versus RHE: a = 0.92 V, b = 1.12 V, c = 1.32 V, d = 1.52 V. Tip scan rate ν = 20 mV/s. d = 2.7 μm; Fitted rate constants: k_1 = 10 m^3 mol^{-1} s^{-1}, k_2 = 30 m^3 mol^{-1} s^{-1} and shown in curves a', b', c', d' (light) for their respective experimental scans (dark curves). For both (a) and (b) PF is shown as dashed grey curve and for both a Au tip (a = 12.5 μm) and Pt substrate (b = 25 μm) were used. (Adapted from Rodríguez-López, *J Phys Chem C*, Vol. 114, (2010): pp. 18645–18655.

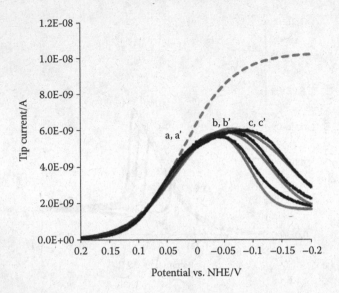

FIGURE 4.20 SI-SECM of Pt oxide using FeEDTA as mediator, pH = 6.8. Substrate oxidation done by holding for 70 s at the indicated potential versus RHE: $a = 1.32$ V, $b = 1.42$ V, $c = 1.52$ V. Tip scan rate $\nu = 50$ mV/s. $d = 2.3$ µm; Solution was 0.89 mM in Fe(III)EDTA in pH = 6.8, 0.1 M phosphate buffer. Fitted rate constants: $k_1 = 10$ m³ mol⁻¹ s⁻¹, $k_2 = 6$ m³ mol⁻¹ s⁻¹ and shown in curves a', b' and c' (light) for their respective experimental scans (dark curves). PF is shown as dashed grey curve. Au tip ($a = 12.5$ µm) and Pt substrate ($b = 25$ µm) were used. (Adapted from Rodríguez-López et al., *J Phys Chem C*, Vol. 114, (2010): pp. 18645–18655.)

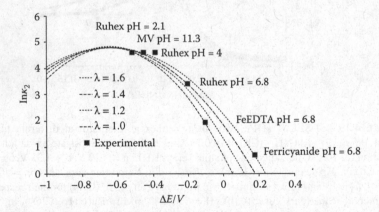

FIGURE 4.21 Dependence of the Pt oxide reduction rate constant on reducing power of mediator. ΔE stands for the difference between the E^0 of the mediator and the peak potential for electrochemical reduction of the oxide at the indicated pH. Values for k_2 as obtained from digital simulations in m³ mol⁻¹ s⁻¹. The graph also shows the parabolic curves obtained by use of Equations 4.22 and 4.23 with different values of λ and adjusted to overlay on the data. (Adapted from Rodríguez-López et al., *J Phys Chem C*, Vol. 114, (2010): pp.18645–18655.)

reorganization energy, λ, e.g., the energy required to transform the nuclear configurations in the reactant and the solvent to those of the product state; Equations 4.22 and 4.23 are relevant [105]:

$$k_{et} = \frac{2\pi}{\hbar} |H_{DA}| \frac{1}{\sqrt{4\pi\lambda RT}} \exp\left(\frac{-\Delta G^*}{RT}\right) \qquad (4.22)$$

$$\Delta G^* = \frac{F(\lambda + \Delta E)^2}{4\lambda} \qquad (4.23)$$

where k_{et} is the rate constant for electron transfer, h is Planck's constant, H_{DA} is the electronic coupling matrix element between the donor and acceptor states, R is the gas constant, and T the temperature [105]; λ in this case is expressed in electron volts. It is tempting to plot the data for the heterogeneous (inner sphere) reaction studied here to this formalism and locate the calculated rate constants in the context of the parabolic relationship to obtain an estimate of λ [106]; SECM kinetic measurements have been previously used to construct such parabolic relationships [107]. Such parabolas are shown in Figure 4.21 for values of λ; our data would fall into the range 1.6 eV > λ > 1.0 eV so an average value of $\lambda = 1.3 \pm 0.3$ eV matches the rate constants fairly well. This value of λ is significantly larger than that attributable to the mediators and suggests that the reduction of the oxide, possibly the place-exchanged oxide (k_2 corresponds to the rate constant for reduction of $\theta_{OX} > 0.5$), is accompanied by a significant rearrangement of the products compared with the reactants. This is consistent with the model proposed by Nagy et al. in which considerable interactions, i.e., induced local fields, need to be overcome in order to reduce the place exchanged oxide [94], and is also consistent with observations of loss of reversibility in cyclic voltammetry upon formation of the place exchanged oxide. The estimated λ value does not include corrections for work terms (e.g., coulombic effects) [1] that may be involved in Equation 4.23, especially considering that the different mediators carry not only different charge, but also opposite sign in some cases; nonetheless, the agreement to the parabolic model is good. Larger (more negative) values of ΔE would be interesting so a possible inverted region could be examined, but could not be accessed due to the interference of parasitic catalytic processes at the Pt substrate [29]. However, less catalytically active surfaces, such as gold or mercury, may yield interesting results (although the generation of more reactive species in water starts to become a problem at potentials more negative than the ones used for MV).

The determination of kinetic parameters for the chemical reaction between a reduced mediator and an oxide can be used for describing electron transfer processes across thin layers of oxide as anticipated in the introduction to this section, Figure 4.17b. The feedback mode of SECM can be used to study the steady-state oxidation of the mediator at the Pt substrate as a function of potential, particularly at very positive (oxidizing) ones. Such measurements are difficult at a single electrode, but straightforward with the SECM, where generating a reducing redox mediator at the tip while inducing strongly oxidizing conditions at the substrate with the formation of a reactive oxide can easily be accomplished. Typically, SECM experiments of PF do not involve bringing the substrate potential to extremely positive values, since for most well-behaved mediators, diffusion-limited conditions of oxidation can be achieved within ~200 mV positive of the mediator E^0. For example, in the case of MV feedback, with $E^0 = -0.45$ V versus NHE, the electrode needs not to be biased further than $E = -0.25$ V versus NHE to observe complete PF; excursion into more positive potential regions where an oxide is formed is rarely carried out, although it is of interest in studying oxide film effects.

Figure 4.22a shows an example of how the kinetic information from SI-SECM experiments can be used to describe the anomalous electrochemical oxidation behavior of FeEDTA at Pt, which has also been reported previously to occur by an EC mechanism producing a nonelectroactive intermediate and low collection at the ring at high rotation rates in rotating ring-disk electrode (RRDE) experiments [108]. In SECM experiments, we found the feedback response of FeEDTA to be sluggish at the less positive potentials and increased following a similar increase in the oxide coverage at the substrate, as shown in Figure 4.22a. The feedback signal reached a maximum at a value of θ_{OX} larger than 0.3 and started decreasing at θ_{OX} slightly larger than 0.4, then kept decreasing until stabilizing at 1.62 V versus RHE (where $\theta_{OX} \approx 1$). Feedback simulations based on the coverage and reactivity with the oxide as the only means of Fe(II)EDTA oxidation show a good agreement with the experimental data.

How does one explain the decrease in the rate of oxidation of Fe(II)EDTA in the potential range of 1.2 V to 1.6 V vs. RHE (potentials well positive of the formal potential of the species)? It cannot be explained by a Marcus inverted region for nonadiabatic electron transfer, since this is not found at metal electrodes; the thin Pt oxide film could be treated as a semiconductor with localized states, but for $\theta_{OX} \leq 1$ this is unlikely [75]. For outer sphere reactions, tunneling effects, as suggested earlier in Equation 4.17, on oxide-covered electrodes have been described. One could then ascribe the changes totally to electron transfer from the Fe(II) EDTA as caused only by tunneling through the oxide film; however, tunneling effects have been typically reserved for thick layers of oxide on metals (e.g., thicker than 6 Å, which is not the case at the potentials shown in Figure 4.22) while chemical effects are proposed to impact the rates of reaction on thinner ones [75].

Thus we prefer to explain the feedback behavior of FeEDTA in Figure 4.22a to the fact that oxidation occurs totally via reaction with the oxide, given the agreement between the simulated feedback from the SI-SECM model and the experimental response. To our knowledge, these chemical effects have not been addressed with the versatility that the SI-SECM technique allows. In this particular case, the oxidation of Fe(II)EDTA would be associated with the presence of oxide at the surface and its reactivity toward it, and the assumption that the higher oxides react more slowly. Notice also that these effects are observed well into the initial states of oxide formation, not only after reaching the first half monolayer of oxide. Since cyanide ion is known to inhibit oxide

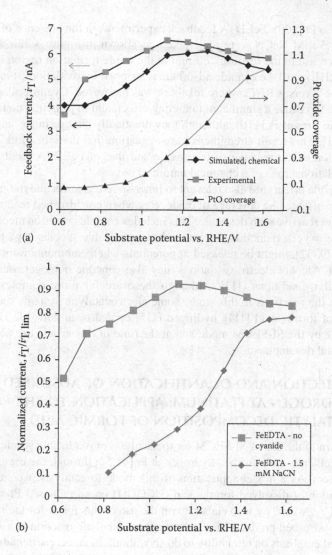

(a)

(b)

FIGURE 4.22 FeEDTA feedback on Pt at anodizing substrate potentials. (a) Comparison of experimental feedback to simulated PF for the Fe(III/II)EDTA system assuming that only oxides are reactive (chemical route) at the Pt substrate and using the PtO coverage values and reaction rates determined by SI-SECM (e.g., Figure 4.20). $E_{tip} = -0.2$ V versus NHE; $d = 2.6$ μm in 0.89 mM Fe(III)EDTA in pH = 6.8, 0.1 M Phosphate buffer. $E°$ (Fe(III/II)EDTA) = 0.12 V versus NHE or a formal potential of 0.44 V versus RHE at pH 6.8. E_{sub} as indicated versus RHE. All points recorded after first holding at the indicated substrate potential for 70 s and then maintaining the potential. (b) Effect of the addition of cyanide to FeEDTA feedback on Pt. All conditions as in (a) except $d = 2.3$ μm for curve with 1.5 mM NaCN; currents are normalized with respect to $i_{T,lim}$, which indicates average maximum feedback limit observed experimentally, curves are normalized to this value to correct for the different distance used. (Adapted from Rodríguez-López et al., *J Phys Chem C*, Vol. 114, (2010): pp. 18645–18655.)

formation on Pt [109], FeEDTA feedback experiments in the presence of cyanide ion, e.g., 1.5 mM NaCN were undertaken, as shown in Figure 4.22b. Note the marked decrease in the feedback current in the oxide formation region (0.8 V to 1.2 V vs. RHE) until the cyanide adsorbate is stripped by oxidation from the surface at ~1.4 V versus RHE, where feedback was recovered. Cyanide adlayers are not expected to cause a significant tunneling effect, and in principle do not block all available sites at Pt [110], although they drastically change inner-sphere processes [111]. This result strengthens our suggestion that there is a link between the formation of the reactive surface oxide and that of FeEDTA feedback, and that this relationship is of a chemical nature.

These oxide effects are also relevant to inner-sphere electrocatalytic processes that may be affected by coverage of oxide, e.g., when nonadsorbed reaction intermediates are reactive with the oxide. This includes complex reaction mechanisms, such as the oxygen reduction reaction, where oxidizable species, e.g., hydrogen peroxide [79,112], might be released at potentials where submonolayers of oxide are present. Alcohol electro-oxidations may also generate homogeneous species under similar conditions [113]. Although these are far more complex mechanisms than the reactions in this study, some electrocatalytic systems such as the oxidation of formic acid [114], hydrogen [115] or hydrogen peroxide [116] may be treatable by the SI-SECM mode and at the time of this article are waiting for experimental development.

4.6 DETECTION AND QUANTIFICATION OF ADSORBED HYDROGEN AT PLATINUM: APPLICATION TO THE CATALYTIC DECOMPOSITION OF FORMIC ACID

We now turn to the use of SI-SECM for the analysis of oxidizable species at electrodes, specifically for adsorbed hydrogen at Pt [41,52] through the use of oxidizing mediators. As a model application of this mode to catalysis, we studied the decomposition of dissolved formic acid (HCOOH) on an unbiased Pt surface to produce $H_{(ads)}$, where we also elaborate on the use of this mode for the interrogation of the adsorbed products of a heterogeneous catalytic reaction at a substrate, with special emphasis on our ability to do so without the direct participation of the substrate in the analytical process. The technique allows examination of a substrate at OC and decouples its role as electrocatalytic surface and means of analysis as is commonly the case in the electrochemical quantification of adsorbed species [1,52]. Such a strategy allows a simple and direct route to determine *in-situ* the surface coverage of reaction intermediates and reveals species that otherwise remain obscured by side processes occurring at the catalytic substrate.

The use of the SECM for the study and design of electrocatalysts is of ongoing interest [26,117–121], which also includes the electrochemical oxidation of formic acid [114]. This process has been widely studied in noble metals because of the potential use of HCOOH for fuel cells [122]. One of the limiting factors for the efficient electrochemical conversion of HCOOH to CO_2 and H^+ at Pt in acidic media is the poisoning and blocking effect of adsorbed intermediates [123–124], such as $CO_{(ads)}$, $HCOO_{(ads)}$ (sometimes called "formate"), $COOH_{(ads)}$, $COH_{(ads)}$, or $HCO_{(ads)}$ [125]. This has been investigated by the use of *in-situ* IR and related methods, which are especially well suited for the detection of carbonyl species [125–129]. Since early studies [130–132], a route termed the "direct pathway" [126,133], i.e., direct oxidation to CO_2 and H^+, was identified as beneficial in terms of avoiding poisoning species. However, the "indirect pathway" (which generates $CO_{(ads)}$) inevitably also comes into play [124].

An important part of the discussion in the same early studies of HCOOH oxidation is the formation of $H_{(ads)}$ by the catalytic decomposition of HCOOH at the Pt surface (a leading route to the "direct pathway"); this discussion was fed by the observation that the OCP of a Pt electrode in contact with solutions of different concentrations of HCOOH would exhibit a transient that stabilized to potentials where $H_{(ads)}$ UPD is possible [130–132,134–138]. The quantification of this adsorbed species and how it related to such changes in the OCP were not fully described [137], although recently these OCP measurements have come back into focus because of the interplay that adsorbed and solution phase species have in HCOOH based catalytic systems [139–140], e.g., in electrochemical oscillators [129,141].

The decomposition of HCOOH to CO_2 and H_2 is thermodynamically favored ($\Delta G^0 = -14$ kcal/mol) [142], and in gas phase studies these are the main products on Pt surfaces [143–147], especially at low temperatures (analogous to low electrochemical overpotentials). In fact, other noble metals such as Pd are known to catalyze this decomposition [147–150], a possible lead that links good electrocatalytic behavior, as it is with Pd and modified Pd electrodes, to gas phase catalytic activity [114,122,132]. The role of HCOOH as a reducing agent in conjunction with Pt surfaces and colloids has been reported [151–153], and recent computational studies have discussed the formation of $H_{(ads)}$ on Pt from HCOOH [154], so the quantification of this species by SI-SECM is clearly of interest. While the detection and quantification of this adsorbate through spectroscopic techniques remain elusive [155–157], this *in-situ* electrochemical approach appears useful for the characterization of $H_{(ads)}$ (or other oxidizable species) on catalysts and electrocatalysts. The adsorbate is formed at a Pt electrode by either of two routes: the electrochemical UPD of hydrogen adatoms achieved by applying a reducing potential to the Pt electrode, Figure 4.23a, or by the catalytic decomposition of a substance, e.g., HCOOH through one possible pathway, e.g., the "direct pathway," at the Pt surface when the electrode rests at OC, Figure 4.23b; the first represents an electrocatalytic route and the second a catalytic route.

In Section 4.2, we described the criteria and process of selection of a suitable mediator for the detection of adsorbed hydrogen; as anticipated, the redox pair formed by the aromatic amine N,N,N′,N′-tetramethyl-p-phenylenediamine, TMPD, and its radical cation were found to be suitable for quantitative interrogation of the adsorbate. Figure 4.24a shows a typical SI-SECM scan of a monolayer of $H_{(ads)}$ on Pt formed by bringing the potential of the electrode to a potential close to that for hydrogen evolution. In this case, the comparison to the PF scan reveals that the interrogation is a fast process; however, it is not completely devoid of kinetic features as it departs slightly from the trace of PF

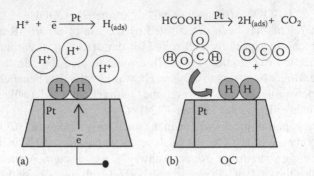

FIGURE 4.23 Electrocatalytic and catalytic formation routes of $H_{(ads)}$ on Pt. (a) Formation of $H_{(ads)}$ by electrochemical UPD of proton in solution or (b) Formation of $H_{(ads)}$ by catalytic decomposition of HCOOH on the Pt substrate at OC through a possible route termed "direct pathway." (Adapted from Rodríguez-López et al., *J Am Chem Soc*, Vol. 132, (2010): pp. 5121–5129.)

at less positive potentials and then shows a sudden increase in reaction kinetics at more positive potentials. Nonetheless, the SI-SECM response is contained within the quantification window, where a low NF background is also characteristic. The quantitative SI of $H_{(ads)}$ on Pt by $TMPD^+$ in acidic media was also studied with different amounts of $H_{(ads)}$ dosed to the substrate electrode. Figure 4.24b shows a typical progression of SI scans performed with a 200 μm diameter glassy carbon interrogator electrode on a 100 μm diameter Pt substrate dosed with $H_{(ads)}$; the amount of $H_{(ads)}$ was adjusted by scanning the substrate to different potentials within the UPD region for hydrogen. A common feature exhibited by these SI scans was a displacement of the peak signal with respect to the dosage. This displacement is related to the shift observed in Figure 4.24a with respect to the PF scans, which is possibly due to the pH sensitive nature of TMPD that may be affected by the discharge of protons after the interrogation process. The SI scans do follow a common trace that ends in a sharp peak whose integration yields ≈2% of a monolayer of $H_{(ads)}$. At lower doses, however, e.g., potential limit of 0.42 V versus NHE, the SI features are less shifted from the PF scan, thus suggesting that pH effects may have a lesser contribution to the shape of the SI scans. A measure of the goodness of the SI scans lies in the quantitative comparison of $H_{(ads)}$ found by SI versus the amount dosed (determined by integration of the charge delivered during the linear sweeps on the substrate). Figure 4.25 shows such a comparison for a series of experiments performed on substrate electrodes of different sizes. There is quantitative agreement of approximately 0.2 V more positive than the onset of hydrogen evolution, with an average standard deviation of 5% of a monolayer for five independent experiments. The disagreement observed for more positive potentials in Figure 4.25, which accounts roughly for 10% of a monolayer of $H_{(ads)}$, was further investigated. The inset of Figure 4.25 shows two SI experiments conducted by taking the substrate to 0.32 V versus NHE in the presence of only traces of O_2 (typical experimental

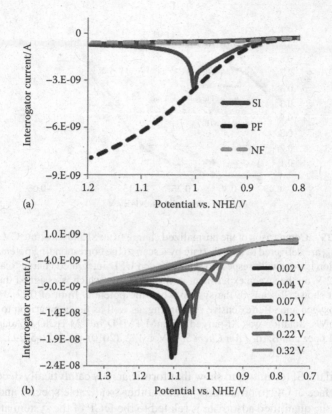

(a)

(b)

FIGURE 4.24 SI-SECM of $H_{(ads)}$ on Pt using 0.5 mM TMPD as mediator in 1 M $HClO_4$. (a) Comparison to NF and PF for the same conditions: Interrogator Au tip ($a = 12.5$ μm), Pt substrate ($b = 12.5$ μm), $L = 0.2$. $H_{(ads)}$ was dosed by taking the Pt substrate through a linear scan from 0.42 V to 0.02 V at 50 mV/s. Interrogator electrode scan rate $\nu = 20$ mV/s. (b) $H_{(ads)}$ on Pt at different dosing. Different doses of $H_{(ads)}$ delivered to the Pt substrate by a linear sweep at $\nu = 50$ mV/s from 0.42 V to the potential limit indicated for each curve. Interrogator glassy carbon electrode ($a = 100$ μm), Pt substrate ($b = 50$ μm), $L = 0.2$. Interrogator electrode scan rate $\nu = 5$ mV/s. (Adapted from Rodríguez-López et al., *J Am Chem Soc*, Vol. 132, (2010): pp. 5121–5129.)

Ar-purging conditions) and upon even more extensive purging with Ar, and its comparison to a NF scan; the results suggest that there is a link between the feature observed at these potentials and a nonnegligible oxygen reduction process, possibly indicative of an adsorbed intermediate derived from this reaction. This peak is only observed upon substrate bias, which is the reason why it is still possible to obtain featureless NF scans (e.g., in contrast with the results for titration of oxides, Sections 4.4 and 4.5). Note, however, that the extensive purging conditions used to prove this point are impractical enough to prevent us from maintaining them over long periods of time, and thus, difficult to implement under our typical experimental conditions in our SECM cells.

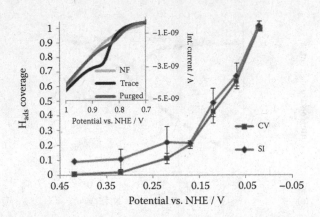

FIGURE 4.25 Comparison of the normalized charge from SI and substrate CV. Different doses of $H_{(ads)}$ are delivered to the substrate by a scan to the corresponding potential limits. Normalization is done with respect to the hydrogen UPD integrated charge to a potential limit of 0.02 V on the sweep experiment at the substrate, which is assigned the value of 1 ML. Inset shows SI scans of the system when the potential limit of 0.42 V was used with trace oxygen and after extensive Ar purging, as well as the comparison to complete NF; $\nu = 5$ mV/s. Solution was Ar-purged 0.5 mM TMPD in 1 M $HClO_4$. (Adapted from Rodríguez-López and Bard, *J Am Chem Soc*, Vol. 132, (2010): pp. 5121–5129.)

We used the SI technique to show that formic acid is catalytically decomposed on a Pt surface at OC to produce $H_{(ads)}$ or another oxidizable species and that the amount of the quantified adsorbate is related to the OCP of the system at the time of interrogation. Figure 4.26a shows a typical OC transient measured in the presence of 12.5 mM formic acid in 1 M $HClO_4$ and 0.5 mM TMPD for a 100 μm diameter Pt substrate aligned to a 200 μm diameter glassy carbon interrogator. In the absence of formic acid, the OCP of the Pt substrate in this solution is typically close to the foot of the TMPD oxidation wave (≈0.7 V vs. NHE); upon injection of HCOOH into the solution, the OCP of the platinum substrate dropped to a limiting potential of ≈0.22 V versus NHE depending on the concentration of HCOOH (in $t \approx 300$ s for this study where [HCOOH] = 12.5 mM) and the amount of trace O_2 in the electrochemical cell. The shape and potential limits of an OCP transient like that shown in Figure 4.26a are consistent with that previously reported [132,139,140], and known since early studies of the decomposition of HCOOH [130–138]. This displacement of the OCP toward more negative potentials is an indicator of the formation of $H_{(ads)}$ or other reduced species, although to our knowledge, a clear quantification of such an intermediate is lacking, especially under the conditions of this study where the Pt substrate is not biased. Figure 4.26a also shows for comparison that in the two-electrode assembly, only the potential of the Pt substrate, which catalyzes HCOOH decomposition, changes with time, while the interrogator shows only a negligible drift. Notice that the measurements here rely on the nondisturbance of the Pt substrate i.e., the two-electrode SI setup is used to probe the catalytic activity of the metal, rather than its electrocatalytic behavior. The interrogator electrode probes the adsorbed species at the Pt surface through the use of the TMPD mediator, without the need for performing any amperometric or other dynamic measurements directly on the Pt substrate electrode.

Figure 4.26b shows a set of SI scans where the substrate was interrogated after introduction of HCOOH into the solution containing TMPD mediator after reaching different OCP values. At HCOOH introduction, the potential is at that of the mediator, but in the presence of HCOOH, the potential changes with time as shown in Figure 4.26a. Qualitatively we note the similarities to the SI scans done for the interrogation of $H_{(ads)}$ from the UPD of hydrogen, Figure 4.25; there is a shift of the peak intensity with respect to the interrogator potential at growing signal intensity. The amount of interrogated adsorbate was larger when the OCP of the substrate is less positive, reaching a limit when the OCP was about 0.22 V.

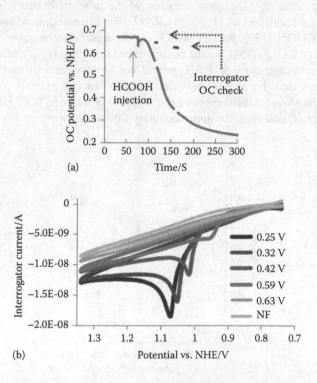

(a)

(b)

FIGURE 4.26 Catalytic decomposition of formic acid (HCOOH) studied by SI-SECM. (a) OCP transient of Pt in the presence of HCOOH. (b) SI-SECM oxidation scans with TMPD as mediator at different OCP limits of Pt in the presence of HCOOH. OCP values at the beginning of the SI-SECM scan are indicated in the graphic. NF was obtained in the absence of HCOOH. For both parts: Solution: 1 M $HClO_4$ + 0.5 mM TMPD + 12.5 mM HCOOH and Ar-purged to traces of O_2. Interrogator glassy carbon electrode ($a = 100 \ \mu m$), Pt substrate ($b = 50 \ \mu m$), $L = 0.2$. Interrogator electrode scan rate $\nu = 5$ mV/s. (Adapted from Rodríguez-López and Bard, *J Am Chem Soc,* Vol. 132, (2010): pp. 5121–5129.)

After an SI scan was performed on the substrate to positive potentials, the OCP restarted from a positive value e.g., 0.6–0.7 V versus NHE, followed by the transient toward less positive potentials; interrogation at different potentials could be carried out numerous times during an experiment, which suggests that the adsorbate forms reversibly on the Pt surface.

Figure 4.27 shows the quantification of the adsorbate (normalized to 1 ML of UPD $H_{(ads)}$) for three series of experiments and corrected for transient effects during the interrogation process [41]; the most relevant experimental finding is the dependence of the adsorbate coverage with respect to the OCP of the Pt substrate at the time of interrogation. As shown in Figure 4.26a, the OCP transient reaches a limiting value near 0.22 V versus NHE which corresponds to the limiting amount of charge used in forming 1 ML of UPD $H_{(ads)}$; the potential window exhibited by Figure 4.27 is wide, and in fact the limiting OCP of 0.22 V versus NHE is barely within the hydrogen UPD region of platinum, although the quantification of the adsorbate strongly suggests that it can be identified as $H_{(ads)}$. Possible larger intermediates, such as physisorbed HCOOH [130], formyl $H \cdot CO_{(ads)}$, "formate"-like species $H \cdot COO_{(ads)}$ [129], or even $CO_{(ads)}$ [158] would occupy more than one $H_{(ads)}$ site, and hence a limiting coverage close to 1 ML would not be attained, although some of these species may be present as intermediates. In the case of physisorbed HCOOH, the slow establishment of the limiting conditions (i.e., formation of a monolayer of adsorbate in approximately 300 s as shown in Figure 4.26a)

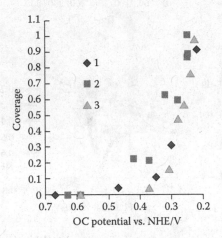

FIGURE 4.27 Plot of $H_{(ads)}$ coverage versus OCP limit at the time of interrogation for the decomposition of formic acid at Pt. Coverage values have been corrected for time uncertainties. All experimental conditions used as in Figure 4.26b. Shown are three independent SI-SECM experiments. Unit coverage of $H_{(ads)}$ was determined from integration of the UPD region at the substrate on a linear potential scan previous to the addition of HCOOH. (Adapted from Rodríguez-López and Bard, *J Am Chem Soc,* Vol. 132, (2010): pp. 5121–5129.)

in an environment that is well in excess of this substance and the potential dependence of the interrogated signal make us discard it as our quantified adsorbate; recent DFT calculations also suggest that HCOOH adsorbs only weakly at the Pt(100) crystal (the same for CO_2 which we will only consider a nonadsorbed product) [154].

The possibility that oxidized TMPD reacts with adsorbed carbon monoxide, $CO_{(ads)}$, a possible product or intermediate of the decomposition of HCOOH, was also assessed. However, SI-SECM experiments with adsorbed CO on Pt and TMPD showed no reactivity. Weak evidence that this species may be formed to some extent lies in the reproducibility of our measurements, especially in the high $H_{(ads)}$ coverage range (OCPs 0.27–0.22 V vs. NHE). Over extended periods of experimentation time, ~2–3 h, quantifications found at the least positive potentials dropped by up to 10%, which may be indicative of a loss in the activity of the Pt substrate due to formation of $CO_{(ads)}$. Even in this scenario, our results contrast with the RAIRS measurements of Ma and Zaera [159], in which exposure of Pt to HCOOH in neutral unbuffered media at OC provided a signal identified as $CO_{(ads)}$ in presumably high coverage almost immediately; the question arises whether the pH or the concentration of HCOOH play an important role in the formation of $CO_{(ads)}$. In the case of the electrochemical oxidation of HCOOH, "formate"-like species and $CO_{(ads)}$ are said to form, even at low potentials, as observed through IR techniques [125–129] (despite certain discrepancies among these studies) [128] and is predicted by some DFT calculations [160]. The slow but reversible route for the decomposition of HCOOH on Pt to yield $H_{(ads)}$ or related oxidizable species (i.e., by TMPD⁺) demonstrated in our work shows an interesting parallel to the gas/solid interface observations. Furthermore, the mechanism by which $CO_{(ads)}$ may be formed also remains an open question, i.e., whether it is a consequence of a parallel path in the decomposition of the "formate"-like species [154] or perhaps a slow inverse water gas shift reaction between $H_{(ads)}$ and weakly bound CO_2 to generate water and CO [161].

Other possible applications of SI-SECM for studying hydrogen on metals were also evaluated [41]. Palladium is known to involve both adsorption and absorption (into the bulk metallic phase) of atomic hydrogen [147–150]. SI experiments with TMPD were conducted to evaluate the possibility of "titrating" this species by use of SI in the chronoamperometry mode (where the mediator is produced by a potential step rather than a scan and is better suited for very large amounts of titrated material). Figure 4.28 shows an example of an experiment in which a bright Pd electrode was dosed with hydrogen at reducing potentials for 3 s in each case; the inset of the figure shows the CV of this type of electrode in acidic media, which shows a poor electrochemical response for the oxidation of hydrogen species. The SI chronoamperometric steps show a very clear and distinct titration response; we envision that our technique could be well suited for estimating the amounts of absorbed hydrogen in confined Pd based materials, such as thin layers [148]. In any case, we make the point that the SI technique can be used for a wider variety of materials and purposes than the main topic discussed in this section.

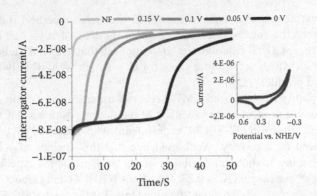

FIGURE 4.28 SI-SECM of adsorbed/absorbed hydrogen at Pd in the chronoamperometry mode. Pd dosed with adsorbed/absorbed hydrogen using linear sweeps to the labeled potential limit versus NHE and holding for 3 s. Interrogator glassy carbon electrode ($a = 100$ μm), Pt substrate ($b = 50$ μm), $L = 0.2$ and $E_{step} = 1.15$ V versus NHE for interrogator electrode. Solution was 0.5 M H_2SO_4 + 0.5 mM TMPD. Inset shows a CV of such "bright" Pd electrode in blank acid. (Adapted from Rodríguez-López and Bard, *J Am Chem Soc,* Vol. 132, (2010): pp. 5121–5129.).

4.7 A CATALYTIC INTERROGATION: THE REACTION OF ELECTROGENERATED BROMINE WITH ADSORBED CARBON MONOXIDE ON PLATINUM

We studied the reaction of bromine, Br_2, with adsorbed carbon monoxide, $CO_{(ads)}$, on polycrystalline platinum in aqueous 0.5 M H_2SO_4 to yield CO_2 and Br^- by SI-SECM [162]. We demonstrate its utility for the evaluation of a new reaction, where Pt not only acts as a support for $CO_{(ads)}$ but also as a heterogeneous catalyst that promotes the oxidation process.

The adsorption and electrochemical oxidation of CO on Pt, both on single-crystal and polycrystalline surfaces, have been widely studied [163–164] by electrochemical techniques [158,165–167] and their combination with Raman and IR spectroscopy [168–170]. A part of the interest in these studies is driven by the identification of $CO_{(ads)}$ as an important blocking intermediate (i.e., a *poison*) in the oxidation of short chain alcohols and organic acids [124–125,171–173], at the anode of fuel cells [122,163–164]. Studies of the electrochemical oxidation of $CO_{(ads)}$ on Pt have traditionally been carried out by scanning or stepping an electrode into the oxidizing potential regime where the generated $OH^{\cdot}_{(ads)}$ radical, or more properly an oxygenated species like $OH_{(ads)}$ or $O_{(ads)}$ on the Pt surface, is proposed to oxidize $CO_{(ads)}$ to CO_2 [166,169,174]. With the SI-SECM technique at hand, we sought to find a tip-generated reactive substance that could achieve this oxidation to provide new strategies for its quantification in catalytic systems, and perhaps for the removal of this strongly adsorbed intermediate. The use of halogens (X_2) is a good starting point since they are good oxidizing agents and their dissociated products, i.e., X radicals, can react with organic species; the reactive Br radical is of particular interest since its precursor, Br_2, can be electrogenerated at a SECM tip [22,24], and its oxidizing properties are comparable to those of $OH^{\cdot}_{(ads)}$ [175].

The SI-SECM strategy in this case includes the bromide ion as precursor (initially present in solution), to produce Br_2, which diffuses to the substrate electrode and reacts with $CO_{(ads)}$ at the Pt surface through Equation 4.24:

$$Br_2 + CO_{(ads)} + H_2O \xrightarrow{\text{Pt}} 2Br^- + CO_2 + 2H^+ \qquad (4.24)$$

Reaction 4.24 regenerates Br^-, which produces a transient PF loop at the interrogator tip as long as there is $CO_{(ads)}$ to react with. This transient PF loop produces an increase in the current at the interrogator tip due to an enhanced diffusive flux of Br^- to the tip. The transient PF that is observed is bounded by the electrochemical window determined by the NF and PF CV scans obtained when the tip and the substrate are aligned and set apart at a small distance,(e.g., 1–3 μm) away when the tip radius $a = 12.5$ μm. The NF scan is obtained through a blank experiment with the Pt substrate at OC and with no $CO_{(ads)}$, while the PF scan is obtained in a blank experiment with the Pt substrate set at a potential where it can reduce Br_2 electrochemically at its maximum rate. The Br^-/Br_2 system, although involving the transfer of 2e, showed good characteristics, i.e., a large electrochemical window with high gain on PF and low background on NF (compare $i_{T,lim,NF} \approx 2$ nA to $i_{T,\infty} \approx 8$ nA) shown in Figure 4.29a. The low background on the NF scan observed with the use of the oxidizing bromine species contrasts with other oxidizing mediators, as discussed in Section 4.2, and probably does not cause appreciable oxidation of the Pt surface, e.g., by suppressing the formation of PtOH or PtO.

Figure 4.29b shows an SI result of the CV mode of a platinum electrode dosed with $CO_{(ads)}$ and its comparison to the NF and PF CV scans. Consumption of the adsorbate causes the response of the interrogator tip to decay to NF as expected from the SI-SECM theory. The SI scans revealed a response that consisted of two peaks, often merging into a single peak with a shoulder when slower scan rates were used. This may be indicative of mechanistic differences at different timescales. More kinetic information is also contained in the displacement of the curve with respect to the PF response, as discussed later. The SI-SECM response, Figure 4.29b, is well bound in the electrochemical window provided by the Br^-/Br_2 couple feedback, which is expected for the SI-SECM response; the small discrepancies observed in the region between 0.9 and 1.1 V can be attributed to small amounts of $CO_{(ads)}$ residues on the interrogator tip rather than a response caused by the substrate.

Independent of shape, the quantification of the amount of adsorbate was consistent in all runs. A quantitative comparison based on the relative charge of $H_{(ads)}$ UPD in surface area measurements and $CO_{(ads)}$ found in our SI-SECM experiments follows. The accepted value for the coverage of UPD hydrogen on polycrystalline Pt is 210 μC/cm² [52]. López-Cudero has reported a typical value of 291 ± 3 μC/cm² for the saturation coverage of CO in the presence of chlorides

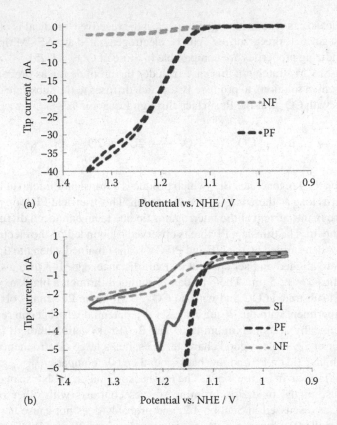

FIGURE 4.29 SI-SECM of adsorbed carbon monoxide at Pt using the Br^-/Br_2 mediator couple. (a) Comparison of NF and PF scans of the Br^-/Br_2 couple on Pt. Solution is 0.5 M H_2SO_4 + 1 mM KBr, $d \sim 2$ μm; both tip and substrate are Pt with $a = 12.5$ μm and $RG \sim 2$. Tip $\nu = 20$ mV/s. NF, substrate at OC; PF, substrate at 0.95 V versus NHE. (b) SI-SECM of $CO_{(ads)}$ on Pt titrated with electrogenerated Br_2. Same conditions as in (a) but substrate in SI scan was predosed with CO saturated 0.5 M H_2SO_4 solution (KBr-free) at a potential of 0.3 V versus NHE for 5 min. SI scan with Pt substrate at OC. (Adapted from Wang et al., *J Am Chem Soc*, Vol. 131, (2009): pp. 17046–17047.)

on polycrystalline Pt [168]. Based on these numbers, the ratio of the $H_{(ads)}$ UPD charge, Q_H, to the $CO_{(ads)}$ charge, Q_{CO}, is $Q_{CO}/Q_H = 1.38(\pm1\%)$; in comparison, the ratio from SI-SECM experiments was found to be $Q_{CO}/Q_H = 0.94(\pm6\%)$. This SI-SECM result is reasonable considering that most θ_{CO} measurements vary between 0.6 and 0.9 depending on the technique[78,158] and comparison standard (in our case, $\theta_{CO} \sim 0.47$, where θ_{CO} is the molar coverage of CO compared with that of H at saturation, if compared with Q_H and assuming 2e transferred per CO). Losses in the saturation coverage have been discussed in terms of equilibration of $CO_{(ads)}$ with CO in solution, although in the case of Pt(111), Offer and Kucernak [158] found that $\theta_{CO,max} \sim 0.68$ (in contrast to previously reported $\theta_{CO,max} \sim 0.75$ values) and that neither spontaneous desorption of CO nor reaction with traces of

O_2 accounted for losses more than 1.24% of a monolayer (ML) per hour. Bromide is known to adsorb on Pt [175] to values up to the range 0.4 to 0.5 ML [78] and strongly adsorbed traces of it may block certain sites for $CO_{(ads)}$. However, this seems unlikely because CO is much more strongly adsorbed than other halides [168] and because the Pt electrode was "pre-cleaned" before the CO adsorption experiments prior to SI-SECM. Traces of bromide adsorption may be responsible for small features in the NF scans, as shown in Figure 4.29b. A last possibility could reside in the selection of the potential of the electrode when CO was introduced, which was set in all experiments shown here to 0.3 V versus NHE; the optimum potential range for adsorption of CO on Pt(100) has been shown to be $E < 0.3$ V versus RHE and for Pt(111) and Pt(110), $E < 0.25$ V [168]. The potential used in our experiments is borderline to these values, yet situated closest to the double layer region of Pt where adsorption of $H_{(ads)}$ and $O_{(ads)}$ is minimized.

Mechanistic insight into the reaction may also explain some of the losses in quantification. In Equation 4.24, we assumed that CO is oxidized to CO_2; since we obtain the transient feedback characteristic of SI-SECM scans, we also assume that Br^- is a product of the reaction that allows for the transient signal. Based on these assumptions, a possible route for the oxidation of $CO_{(ads)}$ by Br_2 is shown in Equations 4.25 to 4.27:

$$Br_2 \xrightarrow{\text{Pt substrate}} 2Br_{(ads)} \tag{4.25}$$

$$2Br_{(ads)} + CO_{(ads)} \xrightarrow{\text{Pt substrate}} OCBr_2 \tag{4.26}$$

$$OCBr_2 + H_2O \longrightarrow 2Br^- + CO_2 + 2H^+ \tag{4.27}$$

The product of Equation 4.26, $OCBr_2$ or "bromophosgene" (an analog of phosgene, $OCCl_2$), has been reported [176–178] (but also debated) [179] to be the product of a photosensitized low yield reaction of Br_2 and CO in the gas phase in the absence of catalyst, where the radical-forming Equation 4.25 proceeds through photochemical excitation. In our experiment, the Pt substrate provides a catalytic surface where Equation 4.25 occurs more readily. The hydrolysis process in Equation 4.27 is thermodynamically allowed and proposed to occur at high altitude in the atmosphere [180] and provides the Br^- necessary for the transient PF as well as a source of oxygen for the production of CO_2.

If CO is not adsorbed on Pt and only dissolved in solution, Br_2 is unable to oxidize it. We studied the reaction between CO and Br_2 in 0.5M H_2SO_4 using a single large Pt (2.1 mm in diameter) electrode. The large Pt electrode was immersed in 1 mM KBr 0.5 M H_2SO_4, Ar-purged, and then saturated with gaseous CO. Then CV was carried out by holding the potential for 20 s at 0.6 V and then scanning from 0.8 V to 1.3 V versus NHE at $\nu = 20$ mV/s. Figure 4.30a shows a comparison between the case with and without CO dosing. The result shows that the presence of CO in solution does not prevent the surface of Pt from oxidizing Br^- to Br_2 (which is an important characteristic

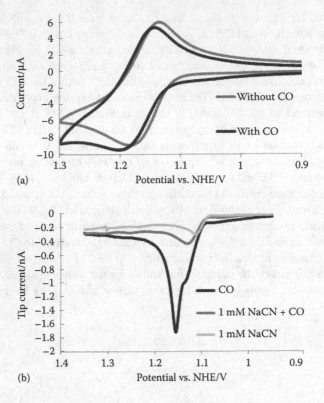

FIGURE 4.30 Supporting experiments for SI-SECM of adsorbed CO on Pt. (a) Effect of CO on the oxidation of Br$^-$ on a Pt macroelectrode ($a = 2.1$ mm) in 0.5 M H$_2$SO$_4$ with 1mM KBr. $\nu = 20$ mV/s. Quiet time 20 s at 0.8 V versus NHE. (b) SI-SECM of CO$_{(ads)}$ on Pt co-adsorbed with CN$^-$. Solution is 0.5 M H$_2$SO$_4$ + 1 mM KBr, $d \sim 2$ μm; both tip and substrate are Pt with $a = 12.5$ μm and $RG \sim 2$. Tip $\nu = 10$ mV/s. Substrate was predosed with CO saturated 0.5 M H$_2$SO$_4$ solution (KBr-free) at a potential of 0.3 V versus NHE for 5 min. SI scans with Pt substrate at OC. Pre-adsorption of cyanide was done by immersion in aqueous 1 mM NaCN in water for 5 min; curve labeled as CO received no cyanide treatment. (Adapted from Wang et al., *J Am Chem Soc,* Vol. 131, (2009): pp. 17046–17047.)

for the interrogator electrode); furthermore, there does not appear to be a significant reaction between Br$_2$ produced at the Pt surface and dissolved CO in solution, since there is only a weak indication of a catalytic effect in the CV response of the Pt electrode compared with the CV in the absence of CO. The presence of a homogeneous catalytic process would be reflected in the shape of the voltammogram showing a larger peak or a plateau current. An SI-SECM experiment involving the coadsorption of CO and cyanide (CN$^-$) on the Pt UME surface was also performed [110]. A cyanide covered Pt UME was used to limit the amount of CO$_{(ads)}$ at Pt in order to verify its reaction with Br$_2$, as shown in Figure 4.30b where the background subtracted responses are plotted for the case where only CN$^-$ is present, when coadsorption is present and when no

CN$^-$ was added. The SI-SECM results show that with CN$^-$, the adsorption of CO is partially blocked, and furthermore, that despite the isoelectronic nature of CN$^-$ and CO, their reactivity with Br$_2$ differs. This suggests that the formation and decomposition of OCBr$_2$ is an important step in the reaction scheme or that CN$^-$ blocks sites for the dissociation of Br$_2$ in a way that CO does not. In Equation 4.26 we have already stated implicitly that Br$_2$ reacts with CO through a Langmuir–Hinshelwood mechanism (both reactants are adsorbed) rather than an Eley-Rideal mechanism (one reactant adsorbed and the other in solution, e.g., CO$_{(ads)}$ + Br$_{2(aq)}$), which draws certain parallelism with the proposed mechanism for oxidation of CO by electrogenerated OH species [181].

A bulk experiment was carried out in order to identify CO$_2$ as the end product of the reaction. A beaker with 10 mmol of Br$_2$ in 100 ml of Ar-purged 0.5 M H$_2$SO$_4$ was bubbled with CO through an inlet where Ar was also continuously passed; Pt black was added as a catalyst. The outlet of the beaker was connected to a container with 100 mL of a saturated and thoroughly Ar-purged solution of Ba(OH)$_2$ (clear when originally prepared with CO$_2$-free water). The reaction between CO$_2$ produced from Equation 4.27 and Ba(OH)$_2$ in solution yields a white precipitate, BaCO$_3$. The control experiment, in which no Pt black was added, showed only cloudiness in the Ba(OH)$_2$ solution after 5 min of CO bubbling. This is attributed to small impurities of CO$_2$ in the CO feed. When Pt black was added to the reaction mixture and the system bubbled for the same amount of time with CO, an abundant precipitate of BaCO$_3$ was observed in the Ba(OH)$_2$ container. The precipitate was filtered and identified as the carbonate through the "carbonate-acid test" where the addition of dilute sulfuric acid releases CO$_2$ bubbles. The addition of a 20 mM solution of AgClO$_4$ in water to the residue of the reaction mixture showed the precipitation of whitish-yellow AgBr, which also supports Br$^-$ as an end product. These experiments show that strongly adsorbed CO on Pt can be oxidized to CO$_2$ by Br$_2$ through the catalytic action of the substrate.

A complete explanation of the complex SI-SECM shape of the scans, Figures 4.29 and 4.30, is beyond the scope of this review. Qualitatively, kinetic effects on the SI-SECM scans might be expected considering the possible interplay of adsorbed species. CO$_{(ads)}$ at high coverage is known to be less strongly adsorbed than at low coverage due to repulsive interactions between neighboring molecules; at saturation coverage, different species of CO coexist (e.g., at Pt(111) approximately 1/3 of the CO adsorbs at atop sites while 2/3 adsorbs in threefold hollow sites) [78]. CO$_{(ads)}$ may indeed have different reactivity toward bromine at different coverage and sites (or may be limited by surface diffusion). Adsorption of bromide has also been shown to modify the reactivity of CO$_{(ads)}$ by making it more stable, i.e., less prone to electrochemical oxidation, although this may be due to the blocking of sites for adsorption/formation of OH$_{(ads)}$ [182]. The chemistry of the Br$^-$/Br$_2$ redox pair itself may play a role in the proposed mechanism, where the formation of the tribromide ion, Br$_3^-$, has been disregarded. The nature of the "bromophosgene" OCBr$_2$ species is also of importance to completely elucidate the mechanism. If this species is released into solution, its overall first-order decomposition rate constant for hydrolysis (Equation 4.27) should be fairly high, $k_{decomp} > 2 \times 10^3$ s^{-1}, for it to feed back the necessary Br^- to detect the reaction with a high CE (calculated using $D \sim 1 \times 10^{-9}$ m^2/s and $k_{decomp} \sim 2D/d^2$, where $d \sim 1$ μm is the tip-substrate distance and the SECM transit time between the electrodes is taken as the half-life for the hydrolysis reaction). If this hydrolysis reaction is localized at the substrate as a result of adsorption of "bromophosgene," then the CE may not be an issue; but the shape of the SI-SECM scan may be dependent on the relationship between the scan rate of the tip (kinetics of generation of Br_2) and the rate of hydrolysis of this product.

4.8   OUTLOOK FOR SI-SECM

The SI mode of SI-SECM has been described and its first applications to the study of electrocatalytic and catalytic systems, including reduction and oxidation reactions, have been developed in this review. SI-SECM has been explored as a tool for the quantification of adsorbed intermediates in these systems; however, adsorbate–mediator kinetic constants can also be determined by comparison of experimental data with digital simulation data obtained with modern software such as COMSOL Multphysics. The adsorbates H, O, and CO, described in this introductory review are small species that are present in large amounts at sub-monolayer coverage at the surface of typical metal electrocatalysts such as Pt or Au. Improvements in the reduction of background currents through the use of relatively inert materials for the interrogator tip should allow access to the detection of larger adsorbates, which typically are present in smaller quantities, or improve the limits of detection for the adsorbates treated here.

There are opportunities for considerable growth in the SI-SECM technique. For example, the identification of sensitive and selective chemical systems for specific adsorbate recognition will expand the types of reactions that SI-SECM can be used to study. As discussed, progress toward this end has been made in the use of TMPD to interrogate adsorbed hydrogen and, in contrast, its inability to interrogate carbon monoxide. Although the differences between these two adsorbed species are quite obvious, it would also be useful to find mediators that are able to resolve the reactivity (e.g., for the same species on different crystallographic planes) and identity of the adsorbed species. The differences in reactivity observed in the case of the titration of Pt oxides formed at different stages of oxidation with different mediators also suggest the possibility of applying similar schemes to solve long-standing issues in electrocatalysis, e.g., the participation of adsorbed oxygen during the OER, or the participation of differently adsorbed hydrogen on the hydrogen evolution reaction.

The elucidation of surface reaction mechanisms in electrochemistry through quantitative techniques is of great interest. Single electrode experiments, where the controlled variable is the potential of the electrode, lack the required versatility to address these challenges on their own. The example of the detection of carbon monoxide through electrogenerated bromine gives a good example of the development of new surface chemistry through the addition of a chemical component to the system. SI-SECM has been introduced experimentally in this review in a tip–substrate configuration that uses electrodes of similar size to circumvent complications arising from OC feedback in the absence of adsorbed species; new strategies need to be developed to limit the response from this effect in systems with substrates much larger than the tip. If this is achieved, then tip sizes can be reduced and access could be gained to a whole new area of analysis of surface domains, maybe even to nanostructures where the complete use of the spatially resolved SECM response can be used to correlate the reactivity of a structure with the type and amount of adsorbates present at it. The study of catalytic processes at insulator materials has yet to be developed and is surely of interest. Advances in the modeling of the SI-SECM response should follow the experimental developments and are of interest not only from a purely pragmatic point of view (i.e., determining rate constants), but also from the access that this technique has given to new spaces in surface chemistry, as exemplified by the reaction of mediators with Pt oxides where no theory, to the best of our knowledge, exists to describe the relationship between the energetics and the rate of these reactions.

REFERENCES

1. Bard, A. J. and Faulkner, L. R. *Electrochemical Methods*, Wiley: New York, 2001.
2. Saveant, J. -M. *Elements of Molecular and Biomolecular Electrochemistry*, Wiley-Interscience: Hoboken, New Jersey, 2006.
3. Abruña, H. D. (ed.). *Electrochemical Interfaces. Modern Techniques for In-Situ Interface Characterization,* VCH: New York, 1991.
4. Bard, A. J. and Mirkin, M. V. (eds.). *Scanning Electrochemical Microscopy,* Marcel Dekker: New York, 2001.
5. Rodríguez-López, J., Alpuche-Avilés, M. A., and Bard, A. J. Interrogation of surfaces for the quantification of adsorbed species on electrodes: oxygen on gold and platinum in neutral media. *J Am Chem Soc*, Vol. 130, (2008): pp. 16985–16995.
6. Bard, A. J., Denuault, G., Lee, C., Mandler, D., and Wipf, D. O. Scanning electrochemical microscopy – a new technique for the characterization and modification of surfaces. *Acc Chem Res*, Vol. 23, (1990): pp. 357–363.
7. Bard, A. J., Fan, F. F., Pierce, D. T., Unwin, P. R., Wipf, D. O., and Zhou, F. Chemical imaging of surfaces with the scanning electrochemical microscope. *Science*, Vol. 254(5028), (1991): pp. 68–74.
8. Wipf, D. O. and Bard, A. J. Scanning electrochemical microscopy. 15. Improvements in imaging via tip-position modulation and lock-in detection. *Anal Chem*, Vol. 64, (1992): pp. 1362–1367.
9. Lee, C., Kwak, J., and Bard, A. J. Application of scanning electrochemical microscopy to biological samples. *Proc Natl Acad Sci, U.S.A.*, Vol. 87, (1990): pp. 1740–1743.
10. Pierce, D. T. and Bard, A. J. Scanning electrochemical microscopy. 23. Retention localization of artificially patterned and tissue-bound enzymes. *Anal Chem*, Vol. 65, (1993): pp. 3598–3604.
11. Horrocks, B. R., Mirkin, M. V., and Bard, A. J. Scanning electrochemical microscopy. 25. Application to investigation of the kinetics of heterogeneous electron transfer at semiconductor (WSe_2 and Si) electrodes. *J Phys Chem*, Vol. 98, (1994): pp. 9106–9114.

12. Borgwarth, K. and Heinze, J. Heterogeneous electron transfer reactions. In *Scanning Electrochemical Microscopy*, A. J. Bard and M. V. Mirkin (eds.), Marcel Dekker: New York, 2001, pp. 201–240.

13. Mirkin, M. V. Theory. In *Scanning Electrochemical Microscopy*, A. J. Bard and M. V. Mirkin (eds.), Marcel Dekker: New York, 2001, pp. 145–199.

14. Amphlett, J. L. and Denuault, G. Scanning electrochemical microscopy (SECM): an investigation of the effects of tip geometry on amperometric tip response. *J Phys Chem B*, Vol. 102, (1998): pp. 9946–9951.

15. Wei, C., Bard, A. J., and Mirkin, M. V. Scanning electrochemical microscopy. 31. Application of SECM to the study of charge transfer processes at the liquid/liquid interface. *J Phys Chem*, Vol. 99, (1995): pp. 16033–16042.

16. Bard, A. J., Mirkin, M. V., Unwin, P. R., and Wipf, D. O. Scanning electrochemical microscopy. 12. Theory and experiment of the feedback mode with finite heterogeneous electron-transfer kinetics and arbitrary substrate size. *J Phys Chem*, Vol. 96, (1992): pp. 1861–1868.

17. Lefrou, C. A very easy kinetics determination for feedback curves with a microdisk SECM tip and rather rapid substrate reaction. *J Electroanal Chem*, Vol. 601, (2007): pp. 94–100.

18. Cornut, R. and Lefrou, C. New analytical approximation of feedback approach curves with a microdisk SECM tip and irreversible kinetic reaction at the substrate *J Electroanal Chem*, Vol. 621, (2008): pp. 178–184.

19. Mirkin, M.V. and Bard, A. J. Simple analysis of quasi-reversible steady-state voltammograms. *Anal Chem*, Vol. 64, (1992): pp. 2293–2302.

20. Wipf, D. O. and Bard, A. J. Scanning electrochemical microscopy. VII. Effect of heterogeneous electron-transfer rate at the substrate on the tip feedback current. *J Electrochem Soc*, Vol. 138, (1991): pp. 469–474.

21. Xiong, H., Guo, J., and Amemiya, S. Probing heterogeneous electron transfer at an unbiased conductor by scanning electrochemical microscopy in the feedback mode. *Anal Chem*, Vol. 79, (2007): pp. 2735–2744.

22. Macpherson, J. V., Slevin, C. J., and Unwin, P. R. Probing the oxidative etching kinetics of metals with the feedback mode of the scanning electrochemical microscope. *J Chem Soc Faraday Trans*, Vol. 92, (1996): pp. 3799–3805.

23. Zoski, C., Simjee, N., Guenat, O., and Koudelka-Hep, M. Addressable microelectrode arrays: characterization by imaging with scanning electrochemical microscopy. *Anal Chem*, Vol. 76, (2004): pp. 62–72.

24. Asscher, M. and Somorjai, G. A. Reactive scattering. In *Atomic and Molecular Beam Methods,* G. Scoles (ed.), Oxford University Press: New York, 1992, Vol. 2, pp. 488–517.

25. Yang, Y. and Denuault, G. Scanning electrochemical microscopy (SECM) study of pH changes at Pt electrode surfaces in Na_2SO_4 solution (pH 4) under potential cycling conditions. *J Chem Soc Faraday Trans*, Vol. 92, (1996): pp. 3791–3798.

26. Sánchez-Sánchez, C. M., Rodríguez-López, J., and Bard, A. J. Scanning electrochemical microscopy. 60. Quantitative calibration of the SECM substrate generation/tip collection mode and its use for the study of the oxygen reduction mechanism. *Anal Chem*, Vol. 80, (2008): pp. 3254–3260.

27. Unwin, P. R. and Bard, A. J. Ultramicroelectrode voltammetry in a drop of solution: a new approach to the measurement of adsorption isotherms at the solid–liquid interface. *Anal Chem*, Vol. 64, (1992): pp. 113–119.

28. Unwin, P. R. and Bard, A. J. Scanning electrochemical microscopy. 14. Scanning electrochemical microscope induced desorption: a new technique for the measurement of adsorption/desorption kinetics and surface diffusion rates at the solid/liquid interface. *J Phys Chem*, Vol. 96, (1992): pp. 5035–5045.

29. Selzer, Y., Turyan, I., and Mandler, D. Studying heterogeneous catalysis by the scanning electrochemical microscope (SECM): the reduction of protons by methyl viologen catalyzed by a platinum surface. *J Phys Chem B*, Vol. 103, (1999): pp. 1509–1517.

30. Burt, D. P., Cervera, J., Mandler, D., Macpherson, J. V., Manzanares, J. A., and Unwin, P. R. Scanning electrochemical microscopy as a probe of Ag^+ binding kinetics at Langmuir phospholipid monolayers. *Phys Chem Chem Phys*, Vol. 7, (2005): pp. 2955–2964.

31. Burshtain, D. and Mandler, D. Studying the binding of Cd^{2+} by ω-mercaptoalkanoic acid self assembled monolayers by cyclic voltammetry and scanning electrochemical microscopy (SECM). *J Electroanal Chem*, Vol. 581, (2005): pp. 310–319.

32. Unwin, P. R. and Bard A. J. Scanning electrochemical microscopy. 9. Theory and application of the feedback mode to the measurement of following chemical reaction rates in electrode processes. *J Phys Chem*, Vol. 95, (1991): pp. 7814–7824.

33. Zoski, C., Luman, C. R., Fernández, J. L., and Bard, A. J. Scanning electrochemical microscopy. 57. SECM tip voltammetry at different substrate potentials under quasi-steady-state and steady-state conditions. *Anal Chem*, Vol. 79, (2007): pp. 4957–4966.

34. Diaz-Ballote, L. F., Alpuche-Aviles, M., and Wipf, D. O. Fast-scan cyclic voltammetry–scanning electrochemical microscopy. *J Electroanal Chem*, Vol. 604, (2007): pp. 17–25.

35. Lie, L. H., Mirkin, M. V., Hakkarainen, S., Houlton, A., and Horrocks, B. R. Electrochemical detection of lateral charge transport in metal complex-DNA monolayers synthesized on Si(111) electrodes. *J Electroanal Chem*, Vol. 603, (2007): pp. 67–80.

36. Slevin, C. J. and Unwin, P. R. Lateral proton diffusion rates along stearic acid monolayers. *J Am Chem Soc*, Vol. 122, (2000): pp. 2597–2602.

37. Mandler, D. and Unwin, P. R. Measurement of lateral charge propagation in polyaniline layers with the scanning electrochemical microscope. *J Phys Chem B*, Vol. 107, (2003): pp. 407–410.

38. O'Mullane, A. P., Macpherson, J. V., Unwin, P. R., Cervera-Montesinos, J., Manzanares, J. A., Frehill, F., and Vos, J. G. Measurment of lateral charge propagation in $[Os(bpy)_2(PVP)_nCl]Cl$ thin films: a scanning electrochemical microscopy approach. *J Phys Chem B*, Vol. 108, (2004): pp. 7219–7227.

39. Zhang, J., Slevin, C. J., Morton, C., Scott, P., Walton, D. J., and Unwin, P. R. New approach for measuring lateral diffusion in Langmuir monolayers by scanning electrochemical microscopy (SECM): theory and application. *J Phys Chem B*, Vol. 105, (2001): pp. 11120–11130.

40. Liu, B., Rotenberg, S. A., and Mirkin, M. V. Scanning electrochemical microscopy of living cell. 4. Mechanistic study of charge transfer reactions in human breast cells. *Anal Chem*, Vol. 74, (2002): pp. 6340–6348.

41. Rodríguez-López, J. and Bard, A. J. Scanning electrochemical microscopy: surface interrogation of adsorbed hydrogen and the open circuit catalytic decomposition of formic acid at platinum. *J Am Chem Soc*, Vol. 132, (2010): pp. 5121–5129.

42. Zhang, M., Bin, S., Cortes-Salazar, F., Hojeij, M., and Girault, H. H. SECM photography. *Electrochem Comm*, Vol. 10, (2008): pp. 714–718.

43. Schnippering, M., Powell, H. V., Zhang, M., Macpherson, J. V., Unwin, P. R., Mazurenka, M., and Mackenzie, S. R. Surface assembly and redox dissolution of silver nanoparticles monitored by evanescent wave cavity ring-down spectroscopy. *J Phys Chem C*, Vol. 112, (2008): pp. 15274–15280.

44. Sheffer, M. and Mandler, D. Scanning electrochemical imprinting microscopy: a tool for surface patterning. *J Electrochem Soc*, Vol. 155, (2008): pp. D203–D208.

45. Kwak, J. and Bard, A. J. Scanning electrochemical microscopy. Theory of the feedback mode. *Anal Chem*, Vol. 61, (1989): pp. 1221–1232.

46. COMSOL Multiphysics v.3.2 software, COMSOL, Inc., Burlington, MA.

47. Celia, M. A. and Gray, W. G. *Numerical Methods for Differential Equations*, Prentice-Hall, Inc.: Englewood Cliffs, New Jersey, 1992, pp. 114–177.

48. Britz, D. *Digital Simulation in Electrochemistry,* 3rd Ed., Lecture Notes in Physics No. 666, Springer: Berlin, 2005, pp. 172–187.

49. Rodríguez-López, J., Minguzzi, A., and Bard, A. J. The reaction of various reductants with oxide films on Pt electrodes as studied by the surface interrogation mode of scanning electrochemical microscopy (SI-SECM). Possible validity of a Marcus relationship. *J Phys Chem C*, Vol. 114, (2010): pp. 18645–18655.

50. Sun, P. and Mirkin, M. V. Kinetics of electron-transfer reactions at nanoelectrodes. *Anal Chem*, Vol. 78, (2006): pp. 6526–6534.

51. Angerstein-Kozlowska, H., Conway, B. E., Barnett, B., and Mozota, J. The role of ion adsorption in surface oxide formation and reduction at noble metals: general features of the surface process. *J Electroanal Chem*, Vol. 100, (1979): pp. 417–446.

52. Woods, R. Chemisorption at Electrodes. In *Electroanalytical Chemistry*, A. J. Bard (ed.), Marcel Dekker: New York, 1976, p. 1–162.

53. Conway, B. E. and Angerstein-Kozlowska, H. The electrochemical study of multiple-state adsorption in monolayers. *Acc Chem Res*, Vol. 14, (1981): pp. 49–56.

54. Pletcher, D. and Sotiropoulos, S. Hydrogen adsorption-desorption and oxide formation-reduction on polycrystalline platinum in unbuffered aqueous solutions. *J Chem Soc Faraday Trans*, Vol. 90, (1994): pp. 3663–3668.

55. Bard, A. J., Fan, F. -R. F., and Mirkin, M. V. Scanning electrochemical microscopy. In *Electroanalytical Chemistry*, A. J. Bard (ed.), Marcel Dekker: New York, 1994, pp. 243–373.

56. Oesch, U. and Janata, J. Electrochemical study of gold electrodes with anodic oxide films. I. Formation and reduction behavior of anodic oxides on gold. *J Electrochim Acta*, Vol. 28, (1983): pp. 1237–1246.

57. Hayden, B. E., Rendall, M. E., and South, O. Electro-oxidation of carbon monoxide on well-ordered Pt(111)/Sn surface alloys. *J Am Chem Soc*, Vol. 125, (2003): pp. 7738–7742.

58. Beltramo, G. L., Shubina, T. E., and Koper, M. T. M. Oxidation of formic acid and carbon monoxide on gold electrodes studied by surface-enhanced Raman spectroscopy and DFT. *Chem Phys Chem*, Vol. 6, (2005): pp. 2597–2606.

59. Spendelow, J. S., Goodpaster, J. D., Kenis, P. J. A., and Wieckowski, A. Mechanism of CO oxidation on Pt(111) in alkaline media. *J Phys Chem B*, Vol. 110, (2006): pp. 9545–9555.

60. Assiongbon, K. A. and Roy, D. Electro-oxidation of methanol on gold in alkaline media: Adsorption characteristics of reaction intermediates studied using time resolved electro-chemical impedance and surface Plasmon resonance techniques. *Surf Sci*, Vol. 594, (2005): pp. 99–119.

61. Kim, T. S., Gong, J., Ojifinni, R. A., White, J. M., and Mullins, C. B. Water activated by atomic oxygen on Au(111) to oxidize CO at low temperatures. *J Am Chem Soc*, Vol. 128, (2006): pp. 6282–6283.

62. Schmid, G. M. and O'Brien, R. N. Oxygen adsorption and double layer capacities; gold in perchoric acid. *J Electrochem Soc*, Vol. 111, (1964): pp. 832–837.

63. Nguyen Van Huong, C., Hinnen, C., and Lecoeur, J. Spectroscopic study of single crystal gold electrodes. Part II. The incipient oxidation of gold electrode. *J Electroanal Chem*, Vol. 106, (1980): pp. 185–191.

64. Marichev, V. A. Contact electroresistance method for in situ studies of metal surfaces in electrolytes: adsorption of hydroxyl ions and submonolayer electrooxidation of gold. *Russ J Electrochem*, Vol. 35, (1999): pp. 434–440.

65. Watanabe, T. and Gerischer, J. Photoelectrochemical studies on gold electrodes with surface oxide layers. Part. I. Photocurrent measurement in the visible region. *J Electroanal Chem*, Vol. 117, (1981): pp. 185–200.

66. DeSilvestro, J. and Weaver, M. J. Surface structural changes during oxidation of gold electrodes in aqueous media as detected using surface-enhanced Raman spectroscopy. *J Electroanal Chem*, Vol. 209, (1986): pp. 377–386.

67. Horányi, G. Indusced cation adsorption on platinum and modified platinum electrodes. *Electrochim Acta*, Vol. 36, (1991): pp. 1453–1463.

68. Lertanantawong, B., O'Mullane, A. P., Surareungchai, W., Somasundrum, M., Burke, L. D., and Bond, A. M. Study of the underlying electrochemistry of polycrystalline gold electrodes in aqueous solution and electrocatalysis by large amplitude fourier transformed alternating current voltammetry. *Langmuir*, Vol. 24, (2008): pp. 2856–2868.

69. Fultz, M. L. and Durst, R. A. Mediator compounds for the electrochemical study of biological redox systems: a compilation. *Anal Chim Acta*, Vol. 140, (1982): pp. 1–18.

70. Mirkin, M. V., Arca, M., and Bard, A. J. Scanning electrochemical microscopy. 22. Examination of thin solid silver(I) bromide films: ion diffusion in the film and heterogeneous kinetics at the film/solution interface. *J Phys Chem*, Vol. 97, (1993): pp. 10790–10795.

71. Pust, S. E., Maier, W., and Wittstock, G. Investigation of localized catalytic and electrocatalytic processes and corrosion reactions with scanning electrochemical microscopy (SECM). *Z Phys Chem*, Vol. 222, (2008): pp. 1463–1517.

72. Liu, B., Bard, A. J., Mirkin, M. V., and Creager, S. E. Electron transfer at self-assembled monolayers measured by scanning electrochemical microscopy. *J Am Chem Soc*, Vol. 126, (2004): pp. 1485–1492.

73. Kiani, A., Alpuche-Aviles, M. A., Eggers, P. K., Jones, M., Gooding, J. J., Paddon-Row, M. N., and Bard, A. J. Scanning electrochemical microscopy. 59. Effect of defects and structure on electron transfer through self-assembled monolayers. *Langmuir*, Vol. 24, (2008): pp. 2841–2849.

74. Baker, B. B. and MacNevin, W. M. The effect of prepolarization of a platinum anode on the current obtained in the controlled-potential oxidation of iron and arsenic. *J Am Chem Soc*, Vol. 75, (1953): pp. 1476–1477.

75. Schmickler, W. and Schultze, J. W. Electron transfer reactions on oxide-covered metal electrodes. In *Modern Aspects of Electrochemistry*, J. O'M. Bockris, B. E. Conway, R. E. White (eds.), Plenum Press: New York, 1986. Vol. 17, pp. 357–410.

76. Moffat, T. P., Yang, H., Fan, F. -R. F. and Bard, A. J. Electron-transfer reactions on passive chromium. *J Electrochem Soc*, Vol. 139, (1992): pp. 3158–3167.

77. Schultze, J. W. and Vetter, K. Influence of the tunnel probability on the anodic oxygen evolution and other redox reactions at oxide covered platinum electrodes. *J Electrochim Acta*, Vol. 18, (1973): pp. 889–896.

78. Markovic, N. M. and Ross Jr., P. N. Surface science studies of model fuel cell electrocatalysts. *Surf Sci Rep*, Vol. 45, (2002): pp. 117–229.

79. *Fuel Cell Catalysis, A Surface Science Approach*, M. T. M. Koper (ed.), John Wiley & Sons: Hoboken, New Jersey, 2009. (Andrzej Wieckowski, Series Editor).

80. Damjanovic, A., Ward, A. T., and O'Jea, M. Effect of the thickness of anodic oxide films on the rate of the oxygen evolution reaction at platinum in sulfuric acid solution. I. Growth at constant current. *J Electrochem Soc*, Vol. 121, (1974): pp. 1186–1190.

81. Damjanovic, A., Birss, V. I., and Boudreaux, D. S. Electron transfer through thin anodic oxide films during the oxygen evolution reactions at platinum electrodes. I. Acid solutions. *J Electrochem Soc*, Vol. 138, (1991): pp. 2549–2555.

82. Kuhn, A. T. and Randle, T. H. Effect of oxide thickness on the rates of some redox reactions on a platinum electrode. *J Chem Soc Faraday Trans*, Vol. 81, (1985): pp. 403–419.

83. Kabanova, O. L. Oxidation of divalent iron and passivation of the platinum electrode. *Zhurnal Fizicheskoi Khim*, Vol. 35, (1961): pp. 2465–2471.

84. Kabanova, O. L. The effect of adsorbed oxygen on the oxidation of ferrocyanide ions on a gold electrode. *Zhurnal Anal Khim*, Vol. 16, (1961): pp. 135–140.

85. Trasatti, S. Electrocatalysis in the anodic evolution of oxygen and chlorine. *Electrochim Acta*, Vol. 11, (1984): pp. 1503–1512.

86. Fierro, S., Nagel, T., Baltruschat, H., and Comninellis, C. Investigation of the oxygen evolution reaction on Ti/IrO_2 electrodes using isotope labeling and on-line mass spectrometry. *Electrochem Comm*, Vol. 9, (2007): p. 1969.

87. McAlpin, J. G., Surendranath, Y., Dinca, M., Stich, T. A., Stoian, S. A., Casey, W. H., Nocera, D. G., and Britt, R. D. EPR evidence for Co(IV) species produced during water oxidation at neutral pH. *J Am Chem Soc*, Vol. 132, (2010): pp. 6882–6883.

88. Angerstein-Kozlowska, H. and Conway, B. E. Real condition of electrochemically oxidized platinum surfaces. I. Resolution of component processes. *J Electroanal Chem*, Vol. 43, (1973): pp. 9–36.

89. Tremilosi-Filho, G., Jerkiewicz, G., and Conway, B. E. Characterization and significance of the sequence of stages of oxide film formation at platinum generated by strong anodic polarization. *Langmuir*, Vol. 8, (1992): pp. 658–667.

90. Dickinson, T., Povey, A. F., and Sherwood, P. M. A. X-ray photoelectron spectroscopic studies of oxide films on platinum and gold electrodes. *J Chem Soc Faraday Trans*, Vol. 71, (1975): pp. 298–311.

91. Wagner, F. T. and Ross Jr., P. N. AES and TDS studies of electrochemically oxidized platinum(100). *Appl Surf Sci*, Vol. 24, (1985): pp. 87–107.

92. Jerkiewicz, G., Vatankhah, G., Lessard, J., Soriaga, M. P., and Park, Y. -S. Surface-oxide growth at platinum electrodes in aqueous H_2SO_4. Reexamination of its mechanism through combined cyclic-voltammetry, electrochemical quartz-crystal nanobalance, and Auger electron spectroscopy measurements. *Electrochim Acta*, Vol. 49, (2004): pp. 1451–1459.

93. Wagner, F. T. and Ross Jr., P. N. LEED spot profile analysis of the structure of electrochemically treated platinum(100) and platinum(111) surfaces. *Surf Sci*, Vol. 160, (1985): pp. 305–330.

94. Nagy, Z. and You, H. Applications of surface X-ray scattering to electrochemistry problems. *Electrochim Acta*, Vol. 47, (2002): pp. 3037–3055.

95. You, H., Zurawski, D. J., Nagy, Z., and Yonco, R. M. In-situ x-ray reflectivity study of incipient oxidation of Pt(111) surface in electrolyte solutions. *J Chem Phys*, Vol. 100, (1994): pp. 4699–4702.

96. Sun, A., Franc, J., and Macdonald, D. D. Growth and properties of oxide films on platinum. I. EIS and x-ray photoelectron spectroscopy studies. *J Electrochem Soc*, Vol. 153, (2006): pp. B260–B277.

97. Xia, S. J. and Birss, V. I. Hydrous oxide film growth on Pt. I. Type I β-oxide formation in 0.1 M H_2SO_4. *Electrochim Acta*, Vol. 44, (1998): pp. 467–482.

98. Noguchi, H., Okada, T., and Uosaki, K. Molecular structure at electrode/electrolyte solution interfaces related to electrocatalysis. *Faraday Discuss*, Vol. 140, (2008): pp. 125–137.

99. Bergelin, M., Feliu, J. M., and Wasberg, M. Study of carbon monoxide adsorption and oxidation on Pt(111) by using an electrochemical impinging jet cell. *Electrochim Acta*, Vol. 44, (1998): pp. 1069–1075.

100. Anderson, A. B., Neshev, N. M., Sidik, R. A., and Shiller, P. Mechanism for the electrooxidation of water to OH and O bonded to platinum: quantum chemical theory. *Electrochim Acta*, Vol. 47, (2002): pp. 2999–3008.

101. Anderson, A. B. O_2 reduction and CO oxidation at the Pt-electrolyte interface. The role of H_2O and OH adsorption bond strengths. *Electrochim Acta*, Vol. 47, (2002): pp. 3759–3763.

102. Marichev, V. A. Reversibility of platinum voltammograms in aqueous electrolytes, ionic product and dissociative adsorption of water. *Electrochem Comm*, Vol. 10, (2008): pp. 643–646.

103. Andrieux, C. P., Savéant, J. -M., and Su, K. B. Kinetics of dissociative electron transfer. Direct and mediated electrochemical reductive cleavage of the carbon-halogen bond. *J Phys Chem*, Vol. 90, (1986): pp. 3815–3823.

104. Saveant, J. -M. *Elements of Molecular and Biomolecular Electrochemistry*, Wiley-Interscience: Hoboken, New Jersey, 2006.

105. Fletcher, S. The theory of electron transfer. *J Solid State Electrochem*, Vol. 14, (2010): pp. 705–739.

106. Miller, C. J. Heterogeneous electron transfer kinetics at metallic electrodes. In *Physical Electrochemistry, Principles, Methods and Applications*, I. Rubinstein (ed.), Marcel Dekker: New York, 1995, pp. 27–79.

107. Barker, A. L., Unwin, P. R., Amemiya, S., Zhou, J., and Bard, A. J. Scanning electrochemical microscopy (SECM) in the study of electron transfer kinetics at liquid/liquid interfaces: beyond the constant composition approximation. *J Phys Chem B*, Vol. 103, (1999): pp. 7260–7269.

108. González, I., Ibáñez, J. G., Cárdenas, M. A., and Rojas-Hernández, A. The electrochemical behavior of the iron(II)-iron(III) system in the presence of EDTA at pH 4.3. *Electrochim Acta*, Vol. 35, (1990): pp. 1325–1329.

109. Huerta, F. J., Morallon, E., Vazquez, J. L., and Aldaz, A. Voltammetric and spectroscopic characterization of cyanide adlayers on Pt(h,k,l) in an acidic medium. *Surf Sci*, Vol. 396, (1998): pp. 400–410.

110. Huerta, F., Morallon, E., and Vazquez, J. L. Voltammetric analysis of the co-adsorption of cyanide and carbon monoxide on a Pt(111) surface. *Electrochem Comm*, Vol. 4, (2002): pp. 251–254.

111. Cuesta, A., Escudero, M., Lanova, B., and Baltruschat, H. Cyclic voltammetry, FTIRS, and DEMS study of the electrooxidation of carbon monoxide, formic acid, and methanol on cyanide-modified Pt(111) electrodes. *Langmuir*, Vol. 25, (2009): pp. 6500–6507.

112. Sánchez-Sánchez, C. M. and Bard, A. J. Hydrogen peroxide production in the oxygen reduction reaction at different electrocatalysts as quantified by scanning electrochemical microscopy. *Anal Chem*, Vol. 81, (2009): pp. 8094–8100.

113. Leiva, E. and Sánchez, C. Theoretical aspects of some prototypical fuel cell reactions. In *Handbook of Fuel Cells*, W. Vielstich, A. Lamm, and H. Gasteiger (eds.), Wiley: Chichester, England, (2003): Vol. 2, pp. 93–131.

114. Jung, C., Sánchez-Sánchez, C. M., Lin, C. -L., Rodríguez-López, J., and Bard, A. J. Electrocatalytic activity of Pd-Co bimetallic mixtures for formic acid oxidation studied by scanning electrochemical microscopy. *Anal Chem*, Vol. 81, (2009): pp. 7003–7008.

115. Zhou, J., Zu, Y., and Bard, A. J. Scanning electrochemical microscopy part 39. The proton/hydrogen mediator system and its application to the study of the electrocatalysis of hydrogen oxidation. *J Electroanal Chem*, Vol. 491, (2000): pp. 22–29.

116. Fernandez, J. L., Hurth, C., and Bard, A. J. Scanning electrochemical microscopy #54. Application to the study of heterogeneous catalytic reactions – hydrogen peroxide decomposition. *J Phys Chem B*, Vol. 109, (2005): pp. 9532–9539.

117. Fernandez, J. L., Walsh, D. A., and Bard, A. J. Thermodynamic guidelines for the design of bimetallic catalysts for Oxygen electroreduction and rapid screening by scanning electrochemical microscopy. M-Co (M: Pd, Ag, Au). *J Am Chem Soc*, Vol. 127, (2005): pp. 357–365.

118. Minguzzi, A., Alpuche-Avilés, M. A., Rodríguez-López, J., Rondinini, S., and Bard, A. J. Screening of oxygen evolution electrocatalysts by scanning electrochemical microscopy using a shielded tip approach. *Anal Chem*, Vol. 80, (2008): pp. 4055–4064.

119. Lin, C. -L., Sánchez-Sánchez, C. M., and Bard, A. J. Methanol tolerance of Pd-Co oxygen reduction electrocatalysts using scanning electrochemical microscopy. *Electrochem Solid-State Lett*, Vol. 11, (2008): pp. B136–B139.

120. Lee, J., Ye, H., Pan, S., and Bard, A. J. Screening of photocatalysts by scanning electrochemical microscopy. *Anal Chem*, Vol. 80, (2008): pp. 7445–7450.

121. Sun, Y. M., Alpuche-Avilés, M. A., Bard, A. J., Zhou, J. P., and White, J. M. Preparation and characterization of Pd-Ti electrocatalysts on carbon supports for oxygen reduction. *J Nanosci Nanotechnol*, Vol. 9, (2009): pp. 1281–1286.

122. Rice, C., Ha, S., Masel, R. I., and Wieckowski, A. Catalysts for direct formic acid fuel cells. *J Power Sources*, Vol. 115, (2003): pp. 229–235.

123. Rodríguez, J. L. and Pastor, E. Consecutive adsorption as studied by electrochemical mass spectrometry: coadsorption, desorption and displacement reactions on platinum. *Electrochim Acta*, Vol. 44, (1998): pp. 1173–1179.

124. Lu, G. -Q., Crown, A., and Wieckowski, A. Formic acid decomposition on polycrystalline platinum and palladized platinum electrodes. *J Phys Chem B*, Vol. 103, (1999): pp. 9700–9711.

125. Chen, Y. -X., Shen, Y., Heinen, M., Jusys, Z., Osawa, M., and Behm, R. J. Application of in-situ attenuated total reflection-Fourier transform infrared spectroscopy for the understanding of complex reaction mechanisms and kinetics: formic acid oxidation on a Pt film electrode at elevated temperatures. *J Phys Chem B*, Vol. 110, (2006): pp. 9534–9544.

126. Chen, Y. -X., Heinen, M., Jusys, Z., and Behm, R. J. Bridge-bonded formate: active intermediate or spectator species in formic acid oxidation on a Pt film electrode. *Langmuir*, Vol. 22, (2006): pp. 10399–10408.

127. Chen, Y. -X., Heinen, M., Jusys, Z., and Behm, R. J. Kinetic isotope effects in complex reaction networks: formic acid electro-oxidation. *Chem Phys Chem*, Vol. 8, (2007): pp. 380–385.

128. Miki, A., Ye, S., Senzaki, T., and Osawa, M. Surface-enhanced infrared study of catalytic electrooxidation of formaldehyde, methyl formate, and dimethoxymethane on platinum electrodes in acidic solution. *J Electroanal Chem*, Vol. 563, (2004): pp. 23–31.

129. Samjeske, G., Miki, S., Ye, A., Yamakata, Y., Mukouyama, H., Okamoto, H., and Osawa, M. Potential oscillations in galvanostatic electrooxidation of formic acid on platinum: a time-resolved surface-enhanced infrared study. *J Phys Chem B*, Vol. 109, (2005): pp. 23509–23516.

130. Breiter, M. W. Oxidation mechanism of formic acid on platinum at low potentials in acidic solutions. *J Electrochem Soc*, Vol. 111, (1964): pp. 1298–1299.

131. Gottlieb, M. H. Anodic oxidation of formic acid at platinum electrodes. *J Electrochem Soc*, Vol. 111, (1964): pp. 465–472.

132. Munson, R. A. Constant current transition time investigations of the electrochemical oxidation of formate-formic acid at a smooth platinum electrode. *J Electrochem Soc*, Vol. 111, (1964): pp. 372–376.

133. Parsons, R. and VanderNoot, T. The oxidation of small organic molecules. A survey of recent fuel cell related research. *J Electroanal Chem*, Vol. 257, (1988): pp. 9–45.

134. Oxley, J. E., Johnson, G. K., and Buzalski, B. T. The anodic oxidation of methanol: open-circuit transient reactions. *Electrochim Acta*, Vol. 9, (1964): pp. 897–910.

135. Schwabe, K. Potential measurements on metal catalysis in aqueous solution. *Z Elektrochem*, Vol. 61, (1957): pp. 744–752.

136. Breiter, M. W. Anodic oxidation of formic acid on platinum. I. Adsorption of formic acid, oxygen and hydrogen in perchloric acid solutions. *Electrochim Acta*, Vol. 8, (1963): pp. 447–456, 457–470.

137. Fleischmann, C. W., Johnson, G. K., and Kuhn, A. T. Electrochemical oxidation of formic acid on platinum. *J Electrochem Soc*, Vol. 111, (1964): pp. 602–605.

138. Eckert, J. Electrochemical behavior of formic acid on platinum electrodes in acidic solutions. *Electrochim Acta*, Vol. 12, (1967): pp. 307–328.

139. Podlovchenko, B. I., Manzhos, R. A., and Maksimov, Y. M. Kinetics and mechanism of interaction between methanol and adsorbed oxygen on a smooth polycrystalline platinum electrode: Transients of the open-circuit potential. *Russ J Electrochem*, Vol. 42, (2006): pp. 1061–1066.

140. Podlovchenko, B. I., Manzhos, R. A., and Maksimov, Y. M. Interaction of HCO-substances with adsorbed oxygen on platinum electrodes: open-circuit transient reactions of HCOOH and CO. *Electrochim Acta*, Vol. 50, (2005): pp. 4807–4813.

141. Tian, M. and Conway, B. E. Electrocatalysis in oscillatory kinetics of anodic oxidation of formic acid: at Pt; nanogravimetry and voltammetry studies on the role of reactive surface oxide. *J Electroanal Chem*, Vol. 616, (2008): pp. 45–56.

142. Papapolymerou, G. A. and Schmidt, L. D. Kinetics of decomposition of formaldehyde, formic acid, methanol, and hydrazine on platinum and rhodium surfaces. *Langmuir*, Vol. 3, (1987): pp. 1098–1102.

143. Columbia, M. R. and Thiel, P. A. The interaction of formic acid with transition metal surfaces, studied in ultrahigh vacuum. *J Electroanal Chem*, Vol. 369, (1994): pp. 1–14.

144. Columbia, M. R., Crabtree, A. M., and Thiel, P. A. Effect of carbon monoxide on platinum-catalyzed decomposition of formic acid in ultrahigh vacuum. *J Electroanal Chem*, Vol. 345, (1993): pp. 93–105.

145. Dahlberg, S. C., Fisk, G. A., and Rye, R. R. Molecular beam study of formic acid decomposition on platinum. *J Catal*, Vol. 36, (1975): pp. 224–234.

146. Tevault, D. E., Lin, M. C., Umstead, M. E., and Smardzewski, R. R. Evidence for production of the hydroxycarbonyl radical in the decomposition of formic acid on platinum. *Int J Chem Kinet*, Vol. 11, (1979): pp. 445–449.

147. Thomas, F. S. and Masel, R. I. Formic acid decomposition on palladium-coated Pt(110). *Surf Sci*, Vol. 573, (2004): pp. 169–175.

148. Baldauf, M. and Kolb, D. M. Formic acid oxidation on ultrathin Pd films on Au(hlk) and Pt(hkl) electrodes. *J Phys Chem*, Vol. 100, (1996): pp. 11375–11381.

149. Brandt, K., Steinhausen, M., and Wandelt, K. Catalytic and electro-catalytic oxidation of formic acid on the pure and Cu-modified Pd(111)-surface. *J Electroanal Chem*, Vol. 616, (2008): pp. 27–37.

150. Burke, L. D. and Casey, J. K. The electrocatalytic behavior of palladium in acid and base. *J Appl Electrochem* 1993, Vol. 23, p. 573–582. Burke, L. D., Kemball, C., and Lewis, F. A. Formation of palladium hydride by reaction of formic acid at palladium electrodes. *Catalysis*, Vol. 5, (1966): pp. 539–542.

151. Ananiev, A. V., Broudic, J. -C., and Brossard, P. The urea decomposition in the process of the heterogeneous catalytic denitration of nitric acid solutions. Part II. Reaction products and stoichiometry. *Appl Catal B*, Vol. 45, (2003): pp. 197–203.

152. Parravano, G. Polymerization induced by catalytic decomposition of formic acid at platinum electrodes. *J Am Chem Soc*, Vol. 72, (1950): pp. 5546–5549.
153. Bickford, D. F., Coleman, C. J., Hsu, C. L. W., and Eibling, R. E. Nobel metal-catalyzed formic acid decomposition and formic acid denitration. *Ceram Trans*, Vol. 23, (1991): pp. 283–294.
154. Yue, C. M. Y. and Lim, K. W. Adsorption of formic acid and its decomposed intermediates on (100) surfaces of Pt and Pd: a density functional study. *Catal Lett*, Vol. 128, (2009): pp. 221–226.
155. Hamada, I. and Yoshitada, M. Density-functional analysis of hydrogen on Pt(111). Electric field, solvent, and coverage effects. *J Phys Chem C*, Vol. 112, (2008): pp. 10889–10898.
156. Kunimatsu, K., Senzaki, T., Tsushima, M., and Osawa, M. A combined surface-enhanced infrared and electrochemical kinetics study of hydrogen adsorption and evolution on a Pt electrode. *Chem Phys Lett*, Vol. 401, (2005): pp. 451–454.
157. Chun, J. H., Ra, K. H., and Kim, N. Y. The Langmuir adsorption isotherms of electroadsorbed hydrogen for the cathodic hydrogen evolution reactions at the Pt(100)/H_2SO_4 and LiOH aqueous electrolyte interfaces. *Int J Hydrogen Energy*, Vol. 26, (2001): pp. 941–948.
158. Offer, G. J. and Kucernak, A. R. Calculating the coverage of saturated and sub-saturated layers of carbon monoxide adsorbed onto platinum. *J Electroanal Chem*, Vol. 613, (2008): pp. 171–185.
159. Ma, Z. and Zaera, F. In situ reflection-absorption infrared spectroscopy at the liquid-solid interface: decomposition of organic molecules on polycrystalline platinum surfaces. *Catal Lett*, Vol. 96, (2004): pp. 5–12.
160. Neurock, M., Janik, M., and Wieckowski, A. A first principles comparison of the mechanism and site requirements for the electrocatalytic oxidation of methanol and formic acid over Pt. *Faraday Discuss*, Vol. 140, (2008): pp. 363–378.
161. Jacobs, G., Patterson, P. M., Graham, U. M., Crawford, A. C., and Davis, B. H. Low temperature water gas shift: the link between the catalysis of WGS and formic acid decomposition over Pt/ceria. *Int J Hydrogen Energy*, Vol. 30, (2005): pp. 1265–1276.
162. Wang, Q., Rodríguez-López, J., and Bard, A. J. The reaction of Br_2 with Adsorbed CO on Pt Studied by the Surface Interrogation Mode of Scanning Electrochemical Microscopy. *J Am Chem Soc*, Vol. 131, (2009): pp. 17046–17047.
163. Vielstich, W. CO, formic acid, and methanol oxidation in acid electrolytes – mechanisms and electrocatalysis. In *Encyclopedia of Electrochemistry*, A. J. Bard, and M. Stratmann (eds.), Wiley-VCH: Weinheim, Germany, 2003, Vol. 2, pp. 466–511.
164. Iwasita, T. In *Handbook of Fuel Cells*. W. Vielstich, A. Lamm, and H. A. Gasteiger (eds.), Wiley-UK: Chichester, 2003, Vol. 2, pp. 603–624.
165. Siwek, H., Lukaszewski, M., and Czerwinski, A. Electrochemical study on the adsorption of carbon oxides and oxidation of their adsorption products on platinum group metals and alloys. *Phys Chem Chem Phys*, Vol. 10, (2008): pp. 3752–3765.
166. Inkaew, P. and Korzeniewski, C. Kinetic studies of adsorbed CO electrochemical oxidation on Pt(335) at full and sub-saturation coverages. *Phys Chem Chem Phys*, Vol. 10, (2008): pp. 3655–3661.
167. Czerwinski, A. and Sobkowski, J. The adsorption of carbon monoxide on a platinum electrode. *J Electroanal Chem*, Vol. 91, (1978): pp. 47–53.
168. Cuesta, A., Couto, A., Rincón, A., Pérez, M. C., López-Cudero, A., and Gutiérrez, C. Potential dependence of the saturation CO coverage of Pt electrodes: the origin of the pre-peak in CO-stripping voltammograms. Part 3: Pt(poly). *J Electroanal Chem*, Vol. 586, (2006): pp. 184–195.

169. Batista, E. A., Iwasita, T., and Vielstich, W. Mechanism of stationary bulk CO oxidation on Pt(111) electrodes. *J Phys Chem B*, Vol. 108, (2004): pp. 14216–14222.

170. Samjeske, G., Komatsu, K. -I., and Osawa, M. Dynamics of CO oxidation on a polycrystalline platinum electrode: a time-resolved infrared study. *J Phys Chem C*, Vol. 113, (2009): pp. 10222–10228.

171. Lu, G. -Q., Chrzanowski, W., and Wieckowski, A. Catalytic methanol decomposition pathways on a platinum electrode. *J Phys Chem B*, Vol. 104, (2000): pp. 5566–5572.

172. Lai, S. C. S., Kleyn, S. E. F., Rosca, V., and Koper, M. T. M. Mechanism of the dissociation and electrooxidation of ethanol and acetaldehyde on platinum as studied by SERS. *J Phys Chem C*, Vol. 112, (2008): pp. 19080–19087.

173. Cao, D., Lu, G. -Q., Wiekowski, A., Wasileski, S. A., and Neurock, M. Mechanisms of methanol decomposition on platinum: a combined experimental and ab initio approach. *J Phys Chem B*, Vol. 109, (2005): pp. 11622–11633.

174. Koper, M. T. M., Jansen, A. P. J., van Santen, R. A., Lukkien, J. J., and Hilbers, P. A. J. Monte Carlo simulations of a simple model for the electrocatalytic CO oxidation on platinum. *J Chem Phys*, Vol. 109, (1998): pp. 6051–6062.

175. Nakanishi, S., Sakai, S. -I., Hatou, M., Fukami, K., and Nakato, Y. Promoted dissociative adsorption of hydrogen peroxide and persulfate ions and electrochemical oscillations caused by a catalytic effect of adsorbed bromine. *J Electrochem Soc*, Vol. 150, (2003): pp. E47–E51.

176. Lissi, E., Simonaitis, R., and Heicklen, J. Bromine atom catalyzed oxidation of carbon monoxide. *J Phys Chem*, Vol. 76, (1972): pp. 1416–1419.

177. Piva, A. The action of bromine on carbon monoxide. *Gazz Chim Ital*, Vol. 45(1), (1915): pp. 219–237.

178. (a) Lenher, S. and Schumacher, H. -J. Bromophosgene. I. Thermal decomposition- a first order wall reaction. *Z physik Chem.* 1928, Vol. 135, p. 85–101. (b) Schumacher, H. -J. and Lenher, S. Bromophosgene. II. Its preparation and properties. *Berichte der Deutschen Chemischen Gesellschaft [Abteilung] B:Abhandlungen*, Vol. 61B, (1928): pp. 1671–1675.

179. Livingston, R. The halogen sensitized oxidation of carbon monoxide. *J Phys Chem*, Vol. 34, (1930): pp. 2121–2122.

180. Francisco, J. S. Energetics for the reaction of CBr_2O with water. *Chem Phys Lett*, Vol. 363, (2002): pp. 275–282.

181. Markovic, N. M., Lucas, C., Grgur, B. N., and Ross, P. N. Surface electrochemistry of CO and H_2/CO mixtures at Pt(100) interface. Electrode kinetics and interfacial structures. *J Phys Chem*, Vol. 103, (1999): pp. 9616–9623.

182. Markovic, N. M., Lucas, C., Rodes, A., Stamenkovic, V., and Ross, P. N. Surface electrochemistry of CO on Pt(111): anion effects. *Surf Sci*, Vol. 499, (2002): pp. L149–L158.

Index

Printed in the United States
by Baker & Taylor Publisher Services